Artificial Intelligence in Drug Discovery

Drug Discovery Series

Titles in the Series:
1: Metabolism, Pharmacokinetics and Toxicity of Functional Groups
2: Emerging Drugs and Targets for Alzheimer's Disease; Volume 1
3: Emerging Drugs and Targets for Alzheimer's Disease; Volume 2
4: Accounts in Drug Discovery
5: New Frontiers in Chemical Biology
6: Animal Models for Neurodegenerative Disease
7: Neurodegeneration
8: G Protein-Coupled Receptors
9: Pharmaceutical Process Development
10: Extracellular and Intracellular Signaling
11: New Synthetic Technologies in Medicinal Chemistry
12: New Horizons in Predictive Toxicology
13: Drug Design Strategies: Quantitative Approaches
14: Neglected Diseases and Drug Discovery
15: Biomedical Imaging
16: Pharmaceutical Salts and Cocrystals
17: Polyamine Drug Discovery
18: Proteinases as Drug Targets
19: Kinase Drug Discovery
20: Drug Design Strategies: Computational Techniques and Applications
21: Designing Multi-Target Drugs
22: Nanostructured Biomaterials for Overcoming Biological Barriers
23: Physico-Chemical and Computational Approaches to Drug Discovery
24: Biomarkers for Traumatic Brain Injury
25: Drug Discovery from Natural Products
26: Anti-Inflammatory Drug Discovery
27: New Therapeutic Strategies for Type 2 Diabetes: Small Molecules
28: Drug Discovery for Psychiatric Disorders
29: Organic Chemistry of Drug Degradation
30: Computational Approaches to Nuclear Receptors

31: Traditional Chinese Medicine
32: Successful Strategies for the Discovery of Antiviral Drugs
33: Comprehensive Biomarker Discovery and Validation for Clinical Application
34: Emerging Drugs and Targets for Parkinson's Disease
35: Pain Therapeutics; Current and Future Treatment Paradigms
36: Biotherapeutics: Recent Developments using Chemical and Molecular Biology
37: Inhibitors of Molecular Chaperones as Therapeutic Agents
38: Orphan Drugs and Rare Diseases
39: Ion Channel Drug Discovery
40: Macrocycles in Drug Discovery
41: Human-based Systems for Translational Research
42: Venoms to Drugs: Venom as a Source for the Development of Human Therapeutics
43: Carbohydrates in Drug Design and Discovery
44: Drug Discovery for Schizophrenia
45: Cardiovascular and Metabolic Disease: Scientific Discoveries and New Therapies
46: Green Chemistry Strategies for Drug Discovery
47: Fragment-Based Drug Discovery
48: Epigenetics for Drug Discovery
49: New Horizons in Predictive Drug Metabolism and Pharmacokinetics
50: Privileged Scaffolds in Medicinal Chemistry: Design, Synthesis, Evaluation
51: Nanomedicines: Design, Delivery and Detection
52: Synthetic Methods in Drug Discovery: Volume 1
53: Synthetic Methods in Drug Discovery: Volume 2
54: Drug Transporters: Role and Importance in ADME and Drug Development
55: Drug Transporters: Recent Advances and Emerging Technologies
56: Allosterism in Drug Discovery
57: Anti-aging Drugs: From Basic Research to Clinical Practice
58: Antibiotic Drug Discovery: New Targets and Molecular Entities
59: Peptide-based Drug Discovery: Challenges and New Therapeutics
60: Drug Discovery for Leishmaniasis
61: Biophysical Techniques in Drug Discovery
62: Acute Brain Impairment Through Stroke: Drug Discovery and Translational Research
63: Theranostics and Image Guided Drug Delivery
64: Pharmaceutical Formulation: The Science and Technology of Dosage Forms
65: Small-molecule Transcription Factor Inhibitors in Oncology
66: Therapies for Retinal Degeneration: Targeting Common Processes
67: Kinase Drug Discovery: Modern Approaches

68: Advances in Nucleic Acid Therapeutics
69: MicroRNAs in Diseases and Disorders: Emerging Therapeutic Targets
70: Emerging Drugs and Targets for Multiple Sclerosis
71: Cytotoxic Payloads for Antibody–Drug Conjugates
72: Peptide Therapeutics: Strategy and Tactics for Chemistry, Manufacturing, and Controls
73: Anti-fibrotic Drug Discovery
74: Protein Degradation with New Chemical Modalities: Successful Strategies in Drug Discovery and Chemical Biology
75: Artificial Intelligence in Drug Discovery

How to obtain future titles on publication:
A standing order plan is available for this series. A standing order will bring delivery of each new volume immediately on publication.

For further information please contact:
Book Sales Department, Royal Society of Chemistry, Thomas Graham House, Science Park, Milton Road, Cambridge, CB4 0WF, UK
Telephone: +44 (0)1223 420066, Fax: +44 (0)1223 420247,
Email: booksales@rsc.org
Visit our website at books.rsc.org

Artificial Intelligence in Drug Discovery

Edited by

Nathan Brown
Benevolent AI, UK
Email: nathan.brown@benevolent.ai

ROYAL SOCIETY OF CHEMISTRY

Drug Discovery Series No. 75
Print ISBN: 978-1-78801-547-9
PDF ISBN: 978-1-78801-684-1
EPUB ISBN: 978-1-83916-054-7
Print ISSN: 2041-3203
Electronic ISSN: 2041-3211

A catalogue record for this book is available from the British Library

The Royal Society of Chemistry is a charity, registered in England and Wales, Number 207890, and a company incorporated in England by Royal Charter (Registered No. RC000524), registered office: Burlington House, Piccadilly, London W1J 0BA, UK, Telephone: +44 (0) 20 7437 8656.

For further information see our website at www.rsc.org

For general enquiries, please contact books@rsc.org

For EU product safety enquiries, please email books@rsc.org or contact Royal Society of Chemistry Worldwide (Germany) GmbH, Römischer Hof, Unter den Linden 10, 10117 Berlin.

Preface

The origins of this book began in early 2018 when I was approached by the Royal Society of Chemistry (RSC) as they were interested in a new book on artificial intelligence. Since a significant number of people had a renewed interest in drug discovery applications of artificial intelligence and machine learning at that time, I took on the not insignificant challenge of putting together a comprehensive book proposal that covered some of the most important areas on which we focus in drug discovery and new applications of artificial intelligence and machine learning to these challenges.

As the project progressed the book evolved and adapted as new and interesting approaches emerged from the literature and conferences. Artificial intelligence and machine learning are experiencing a fast moving renaissance at the moment and it was not an insignificant challenge to both set the scene of the field for those new to the area, but also provide enough substance regarding new methods to appeal to those heavily involved in the field. What I believe has resulted is a comprehensive and up-to-date overview of where we are, where we are going, and, not least of which, where we have been. My hope is that this provides a useful opportunity for everyone to become more involved in this area and contribute not only to the development of new applications but also to how we successfully apply and adapt those methods to demonstrate impact on drug discovery as a field.

This book would not exist were it not for the huge amount of work from the contributing authors to this book. I am particularly excited by the balance struck between experienced and early career researchers, reflecting the fast moving and partly nascent aspect of the new advances in artificial intelligence and machine learning. A big thank you to all of the authors and co-authors of each and every chapter, and I hope that the wisdom these chapters impart will inspire new and exciting work to be published in the field.

Drug Discovery Series No. 75
Artificial Intelligence in Drug Discovery
Edited by Nathan Brown

Published by the Royal Society of Chemistry, www.rsc.org

A huge thank you to the publishers, in particular Drew Gwilliams and Katie Morrey, for doing their best to keep me on track, remind me of various deadlines, and supporting all of the work as editor where I have needed their help.

Lastly, I would like to thank my partner, Lami, for supporting me through this process, and our two cats – Dexter and Sophie – for largely distracting me from writing by complaining for more food or sitting on my laptop. Although I have written and edited a number of books in the past, no book project is like any other and is fraught with unforeseen challenges that require a concerted effort to overcome when they arise and this book is no exception. With a huge amount of support from the contributing authors, publishers, and my friends and family, we finally have a manuscript that is both timely and holistic in scope.

Nathan Brown
London, United Kingdom

Contents

Drug Discovery Series No. 75
Artificial Intelligence in Drug Discovery
Edited by Nathan Brown

Published by the Royal Society of Chemistry, www.rsc.org

Section 1: Introduction to Artificial Intelligence and Chemistry

CHAPTER 1

Introduction

NATHAN BROWN

BenevolentAI, 4-8 Maple Street, London W1T 5HD, UK
Email: nathan.brown@benevolent.ai

1.1 Introduction

Drug discovery applies to a vast range of technologies in the interest of ushering new chemical entities of disease relevance into the clinic to meet, as yet, unmet patient needs. While many of the technological methods use experiments in so-called "wet" laboratories, the development and application of computational methods, often called *in silico* as opposed to *in vitro* or *in vivo*, have been in wide usage for many decades. Recently, however, a renaissance of Artificial Intelligence, and specifically Machine Learning, approaches have led the vanguard of novel applications to speed drug discovery not only in efficiency gains but also in the development of improved medications for patients.

This book covers a number of different and new approaches to applying artificial intelligence and machine learning methods in drug discovery, sometimes with entirely new algorithms, but often-times building on years of research and applied anew aided by concomitant algorithmic improvements, but also significant improvements in software and hardware that allows hitherto inaccessible quantities of computational power.

The first chapters of this book consider chemical data and learning from unstructured data sources, such as those found in the literature and in patents. One of the most challenging aspects of using Artificial Intelligence in drug discovery is teasing out insights that have been discovered but often published in a way that is not amenable to further analysis – essentially

Drug Discovery Series No. 75
Artificial Intelligence in Drug Discovery
Edited by Nathan Brown

Published by the Royal Society of Chemistry, www.rsc.org

obfuscating this data through publication in formats that cannot be easily read by machines. The first of these chapters looks at the area of chemical topic modelling, which is an unsupervised learning method to extract meaning from natural language, using a methodology called natural language processing. The data extracted permits the clustering of documents based on the co-occurrences of words in these publications thereby permitting an easier approach to grouping documents based on subject potentially obviating the need for manual curation.

Later chapters of the book cover predictive modelling, specifically making use of ligand and protein structure data, respectively. However, it can often be challenging to entirely deconvolute these two data types, since aspects of each are at least implicit, if not explicit, in the other. Ligand-based predictive modelling has a long and distinguished history in drug discovery, going right back to the pioneering work of Corwin Hansch and Toshio Fujita of Quantitative Structure-Activity Relationships (QSARs) back in the 1950s and 1960s. These QSAR equations were derived manually, originally to be used to predict certain molecular properties using mathematical models. In more recent decades, the processes have moved towards using large scale data sources and libraries of molecular descriptors to automate the generation of predictive models using more modern machine learning algorithms. However, it should be recognised that the field of Machine Learning arose contemporaneously with QSARs in the late 1950s.

One of the most important aspects of predictive modelling is not using the predicted values of the endpoint independently, but rather incorporating that with consideration of whether that prediction may be reliable based on the data that are available, or even when the data available is simply insufficient to make any predictions and perhaps the best recommendation is to generate new data to inform our data space. Krstajic offers some important advice in this area with his chapter on the importance of defining the "don't know" answer. This advice is two-fold in importance. Firstly, it encourages diligence in application of our predictive models to drug discovery. Secondly, however, it offers an honesty in modelling to ensure that scientists from medicinal chemistry and computational methods understand more clearly where the data is limited and offer new insights and priorities for chemical synthesis and testing to bolster those datasets.

As protein structure data becomes more prevalent due to protein crystallography becoming a routine assay type in drug discovery, new methods arise that incorporate these more sizable datasets to offer more protein binding site contextual information rather than that merely implied by ligand structure data alone. Methods included in the structure-based predictive modelling chapters are predicting protein-ligand molecular recognition, approaches to using convolutional neural networks in virtual screening, and applying machine learning methods in enhancing the impact of molecular dynamics simulations.

One of the most significant challenges in synthetic organic chemistry, let alone medicinal chemistry and drug discovery, is the design and planning of new chemical syntheses. Given a molecular target, what series of reactions and indeed conditions can be optimised to minimise materials, effort and time to produce the desired results in appropriate yield for its intended purpose in the laboratory? The challenge of synthesis planning has been something of a holy grail in the area of applying artificial intelligence methods to chemistry and drug discovery over many decades, going back to the work of E.J. Corey and others with their retrosynthetic planning working backwards from the desired product to decide which steps should make up parts of the synthesis. However, in recent years, more modern deep learning artificial intelligence, in addition to symbolic artificial intelligence methods have come to the fore. These new methods take advantage of the vast repositories of reaction data held in public databases, proprietary data held by publishers, and internal data sources at chemical companies to rapidly synthesise options of synthetic routes that have been demonstrated to be competitive with human experts. While a significant amount of research remains outstanding to develop these methods into something that is capable of working competitively with human experts, this area of research is one of the most researched areas of computational sciences applied to chemistry in recent years.

The last chapters of this book bring together many of the topics already covered earlier in the book but brings to bear this combination of methods to the more holistic approach of molecular design. The field of drug discovery is itself a multi-objective or multi-parametric challenge. Approved drugs must satisfy requirements of safety and efficacy at the intended dose, but this is itself a convolution of many different requirements and vast arrays of assays that must be considered in the eventual recommendation of not only a drug for human benefit, but also the nomination of a clinical candidate. While methods for molecular design incorporate many different approaches, the heart of the challenge is molecular graph generation: being able to recommend what to make and how to make it to satisfy the various constraints identified as important for the disease area being considered. A number of artificial intelligence and deep learning approaches have been investigated in recent years, including recurrent neural networks, junction tree variational autoencoders and other deep generative models, riding the resurgent wave of artificial intelligence methods, it is important to reflect on the challenges involved in molecular design, independent of the method used to suggest said designs. Medicinal chemistry and drug discovery have a long history of designing new molecules both manually and using automated or semi-automated ways to suggest novel designs. A significant approach is that of matched molecular pair analysis, which has historically been considered as a relatively manual approach investigating molecular replacements around a chemical series of interest. However, in more recent years, computational methods have been developed to abstract the wealth of information from data generated in various organisations to automate a data-driven process to matched molecular pairs giving rise to a vast database of molecular transforms that have concomitant levels of statistical significance in terms of modulating certain properties of interest.

As research progresses in the area of molecular design, many of the emerging methods, together with older and more established approaches, will likely combine into various hybrid systems that permit reliable and reproducible molecular design at scale. Some of the most significant challenges include those in optimising not only potency but also ADMET properties, in addition to the design of molecular structures that can indeed be made effectively and efficiently in the laboratory to test the virtual hypotheses being asked. As we draw closer the theoretical worlds of design and the practical efforts of synthesis and testing of those designed compounds, so will the modern laboratory change and adapt to facilitate more close-coupled drug discovery and development and thereby enhance efficiency and optimisation processes, which is touched on in the last main chapter of this book with a practical vision of how we can apply these methods to democratise the discovery of new chemicals.

CHAPTER 2

The History of Artificial Intelligence and Chemistry

NATHAN BROWN

BenevolentAI, 4-8 Maple Street, London W1T 5HD, UK
Email: nathan.brown@benevolent.ai

2.1 Artificial Intelligence in History

Although relatively new in name, the concepts of artificial intelligence have had a long history throughout millenia of mythology and human progress.[1] The concepts of artificial intelligence, robots and autonomous robots were highlighted by Greek poets from almost three thousand years ago, from the likes of Homer and Hesiod. Talos, mentioned by Hesiod, is an early example of a robotic creature made of bronze created by Hephaestus, the Greek god of blacksmiths and metalworking, amongst other trades. Other examples include those of Galatea and Pandora. The sacred mechanical statues which were built in Egypt and also in Greece, were thought to have the capacity for emotion and wisdom. Trismegistus wrote: "they have sensus and spiritus ... by discovering the true nature of the gods, man has been able to reproduce it".

Automata and mechanical beings continued in their development for many centuries, from Yan Shi presenting the King Mu of Zhou with mechanical men in around the tenth century BCE, to Al-Jazari creating an orchestra that could be programmed to play music by a group of mechanical human beings. In the 1500s, Paracelsus, the Swiss physician, claimed to have created an artificial man from magnetism, sperm, and alchemy, but his claims are somewhat in dispute. Paracelsus, of course, also being the scientist who coined the phrase: "*Sola dosis facit venenum*" (Only the dose makes the poison). This being the

Drug Discovery Series No. 75
Artificial Intelligence in Drug Discovery
Edited by Nathan Brown

Published by the Royal Society of Chemistry, www.rsc.org

foundation of toxicology in drug research. Paracelsus described his creation of an homunculus, or small human being, as such:

> That the sperm of a man be putrefied by itself in a sealed cucurbit for forty days with the highest degree of putrefaction in a horse's womb, or at least so long that it comes to life and moves itself, and stirs, which is easily observed. After this time, it will look somewhat like a man, but transparent, without a body. If, after this, it be fed wisely with the Arcanum of human blood, and be nourished for up to forty weeks, and be kept in the even heat of the horse's womb, a living human child grows therefrom, with all its members like another child, which is born of a woman, but much smaller.

In the early 17th century, Descartes suggested that animals were complex machines, but did claim that mental phenomena were of a different substance.[2] However, Hobbes countered this by suggesting a mechanical and combinatorial theory of cognition, which led him to conclude that "... reason is nothing but reckoning".

Mathematical calculators were subsequently invented and developed by the likes of Pascal and Leibniz, introducing first addition and subtraction, with the later additions of multiplication and division by Leibniz. Leibniz also developed binary representations around that time that became the foundation of modern computation systems.

Of course, theories of artificial intelligence were not only within the realms of science, technology, and philosophy, but also the arts and literature. Swift's *Gulliver's Travels* in 1726, imagined a world that included the Engine, that was able to assist people in the ability to write "Books in Philosophy, Poetry, Politicks (sic), Law, Mathematicks (sic), and Theology, with the least Assistance from Genius or study". The Engine itself being a parody of the *Ars Magna*, and inspiration for Leibniz's mechanism. One of the most significant cultural references to artificial intelligence and artificial life is from Shelley's *Frankenstein*, or the *Modern Prometheus*, which saw Dr Frankenstein create his own homunculus, after Paracelsus, and considered the ethical challenges raised when attempting to create sentient beings. Later, Butler would propose that the theories of Darwin and evolution, would also apply to machines and it would be only a matter of time before they would become conscious and eventually replace the human race.[3]

Moving into the twentieth century, further advances were made, particularly in formal logic and the concepts of thought and automata. Russell and Whitehead published their *Principia Mathematica*, which laid out the concepts of formal logic over three volumes.[4] However, the arts were close to follow, with the first use of the term "robot", by the playwright Capek in his play *R.U.R.* or *Rossum's Universal Robots*.[5]

In 1931, Kurt Gödel demonstrated that sufficiently powerful systems were capable of allowing the formulation of true theorems, meaning that they were able to derive all possible theorems from the axioms. To achieve this,

Gödel built a universal, integer-based programming language, which led him to be called the father of theoretical computer science.

Throughout the 1940s and 1950s, a lot of advances were made in the foundational science of what we today call artificial intelligence. In 1944, the concept of Game Theory was introduced that would be hugely important in the progression of artificial intelligence, in the paper from von Neumann and Morgenstern: *Theory of Games and Economic Behaviour.*[6] Around the same time, Bush published a highly prescient vision of the future where computers were seen as assistants rather than supplanting humanity, in his article *As We May Think.*[7] As the 1950s progressed, many advances took place, with Turing's test: a proposed measure of the intelligence of machines as compared with humans. Asimov published his *Three Laws of Robotics*, defining some of the ethical considerations that should be undertaken by those involved in developing robots. Shannon then published a paper reframing the task of playing a game of chess as a search and optimisation problem.

In 1956, the field of artificial intelligence was finally christened with this name at the Dartmouth Workshop, organised by John McCarthy, Marvin Minsky, Nathan Rochester and Claude Shannon, with McCarthy coining the term artificial intelligence for the entire field.

During the 1960s, a terrific amount of work was conducted into the research of natural language processing (NLP) and tasks such as playing checkers and chess. In addition, the Dendral system was developed to interpret mass spectra on organic chemical compounds. This was not only the first application in chemistry, but also the first successful knowledge-based program for scientific reasoning.

2.2 The Winters of Artificial Intelligence

Although artificial intelligence has had an illustrious existence, it has experienced a number of so-called "winters", where expectations were not met sufficiently with ambition and promise, and this led to a reduction of government and research council funding. The first AI winter took place in the early 1970s, due to a number of unfulfilled promises and evaluation of the field and its progress. In the late 1960s, during the Cold War, a lot of important tasks related to automated machine translation were scuppered due to unforeseen challenges in word-sense disambiguation—understanding the intended definition based on sentence context and structure—which was hugely important in the intelligence community to process vast quantities of information coming from the Soviet Union. This challenge remains to a large extent to this day, but with a number of advances making the task somewhat successful with tools such as Google Translate.

The first AI Winter was also reinforced with the *Lighthill Report* of 1973, reviewing the state of artificial intelligence research in the United Kingdom. The *Lighthill Report* led to a drastic reduction of funding thanks to the report stating the utter failure of AI to achieve its "grandiose objectives". Many of the challenges tackled in AI were seen as toy challenges and the methods would not scale to real world problems due to combinatorial explosion and intractability, which could otherwise be tackled better in other sciences and disciplines. The report led to the complete dismantlement of AI research in England, and research only continued in a few universities: Edinburgh (one of the prime establishments for AI research, Essex, Sussex). This cut in funding led to what James Hendler coined a "bow-wave effect that led to funding cuts across Europe".[8]

2.3 Chemistry Finding Artificial Intelligence

The first recognised artificial intelligence application to chemistry was the Dendral (*Dend*ritic *Al*gorithm) system, begun in 1965. The primary aim of the Dendral system was to study the formation of hypotheses and scientific discoveries. A path to achieving this was the development of an expert system to analyse mass spectrometry data permitting their interpretation. The research was primarily conducted at Stanford University by Edward Feigenbaum, Bruce G. Buchanan, Joshua Lederberg, and Carl Djerassi, with a large team of researchers.[9]

Dendral is not only recognised as the first formal application of artificial intelligence to chemistry, but also as the first expert system since it was capable of automating some tasks of organic chemists, such as problem solving and thereby improving decision making. Dendral was the precursor to many derivatives, MYCIN, MOLGEN, PROSPECTOR, XCON, and Steamer. However, the Dendral system itself consisted of two modules, Heuristic Dendral and Meta-Dendral, and was written in Lisp, which was recognised as the primary programming language for artificial intelligence, since it was seen as highly flexible and easy to rapidly code solutions. Lisp also introduced a number of key concepts to the field of computer science, such as tree data structures, automatic storage management, dynamic typing, conditionals, higher-order functions, recursion, the self-hosting compiler, and the read–eval–print loop.

Heuristic Dendral used a knowledge base derived from expert chemical knowledge, which was then used, in concert with experimental and mass spectrometry data, to identify feasible chemical structures. This algorithm did, as a side effect, make significant contributions to graph theoretic algorithms that could be translated into applications in other domains. Conversely, Meta-Dendral considered a learning system that could learn from chemical structures and their associated mass spectral data, to develop rules that could then be used in Heuristic Dendral to improve its predictions.

2.4 Synthesis Planning

In recent years, artificial intelligence has found significant applications to the challenge of synthetic tractability and synthesis planning. Chemical synthesis systems have later been accelerated in their potential impact by advances in both symbolic AI and deep learning . Indeed, these recent advances have demonstrated their effectiveness through a variation of the Turing test, where expert human organic chemists were pitted against the computer software systems and able to compete well with synthesis plans designed by the human experts.

Although synthesis planning has recently been vastly improved by the advent of new algorithms, the history of artificial intelligence and chemistry goes back many decades to the pioneering work of E. J. Corey. Corey developed a knowledge base of chemical reactions that could be applied in performing a retrosynthetic analysis of target molecules, to determine a potential synthetic route. The analysis conducted would perform a logical process to predict which steps can be best conducted synthetically to achieve the target molecule. Through this process, a tree-based data structure is generated and expanded that permits optimization of the route to suggest back to the user an optimal route.[10]

Corey was not the only chemist working in this area of logical analysis of chemical reactions. Ugi and co-workers also dedicated a significant portion of their time to researching possible new chemical reactions using an extant knowledge base of known reactions. Rather than merely reproducing what is already known in the knowledge base, Ugi's system was able to interpolate within what was known to reveal hitherto unknown suggestions with evidence to support their suitability.[10,11]

Although recent advances in applying artificial intelligence to challenges in synthetic organic chemistry, it is all too easy to forget the advances of those that came before us. A huge amount of significant work was conducted in the laboratories of both Corey and Ugi decades ago which led not only to advances in the chemical sciences, but also to advance algorithmic development in the computational sciences. In this way, it is an important message to ensure one has fully considered the prior art in a field before applying new algorithms before assuming the state of the art fully covers these advances.

2.5 Predictive Modelling of Properties

One of the most important and also valued tasks in drug discovery is the prediction of properties relevant to optimising a small molecule to the clinical candidate stage. The ability to predict effectively a wide range of properties of a small molecule prior to synthesis and testing is hugely valuable, not only in time and cost, but also—possibly even more significantly—optimising through the vast swathes of drug-like chemistry space that are feasible to home-in on the most relevant ones to make and test. It is important to highlight here that the synthesis and testing of a molecular design is first and foremost to drive forward a project in terms of the parameters being optimised. However, it is also important to realise that the synthesis and testing of a compound can also contribute to improving the prediction of properties, thereby having the side-effect of indirectly improving a project, while improving the predictive models in concert with the exploration and exploitation of the chemistry space under consideration.

Reading the recent artificial intelligence and deep learning literature could lead to the conclusion that very little work has been done in this area in the previous years, however, this is far from the truth. One of the earliest attempts to reconcile the properties of molecules compared to its chemical structure go back almost to the beginnings of atomistic theory in the mid-19th century. Alexander Crum Brown formalised the fundamental equation of a structure–activity relationship or model in his work of 1869.[12] Crum Brown posited in this work that the physiological action of a compound is merely a function of its chemical constitution: essentially a mathematical equation that we have spent the past century and a half optimising its definition in a multitude of applications.

The work of Crum Brown paved the way for a significant amount of work almost a century later in the discipline of what was coined mathematical chemistry[13] and the subsequent advances in the definition of Quantitative Structure-Activity Relationship (QSAR) equations by Hansch and Fujita.[14] It is interesting to note that QSAR models emerged around the same time as machine learning and artificial intelligence and the related fields have shared a number of advances over the past 60 years.

As statistical learning methods improved to derive predictive models from dependent and independent data, so too did the chemical and chemoinformatics fields in molecular representation methods. One of the most significant challenges in building predictive models for molecular properties is the molecular representation problem of how best to convert the molecular structure into a format that is amenable to analysis through statistical and machine learning algorithms.[15] While it would be impossible to consider all the different types of molecular descriptors that have been reported over the years—to that the reader is referred to the seminal work of Mauri, Consonni, and Todeschini[16] that is an extensive compendium of many molecular descriptors that have been used in a variety of applications—it is perhaps useful to outline a few of the fundamental types of molecular descriptor that are used. Molecular descriptors can roughly be partitioned into knowledge-based and information-based descriptors. Knowledge-based descriptors are based on pre-existing knowledge such as molecular weight—the sum of sums of the empirical formula—to what are predictive models in their own right, such as calculated logP, or the calculated octanol–water partition coefficient. Quite often the latter is calculated using existing data to derive a computational model, where the descriptors in those models are frequently atomic or fragment contributions derived from the experimental data. Information-based methods tend to encode the information present in a chemical structure, without any explicit knowledge, to generate a molecular descriptor that is abstract but frequently holistic in its nature. Molecular fingerprints are such a class of these molecular descriptors and have been demonstrated on a number of occasions to be an effective molecular representation of chemical structures for model generation.[17,18] Typically, molecular descriptors fall on a trade-off surface in appropriateness to predictive modelling based on the mode in which the subsequent models are to be applied.[19] Knowledge-based descriptors are typically better for generating diagnostic models that enhance interpretability, while information-based descriptors are highly effective at generating accurate predictions, but are significantly more challenging to interpret.

2.6 Summary

The fields of chemistry and artificial intelligence have a long and intertwined history. Indeed, chemistry led to the foundations of a number of different techniques and methods that have since found favour in artificial intelligence. From Crum Brown's definition of a predictive modelling in the mid-19th century onwards, together with formalisation of many aspects of mathematical graph theory, through early research in QSAR models and symbolic artificial intelligence in the 1960s, chemistry has been heavily involved in many advances of artificial intelligence. A number of these advances have historically been due to the primary data structure in chemical information systems being graph representations. Therefore, it logically follows that many of the algorithms developed for chemical information storage, processing and analysis have been implicitly linked with advances in graph theory.

This chapter has attempted to demonstrate and perhaps correct some of the assumptions about artificial intelligence and machine learning and its connection to drug discovery. Both artificial intelligence and machine learning have a long and illustrious relationship with drug discovery and many significant advances have been made thanks to numerous endeavours over many decades. It is important in this situation, particularly when new methods that attempt to disrupt a field and occur frequently in relation to where they sit on the Gartner hype cycle, that the methods are already in common usage in many drug discovery projects throughout the world and have played a significant role in advancing small molecule drug discovery to advances and success in the clinic.

References

1. A. Mayor, *Gods and Robots: Myths, Machines, and Ancient Dreams of Technology*, Princeton University Press, 2020.
2. P. McCorduck and C. Cfe, *Machines Who Think: A Personal Inquiry into the History and Prospects of Artificial Intelligence*, CRC Press, 2004.
3. S. Butler, *Erewhon, or, Over the Range*, DOI: 10.5479/sil.1036699.39088016476525, 1872.
4. I. Grattan-Guinness, A.N. Whitehead and Bertrand Russell, principia mathematica, first edition (1910–1913), *Landmark Writings in Western*

Mathematics 1640–1940, 2005, pp. 784–794. DOI: 10.1016/b978-044450871-3/50142-x.
5. K. Capek, *Rur and War with the Newts*, S.F. Masterworks, 2011.
6. L. O. Kattsoff, *Theory of Games and Economic Behavior*, ed. J. von Neumann and O. Morgenstern, Princeton University Press, Princeton, N. J., 625 pp. $10.00, *Social Forces*, 1945, pp. 245–246.
7. V. Bush, *The Atlantic*, 1945, Available at: https://www.theatlantic.com/magazine/archive/1945/07/as-we-may-think/303881/ (Accessed: 7 May 2020).
8. J. Hendler, Avoiding Another AI Winter, *IEEE Intell. Syst.*, 2008, 2–4.
9. J. Lederberg, How DENDRAL was conceived and born, *Proc. ACM Conf. History Med. Informatics*, 1987, DOI: 10.1145/41526.41528.
10. D. A. Pensak and E. J. Corey, LHASA—Logic and Heuristics Applied to Synthetic Analysis, *ACS Symp. Ser.*, 1977, 1–32.
11. J. Dugundji and I. Ugi, An algebraic model of constitutional chemistry as a basis for chemical computer programs, *Fortschritte Chem. Forschung*, no date, 19–64.
12. A. C. Brown, A. Crum Brown and T. R. Fraser, XX.—On the Connection between Chemical Constitution and Physiological Action. Part II.—On the Physiological Action of the Ammonium Bases derived from Atropia and Conia, *Trans. Royal Soc. Edinburgh*, 1869, 693–739.
13. A. T. Balaban, Reflections About Mathematical Chemistry, *Found. Chem.*, 2005, 289–306.
14. C. Hansch, P. P. Maloney, T. Fujita and R. M. Muir, Correlation of Biological Activity of Phenoxyacetic Acids with Hammett Substituent Constants and Partition Coefficients, *Nature*, 1962, **194**, 178.
15. T. Hastie, R. Tibshirani and J. Friedman, *The Elements of Statistical Learning: Data Mining, Inference, and Prediction*, Springer Science & Business Media, 2013.
16. A. Mauri, V. Consonni and R. Todeschini, Molecular Descriptors, *Handbook Computat. Chem.*, 2017, 2065–2093.
17. N. Brown, B. McKay and J. Gasteiger, Fingal: A Novel Approach to Geometric Fingerprinting and a Comparative Study of Its Application to 3D-QSAR Modelling, *QSAR Combinatorial Sci.*, 2005, 480–484.
18. P. Gedeck, B. Rohde and C. Bartels, QSAR–how good is it in practice? Comparison of descriptor sets on an unbiased cross section of corporate data sets, *J. Chem. Inf. Model.*, 2006, **46**(5), 1924–1936.
19. N. Brown and R. A. Lewis, Exploiting QSAR methods in lead optimization, *Curr. Opin. Drug Discovery Dev.*, 2006, **9**(4), 419–424.

Section 2: Chemical Data

CHAPTER 3

Chemical Topic Modeling – An Unsupervised Approach Originating from Text-mining to Organize Chemical Data

NADINE SCHNEIDER,*[a] NIKOLAS FECHNER,[b] NIKOLAUS STIEFL[a] AND GREGORY A. LANDRUM[c]

[a] Novartis Institutes for Biomedical Research, Global Discovery Chemistry/CADD, Novartis Pharma AG, CH-4002 Basel, Switzerland; [b] Novartis Institutes for Biomedical Research, Informatics/Chemistry Information Systems, Novartis Pharma AG, CH-4002 Basel, Switzerland; [c] KNIME AG, Hardturmstr 66, 8005 Zurich, Switzerland
*Email: nadine-1.schneider@novartis.com

3.1 Introduction

A key step in extracting and understanding the information contained in data is structuring the data. This is especially the case when dealing with large data collections that lack obvious or pre-defined organizational attributes. When approaching this problem, humans typically prefer high-level semantic structures over purely descriptive attributes of the data. For instance, one would rarely describe a book by stating its percentage of verbs or adjectives instead of trying to summarize the plot or the author's intention. In chemistry, we tend to describe large sets of molecules using the types of compounds it contains, *e.g.* peptides, nucleotides or kinase inhibitors, and

Drug Discovery Series No. 75
Artificial Intelligence in Drug Discovery
Edited by Nathan Brown

Published by the Royal Society of Chemistry, www.rsc.org

not by the number of atoms and bonds the molecules contain. Medicinal chemists often think in chemical series – sets of structurally related compounds – or in protein targets for which a compound was designed. Being able to automatically organize sets of molecules in a similar way would allow chemists to more naturally explore and retrieve information and knowledge from those sets. The typical tools applied to help with this challenge are unsupervised machine-learning/statistical approaches like clustering, principal component analysis (PCA) and, more recently, autoencoder-based approaches[1] or probabilistic methods like Gaussian Mixture Models (GMM).[2] All of these unsupervised methods reduce the dimensionality of the data in order to try and enable a better understanding of the data and, ideally, the discovery of a hidden underlying structure.

In this chapter, we introduce and discuss an unsupervised approach called "chemical topic modeling" which we adopted from the text-mining field.[3] The probabilistic framework used in topic modeling provides an intuitive and meaningful way to organize large data sets. In the first part of the chapter the underlying probabilistic approach – in our case Latent Dirichlet Allocation (LDA)[4,5] – is described at a high level to introduce the basic concepts and ideas behind topic modeling and its extensions. Furthermore, the relation of this approach to other well-known machine-learning methods will be summarized. The second part of the chapter brings in the transformation and application of LDA to handle chemical data. In concrete examples the advantages of chemical topic modeling are shown and further applications are discussed. The source code for chemical topic modeling along with the Jupyter notebooks containing the experiments discussed in this chapter can be found in GitHub.[6]

3.2 Topic Modeling and LDA

The wealth of prior knowledge and understanding of context that a human applies to create a semantic structure when describing a book or set of books is not at all straightforward to put into an algorithmic framework. However, one can assume that the words found in the books are not generated totally arbitrarily, but come into existence within a context. For instance, even though a fantasy novel and an engineering textbook may have been written in the same language, drawing on the same underlying vocabulary, the distributions of words used in each will likely be quite different. Taking the example one step further, the distribution of words used will often differ in a single book depending on the chapter or topic covered. Thus, at least to some degree, it is possible to draw conclusions about the context of a text from its statistical composition, and books from the same genre would be expected to be covered by similar word distributions. Given a set of arbitrary texts, while the genres themselves might not be explicitly defined, the word distributions are and can be used to organize those books in a way that implicitly captures the genres. This interplay of word distributions and context allows interpretable connections to be made and provides the foundation of topic modeling.

There exist several mathematical and algorithmic frameworks for constructing such topic models. They usually share the common idea that the data was generated in a context and that information about the context is latently present in the data. When provided with enough data the most common contexts can be reconstructed.

The general idea that texts are created within a context lies at the center of one of the most commonly used and successful topic modeling approaches – Latent Dirichlet Allocation.[4,5,7–9] Imagine an author writing an English novel. The words used will be drawn from the known English vocabulary. Depending on the context (or topic) each word from the vocabulary has a different likelihood to be used in a text. One way of describing the context computationally is as a specific unordered collection of words where each word can appear multiple times – a so-called "bag-of-words". Usually the author has a pre-defined set of topics in mind from which he or she creates the novel, in our formulation these topics provide the bags-of-words the author uses when writing. In the LDA approach this is modelled fairly directly as a stochastic generative process: The model assumes a fixed set of bags-of-words (topics), each of which has a specific distribution of words (bag-of-words). When creating a text in this model, first a distribution over the bags-of-words is randomly defined. Then, for each word that is to be added to the text, one of the topics is randomly drawn from this distribution, and a word is randomly drawn from the bag-of-words corresponding to that topic.

Obviously, such a generative model is more of a thought experiment than something that would actually be applied. However, it provides a powerful framework for topic modeling when conceptually inverted: Instead of starting with a distribution over topics and generating a collection of texts, the real starting point now is the collection of documents that is assumed to originate from the stochastic process described above. Given a large number of such documents, it is possible to mathematically infer, or at least approximate, the most likely distributions of topics and words (within the topics) that would have been used to generate said documents. Those distributions further allow the estimation of which topics have most likely contributed to a given document and to which extent they have contributed. Put differently, LDA enables the description of each of the documents in terms of how relevant each of the inferred topics are to them. Figure 3.1 provides a schematic illustration of the different facets of topic modeling.

3.2.1 The Mathematical Framework of LDA

The mathematical framework of LDA relies on two sets of distributions, one set with distributions over topics for the corpus (collection) of documents (document–topic distribution) and one set with distributions over words for the topics (topic–word distribution). Here, we will only introduce the ideas

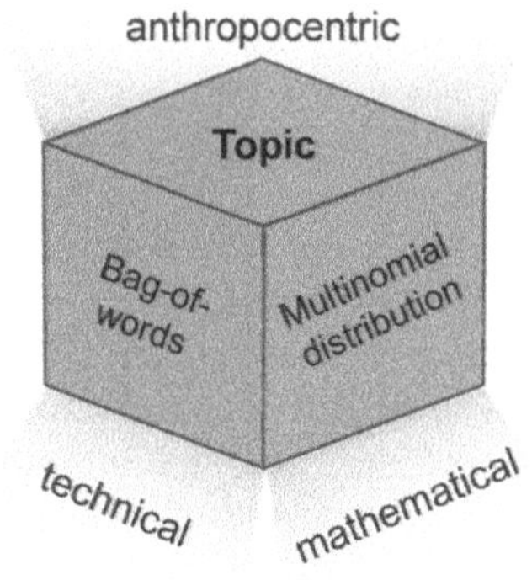

Figure 3.1 The different facets of a "topic" in topic modeling. Looking from the anthropocentric perspective, the general meaning of a text/document is usually described as a "topic". From the technical topic modeling perspective it is described as a bag-of-words and mathematically it can be formulated as multinomial distribution that was drawn from a Dirichlet distribution.

on a very high level based on the original publication by David Blei and coworkers,[4] from which we have also adopted the mathematical notations.

Since each document–topic and topic–word distribution is a multinomial distribution over either the topic set or the word vocabulary, these distributions are assumed to be drawn from two Dirichlet distributions Dir(α) and Dir(β). The Dirichlet distribution is the conjugate prior to the multinomial distribution and hence samples drawn from it correspond to parameterizations of multinomial distributions. The parameterization of these two Dirichlet distributions are the main non-data-driven influences on an LDA topic model. The document-topic Dirichlet prior Dir(α) is parameterized by the number of topics K that is assumed to exist and a Dirichlet parameter vector α that controls the topic mixture of documents (*i.e.* documents are assumed to belong to only a few very pronounced topics or many equally relevant topics). Likewise, the Dirichlet prior Dir(β) is parameterized by the size of the word vocabulary and the number of topics and an analogous parameter β that in this case controls the mixture of words per topic (*i.e.* topics are based on a few very relevant words or on a more diverse word collection).

The introduced stochastic generative process above, can be reintroduced in this framework under the assumption that for each document that is supposed to be generated, a multinomial distribution over the topics (*i.e.* bags-of-words) is sampled from the Dirichlet distribution Dir(α). This distribution describes how the different topics are contributing to this document. Likewise, each of these topics is assumed to be a multinomial distribution over words that was sampled from the Dirichlet distribution Dir(β). In contrast to document-topic distributions, these topic-word distributions are sampled for the whole document corpus and not specifically for each document. A document is then generated word-by-word by drawing first a topic from the document-topic distribution for this document and then drawing the word itself from the topic-word distribution for the selected topic. Here, it is important to understand that words are not unique for topics since this has an important implication for reconstruction of document-topic and topic-word distributions later on.

The framework of assigning topics to word positions in the documents has an important implication: it allows for each word to have been sampled from a different topic (*i.e.* bag-of-words). Consequently, the obtained document would not have been generated from one topic only but from a mixture of topics. This is an important attribute that makes an LDA model a mixed-membership model and forms a key distinction from other data organization concepts like cluster membership assignments.[4]

Of course, computing the probability distribution over documents when given topic and word distributions is usually not of utmost interest. In real world scenarios the documents are the accessible observations whereas the topic and word distribution must usually considered to be hidden variables that are not directly obtainable. Thus, the posterior distribution of the topics over the documents and the words within the topics given the observed document corpus would be more relevant to obtain. This inference of the posterior distribution of the document-topic distributions and the topic-word distributions is the central component of topic modeling using LDA.[4] Computing this posterior distribution, as shown by Blei,[4,7] is only tractable using approximate inference techniques, given a corpus of documents and two defined Dirichlet prior distribution Dir(α) and Dir(β).

3.2.2 Advanced Topic Modeling Extensions

The probabilistic model of the Latent Dirichlet Allocation is based on the assumption that the topics that comprise the documents are infinitively exchangeable and mutually largely independent.[4] However, as Blei and Lafferty[10,11] argue, these assumptions are not necessarily true for real-word texts in which the co-occurrence of specific topics within the same document is likely not independent and also topics may evolve over time. To address these limitations, they introduced modifications to the LDA that make use of log-normal distributions instead of Dirichlets. These log-normal distributions, which form the foundation of the *Correlated Topic Models*,[10] are better suited to capture covariance between topics. Moreover, log-normal distributions can also be used to chain the parameterizations of topics at discrete, sequential time frames and hence to model the evolution of topics over (discrete) time, which is introduced as *Dynamic Topic Models* in Blei and Lafferty.[11] Another extension to the original topic modeling idea is to incorporate additional information about documents into the topic model in order to form a *Supervised Topic Model.*[12] Here, the idea is not to learn the topics directly in a supervised machine-learning fashion, but to take into account an existing response variable that is available for each document in order to identify latent topics that are connected to the response variable (*e.g.* can be used to predict those). To achieve this, McAuliffe and Blei[12] suggested an extension to LDA in which the response variable is jointly modeled with the topics and used to inform a maximum likelihood estimation of the parameters of the topic model.

Although all these extensions to LDA might also provide interesting opportunities for chemical topic modeling, the utility of these extensions for this purpose has not yet been thoroughly investigated and hence will not be discussed further in this chapter.

3.2.3 Topic Modeling and Its Relation to Other Machine Learning Methods

As introduced earlier, topic models are an approach to infer the latent structure in textual data. The identification of such latent structure normally results in a lower-dimensional representation of the data. This type of dimensionality reduction is a frequently re-occurring concept in many machine learning techniques. Some are very directly inter-related, both in terms of the mathematical framework and methodological evolution, for instance latent semantic analysis (LSA) or latent semantic indexing (LSI),[13] an early approach to topic modeling, is based on singular value decomposition and hence bears some close conceptual similarities to factor analysis and principal component analysis. The close relationship between factor analysis, latent semantic analysis and LDA as a Bayesian extension to probabilistic LSA (pLSI)[14] has been discussed and compared by Péladeau and Davoodi.[15]

The reduction of data dimensionality also plays a central role in many modern deep learning architectures. For instance, Hinton and Salakhutdinov[16] describe in one of the early "deep" learning publications on "Reducing the Dimensionality of Data with Neural Networks" how a network of stacked Restricted Boltzman Machines (RBM) is used to construct a deep autoencoder to achieve this goal and compare the results to PCA. In this case the connection between topic models and RBMs as a kind of neural network is only indirectly present by both approaches sharing the objective on data dimensionality reduction. However, in a later publication, the same authors[17] introduce an extension to RBMs to be used directly for topic modeling and compare it to LDA.

The dimensionality reduction perspective is of course only one possible point of view on topic models. Another interesting angle and analogy is discussed by Gerlach *et al.*[18] who compared the ideas around topic modeling with approaches from network analysis and community detection and shows that adopting approaches from those domains for topic modeling can have advantages over LDA.

3.2.4 Topic Modeling in Different Scientific Disciplines

Apart from the development and usage of topic modeling in the natural language processing field, topic modeling has been introduced and successfully employed in many other different disciplines like digital humanities or computational social sciences.[19,20] In these areas it allows scientists to explore the meaning and the structure of large text corpora over a long time frame. Before these methods were available it was a rather time-consuming work limited by the number of books or articles that humans can read to retrieve the meaning of information from them.

Furthermore, due to the ability of topic modeling to organize large amounts of data in a very intuitive way it has also gained prominence in the area of natural sciences. Here the focus is not so much on text as on other types of data like, for example, proteomics,[21] metabolomics,[22–24] transcriptomics[25] or genomics data.[26,27] For example, Rogers *et al.*[24] showed that LDA can be applied to mass spectrometry fragmentation spectra of metabolomics data sets to extract biochemically relevant features in an unsupervised manner. These features or substructures can be used then to support *de novo* structural annotation of molecules which is a real challenge in that area.[22–24] Regarding genomics data, Chen and coworkers,[26] for example, showed that the topic models could retrieve the correct grouping into serotypes of bacteria and that these models are more successful in dividing patient cancer data into specific subtypes than conventional clustering methods. A more detailed overview of the use of topic modeling in the context of biological data can be found in the review article by Liu and colleagues.[28] With this background, introducing topic modeling to the field of chemistry seems to open new opportunities to organize chemical data sets. Throughout the rest of this chapter the application of topic modeling to chemical data will be discussed and demonstrated.

3.3 Chemical Topic Modeling

In drug discovery and especially in medicinal chemistry, novel chemical matter in the form of large screening libraries of compounds is of major interest. Over the past decades proprietary, commercially accessible and public compound databases like ZINC,[29] PubChem,[30] ChEMBL,[31,32] and also virtual libraries like GDB17[33] have grown to encompass billions of molecules. Furthermore, novel experimental technologies, like DNA encoded libraries[34] are novel sources of very large subsets of chemical space. They allow the generation and screening of billions of novel molecules. Similarly, automated synthesis labs that use sets of well-established reactions in combination with large numbers of diverse building blocks[35] provide access to enormous quantities of novel chemical matter. Common tasks like exploring those sets to understand which areas of the chemical universe are covered, re-organizing the sets to build smaller, more focused sub-libraries or triaging the results from screening campaigns on these sets require approaches that can deal with these massive amounts of data. To allow humans, typically medicinal or computational chemists, to understand and interrogate this data, we need intuitive and sophisticated ways to organize and browse it. In the past methods like K-Means[36] or other clustering approaches (*e.g.* see ref. 37–40) were typically applied for these tasks. These methods are powerful, but they have their inherent weaknesses.[41] For example, they typically require the definition of a similarity measure, not without controversy for chemical compounds. In addition, their reliance on a pairwise similarity/distance matrix places severe limits on the data set size they are able to manage. Furthermore, the results of many of these approaches are often hard to interpret and may be critically sensitive to changes in the starting conditions. Finally, depending on the chosen parameters especially "single-membership" clustering approaches could end up with very large diverse clusters or many disconnected singletons. Given the success that topic modeling has had in many other areas as a tool for organizing large amounts of data in an intuitive way it is an interesting alternative to clustering algorithms in chemistry.

The basic idea of chemical topic modeling is to regard molecules as the "documents" in the topic model. When working with text, documents are split into words or groupings of words (n-grams) to build the basis – the bag-of-words – for the model. As this transformation is not immediately obvious for molecules, different methods will be discussed in Section 3.3.1. In general, molecules need to be decomposed into smaller subgraphs or substructures to construct a "bag of molecular fragments". After decomposing the molecules into fragments, a matrix – containing the molecules as rows and the counts of each fragment as columns – can be created and used as the input for the topic model. Application of LDA to this matrix results in two new matrices: "the topic model" which is a probability distribution for topics over fragments and "the hidden thematic structure of a data set" which is a probability distribution for molecules over topics. The overall principle of chemical topic modeling is outlined in Figure 3.2 and more details of the process are discussed in Section 3.3.2 of this chapter.

3.3.1 Feature Representation for Chemical Topic Modeling

In Section 3.2 the idea of a topic model was introduced for the classical application in the natural language processing field. The representation of the input features for the model there is rather obvious and simple: documents are tokenized into words. In chemistry the featurization of molecules is more complex and still an active field of research in cheminformatics (see for example ref. 42,43). Nevertheless many different methods were developed and successfully applied over the past decades like Atom Pair fingerprints,[44] Topological Torsions[45] or Morgan/ECFP[46,47] for representing the molecular topology (see below). These fingerprint representations of molecules are often used to calculate chemical similarity or as input for machine learning models. Hull and coworkers[48–50] already transformed molecules to an input format suitable to apply Latent Semantic Indexing (LSI)[13] – a method originating from the text-mining field and which is related to topic modeling

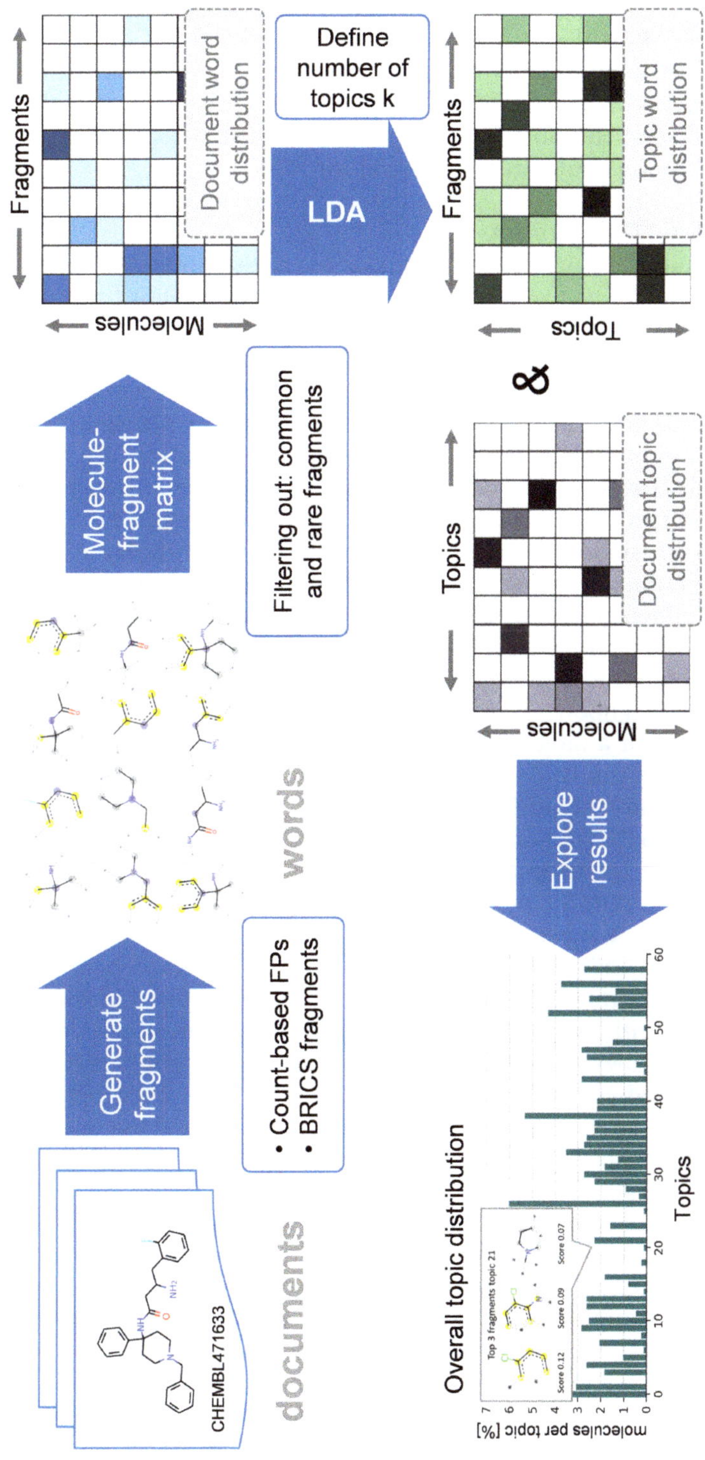

Figure 3.2 Chemical topic modeling workflow. To make the connection to chemical topic modeling, the terms used in the context of topic modeling of text documents are shown as grey text. Reproduced from ref. 3 with permission from American Chemical Society, Copyright 2017.

(see Section 3.2.3). They used Atom Pairs[44] and Topological Torsions[45] to represent the molecules.

To transform a molecule to an input format that is compatible with a topic model the molecule also needs to be tokenized. This means that the molecule is deconstructed into smaller pieces so-called substructures or fragments. Different techniques can be used for this:

- **Chemical 2D fingerprints**
  Chemical 2D fingerprints mainly come in two different flavors: *path-based fingerprints* like Daylight,[51] Atom Pairs[44] or RDKit fingerprints,[52] or circular fingerprints like Morgan/ECFP fingerprints.[46,47] The RDKit fingerprints are created for example by generating all possible paths of certain lengths starting at each heavy atom in the molecular graph (see Figure 3.3 top). For each atom and bond within the path, different properties (*e.g.* element, bond order, hybridization, *etc.*) are hashed to allow discrimination of different chemical substructures/environments from each other. *Circular fingerprints* like the Morgan/ECFP fingerprint are constructed by iterating several times over each heavy atom in the molecule and including information about the neighboring atoms up to a predefined radius in each iteration (see Figure 3.3 bottom). As for RDKit fingerprints, different chemical properties depending on the exact definition of the fingerprint type are stored for the central atom and its neighbors, providing a rather detailed description of the subgraphs of a molecule. A more detailed description of chemical 2D fingerprints can be found in.[53]
- **Fragmentation algorithms**
  In contrast to the rather generic substructures that are generated for a molecule by using a fingerprinting algorithm, fragmentation algorithms like BRICS[55] or RECAP[56] cut the molecule at certain bonds to create smaller substructures. These approaches typically do not break ring bonds, thereby preserving chemical ring systems which often are important parts of the molecule. For defining the bond-breaking rules these algorithms sometimes make use of chemical reaction types/principles/rules so that the resulting fragments might be synthetically re-combined to yield new molecules. A big difference between the fragments derived from 2D chemical fingerprints and the ones created by a fragmentation algorithm is that the latter do not result in overlapping substructures.

The many different methods to generate fragments or tokens from a molecule allow us to explore and compare the influence of the representation on the resulting topic model.[3] In the natural language processing field researchers experimented with different tokenization methods by creating n-grams of the documents instead of using single words.[57] This is related to the overlapping fragments for molecules tokenized using 2D chemical fingerprints. It can also be compared with the convolution of

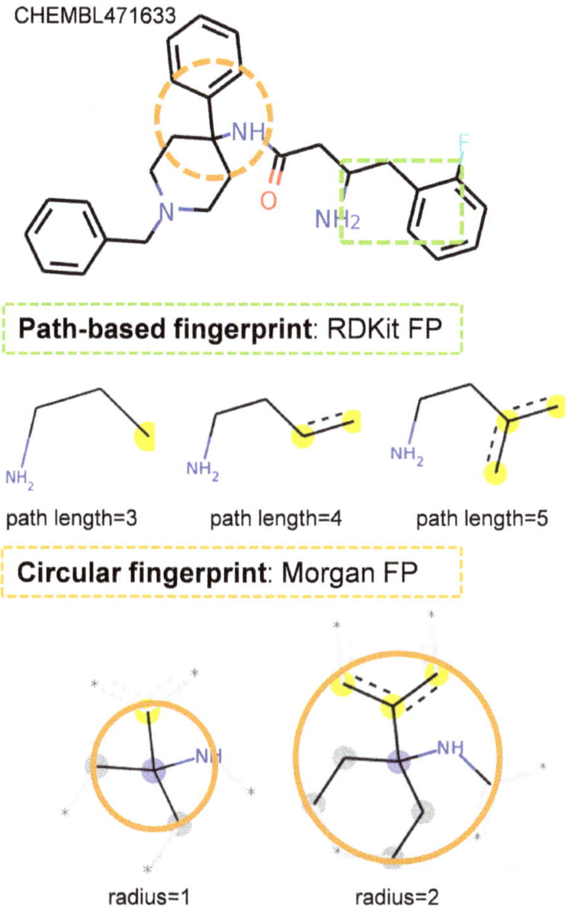

Figure 3.3 Fragment generation by using 2D chemical fingerprints. Top: example molecule from ChEMBL[31,32] (CHEMBL471633) to show circular (orange) and path-based (green) fingerprint fragment generation. Middle: RDKit FP fragments with a path length between three and five bonds. Linear and branched paths are possible. Bottom: Morgan/ECFP FP fragments with radius 1 and 2. Grey highlights: atoms in aliphatic rings. Yellow highlights: atoms in aromatic rings. Grey lines: bonds to atoms which are not part of the fragment. The fragment visualization was done using RDKit.[54]

images in deep convolutional neural nets, an approach which is known to improve the performance in image recognition.

Typically, in cheminformatics applications 2D chemical fingerprints are encoded in the form of bit vectors. Here, only the information whether a feature/substructure is present or not is available. For topic modeling it is important to keep the exact number of occurrences of each feature in a molecule.[3] This information is used to prepare the input matrix for the topic model after the fragmentation of the molecules. In this matrix, the rows represent the molecules M in the data set and the columns represent the different fragments F. Each cell in the matrix captures the number of occurrences of the respective fragment in the molecule (compare Figure 3.2).

When working with chemical 2D fingerprints it is often useful and sometimes mandatory to fold the resulting fingerprint vectors to a certain size like 2048 or 4096 bits to be able to use them as input for machine learning models. Depending on the size of the fingerprint and the fingerprint type, the folding can lead to bit collisions that cause different fragments to be assigned to the same bit.[58] This introduces a certain amount of noise into the models which usually can be ignored if the data set is large enough but which complicates interpretation/description of models. In the case of chemical topic modeling, where a primary goal for the model is to offer a way to explore, visualize and interpret the data, these collisions would be misleading and should be avoided to the extent possible. To overcome this, fingerprints of unrestricted length are generated.[3] This may result in a rather large and sparse input matrix depending on how many very specific fragments occur in the different molecules of the data set. In the natural language processing field a common approach to reducing the size of the input matrix is TF–IDF filtering (Term Frequency–Inverse Data Frequency).[59] It is a way to reflect how important terms are in a collection of documents. In general, terms that have very high or very low document frequencies are usually ignored in the final vocabulary.[8] In chemical topic modeling a similar approach was applied: fragments occurring in more than 10% of the molecules or those that are found in less than 1% of the compounds (0.1% for the larger data sets) were ignored for the final input matrix[3] (compare Figure 3.2).

3.3.2 Creating and Interpreting a Chemical Topic Model

Following the chemical topic modeling workflow in Figure 3.2 the resulting pre-filtered molecule-fragment matrix ($M \times F$) is used as input for the LDA algorithm. The only parameter which needs to be selected before building the chemical topic model is the number of expected topics K (the Dirichlet priors α and β can be changed as well, otherwise default values of $1/K$ are taken). Since a (chemical) topic model is an unsupervised approach there is no optimal value for K and the model builder can freely choose depending on the desired granularity of the final model: a larger number of topics usually leads to more specific topics while a small number of topics forces the model to more strongly generalize. Extreme scenarios – either very few or very many topics – can lead to topics that are hard to interpret. Usually choosing a slightly larger number of topics than one would expect/desire might give more appropriate results than overly restricting the model. Furthermore the runtime for fitting the model increases linearly with the number of topics[3] (for an analysis of the time complexity of the LDA algorithm see ref. 4).

The LDA algorithm results in two different matrices: a probability distribution over fragments for each topic – the fitted model – and a probability distribution over topics for each molecule of the data set (the applied model). These matrices are normalized to contain a probability for a fragment given a topic and a probability for a topic given a molecule, respectively. Both can be used to explore, visualize and interpret the chemical topic model. The topic–fragment matrix (see Figure 3.2) can be used to identify fragments that are associated with a certain topic (Figure 3.4 top). This information can also be used to highlight a topic for a molecule by emphasizing the respective fragments in the molecule (Figure 3.4 bottom). This is a distinct advantage of chemical topic modeling compared to most other clustering methods. Here, for clustering, interpretation is difficult because clusters are defined solely by the compounds that compose them and their respective similarities. Figure 3.4 exemplifies a visualization of a topic model. Besides the most important fragments for a certain topic, the most likely topic k_{max} for a molecule can be extracted from the document-topic

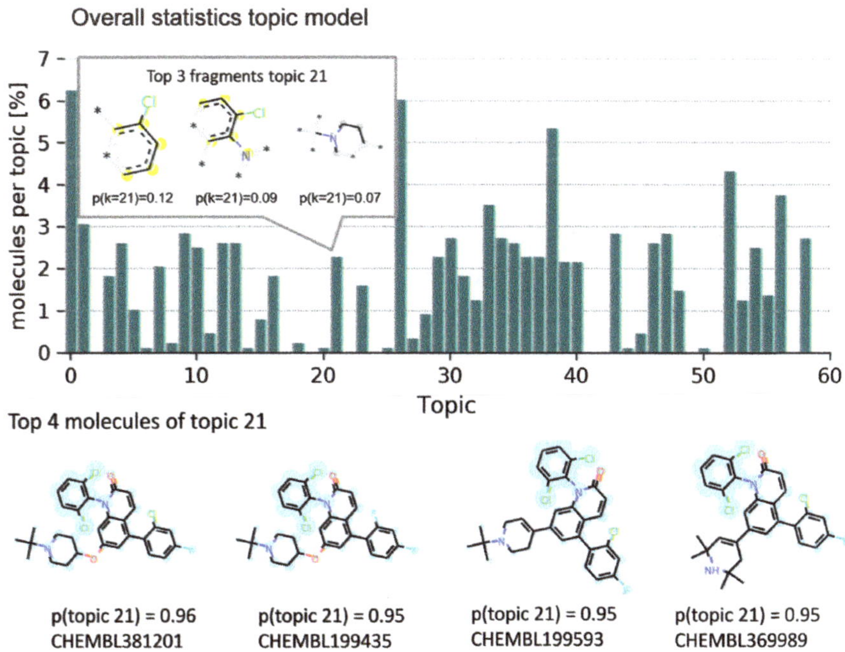

Figure 3.4 Visualization for an example topic model (60 topics). For molecular fragmentation the Morgan FP algorithm was used. Top: for each molecule the most probable topic is determined and the molecule is assigned to this topic. The distribution of this assignment is depicted in the histogram indicating the fraction of molecules associated with a certain topic. Top left: top three fragments of topic 21 are shown along with their probability to be associated with this topic. Bottom: top four molecules of topic 21 ranked based on their topic probability; within the molecules fragments with a high probability for topic 21 are highlighted in light blue (the larger the highlight radius the higher the probability). Reproduced from ref. 3 with permission from American Chemical Society, Copyright 2017.

distribution and is used to plot the distribution of the molecules in the different topics (see histogram Figure 3.4 top). Furthermore for each topic the molecules with the highest probability for that topic can be analyzed and visualized (Figure 3.4 bottom). The example in Figure 3.4 nicely demonstrates that the molecules in the same topic are structurally related to each other. Highlighting the fragments with a high probability for that topic in the molecules easily allows identification of the major structural motif(s) in the topic. The probabilities for the molecular fragments the topic model provides and the topics themselves constitute a unique advantage over other techniques and a good starting point for large-scale data exploration. This is also discussed in Section 3.4 of this chapter.

3.3.3 Evaluation of a Chemical Topic Model

As with all other unsupervised learning algorithms, the quantitative evaluation of chemical topic model quality is difficult. For topic modeling and other unsupervised probabilistic models, perplexity is often used as a measure of model quality.[4,57,60] Usually the perplexity is calculated on a held-out test set to probe how well the model performs on unseen documents. Lower values of perplexity indicate better generalization performance of the model. In general, perplexity is difficult to interpret and sometimes anti-correlates with human judgement.[60] An alternative approach to evaluate the performance of unsupervised methods is to combine them with a supervised learner. Since unsupervised methods are often used for representation-learning tasks, the latent variables derived from these methods can be used as input for a related learning task. The performance of the supervised model gives some insight about the quality of the unsupervised model (see for example ref. 61,62). Another method to probe if an unsupervised model is working as expected (which already assumes an expectation or ground truth) is to pre-label the data and estimate how well the approach is able to organize/categorize the data according to the pre-defined labels. The challenge/shortcoming of this evaluation method is that a ground truth categorization with labels needs to be available. If this is available, different metrics of how well the model recapitulates the base categorization can be calculated such as adjusted mutual information (AMI)[63] or homogeneity or completeness of clusters.[64] A similar approach was used to evaluate chemical topic models.[3] Here, labels were chosen for the molecules in a data set by relying on the protein target these molecules were designed for. The data set of molecules for this was compiled from the ChEMBL database[31,32] which provides experimental data like protein targets for around 1.6 million molecules. The aforementioned data set was originally created by Riniker and coworkers to evaluate ligand-based virtual screening techniques[65] and was slightly modified for the evaluation of chemical topic modeling.[3] It contains 880 molecules belonging to 36 chemical series (sets of molecules that are structurally related) that were measured against, and presumably designed to interact with, a certain

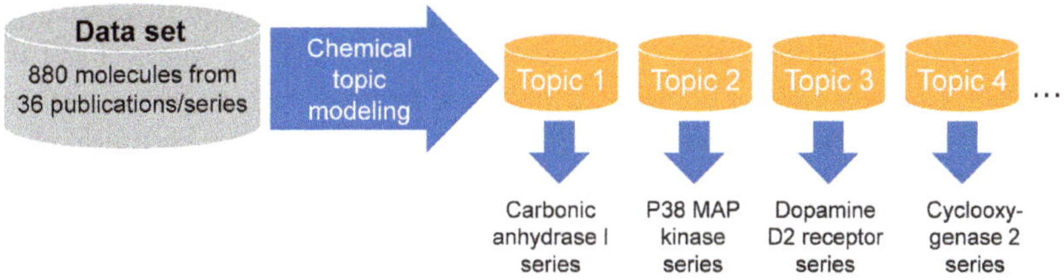

Figure 3.5 Evaluation of chemical topic modeling using pre-defined labels: comparing the grouping resulting from the chemical topic modeling to the original chemical series labels (ground truth labels).

protein. After creation of a chemical topic model using the workflow shown in Figure 3.2, the most likely topic k_{max} is extracted for each molecule in the data set. The grouping of the molecules based on k_{max} is compared to the grouping based on the protein target (see Figure 3.5). For each protein target, the predominant topic – based on the number of molecules assigned to this topic – is determined and the recall and precision are calculated for each target. In this context, this means how many molecules of a series were assigned correctly to this major/predominant topic (recall) and how many molecules assigned to that topic belong to the correct protein target (precision). This approach allows us to quantitatively evaluate how well the inferred topics match the concepts of medicinal chemists. Additionally, the approach can be used to explore and optimize different parameters important for the topic model, for example the number of topics and the fragmentation methods.[3]

In summary, extensive evaluation of topic modeling on chemical data resulted in the following findings (more detailed results can be found in ref. 3):

- Chemical topic modeling allows us to organize molecules in a sensible way that corresponds to human intuition (chemical series).
- The fingerprint-based fragmentation methods provide better results compared to the rule-based fragmentation, most likely due to the fact that they contain overlapping fragments (discussed above).
- Choosing a larger number of topics than anticipated leads to more reasonable topics.

3.4 Exploring Large Data Sets with Chemical Topic Modeling

In many different fields, topic modeling is used to organize large collections of documents or other data (compare introduction and Section 3.2.4). As topic modeling has proven to be useful for chemical data, the next step is to use it as a tool to explore large chemical data sets like the ChEMBL database.[31,32] Here one of the major challenges is efficiently training such large models and managing the size of the matrices that result by applying it to such large data sets. Online learning can help to create large models and is a commonly used technique in different machine-learning algorithms in the era of big data. Here, in the training phase not all data is presented to the model at once, instead the topic model is trained using sequential mini-batches of data.[8] In addition, in a true online-learning scenario the model can dynamically adapt to new patterns in the data. Chemical topic modeling is based on an online-LDA method so that it is applicable to large chemical data sets. Building a 100-topic model on the 1.6 million molecules of the ChEMBL23 data set could be achieved in about an hour on a typical computational chemistry workstation.[3] As mentioned above, the runtime increases linearly with the number of topics. To handle the input molecule-fragment matrix ($M \times F$) (*e.g.* on ChEMBL23: ~1.6M rows× ~3.5K columns (number of fragments after filtering rare and common fragments as described above)) a minimum demand on memory of at least 16 GB exists. A reasonable alternative for decreasing runtime and memory requirements during training is to build the model on a smaller random sample of the data set (*e.g.* only 10%). Experiments show that the models obtained using 100% of ChEMBL23 compared to those built on a 10% random sample of ChEMBL23 are similar.

As mentioned before, massive amounts of data can often be challenging to explore in a reasonable and intuitive way. Chemical topic models built on smaller data sets, *e.g.* to extract chemical series (see Section 3.3.3), can be easily explored by examining the most likely fragments and molecules of the topics. With larger, more diverse data sets like ChEMBL23 or the proprietary screening collections of pharmaceutical companies we need alternative metrics to guide the browsing of topic models. Unlike in the text-mining field, where it is generally possible to use most-likely words to quickly get a thematic overview of topic models built even on large corpora like all Wikipedia articles,[7] more quantitative measures are necessary to help explore and understand a chemical topic model based on millions of compounds. To underpin this, Figure 3.6 shows the most likely fragments of eight different topics of a 100-topic model of ChEMBL23. For the first six topics, studying the most likely fragments leads us to an idea of the types or classes of molecules found in these topics. In contrast, the last two topics are more difficult to interpret without additional investigation of their most likely molecules.

For guiding the exploration of larger chemical topic models the probabilistic framework of topic modeling creates an excellent starting point by providing the different probability distributions for topics and fragments (see Section 3.3.2). Using these probability distributions, different values can be calculated to provide information about the specificity, size or composition of a topic. We can use these concepts to rank topics, relate them to each other, and identify the most relevant topics (see, for example, also ref. 66,67). Different simple metrics based on the molecule–topic probability distribution and on the topic–fragment distribution can be calculated as follows.

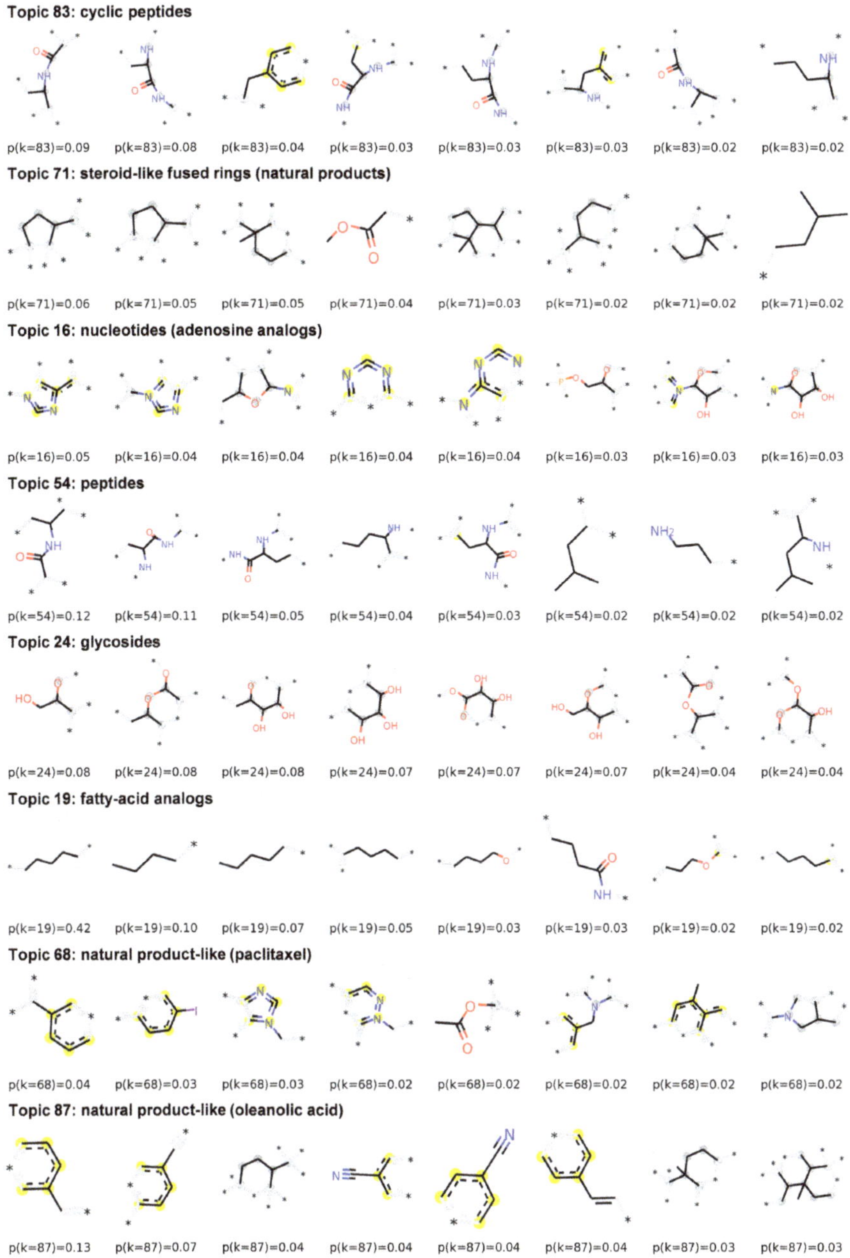

Figure 3.6 Most likely fragments of eight different topics of a 100-topic for ChEMBL23. For each fragment the normalized probability for a certain topic is shown. The names of the topics were manually assigned after investigation of the most likely fragments and molecules of the topics.

Metrics based on the document–topic distributions:

- **Topic size:** calculated as the number of molecules for which the topic is the most likely topic.
- **Topic specificity:** this is related to the number of molecules for which the topic has a high probability. Topics which have a high percentage of molecules with low probability are usually less specific in a sense that they exhibit different types/classes of molecules. To evaluate the topic specificity the *fraction of highly probable molecules* can be calculated, *i.e.* the fraction of all molecules that have a probability for a certain topic higher than a certain threshold (*e.g.* 0.6) is compared to all molecules assigned to that topic (topic size).
- **Number of topic memberships:** as mentioned before, topic modeling leads to a mixed-membership model where each molecule can belong to more than just one topic. This value also provides an interesting characteristic for identifying molecules that have either a strong membership to one (or a few) topics and others that are spread across a larger number of topics. The latter are usually larger heterogeneous molecules.

Metrics based on the topic–fragment distributions:

- **Number of relevant fragments:** in general, this measures how many fragments are necessary to describe a topic. It can be calculated as the number of fragments that have a probability higher than a basic uniformly distributed probability for a certain topic. This value can be normalized within the model and used to identify the topics with larger and lower numbers of relevant fragments.
- **Concentration of the fragment probability distribution:** the distribution of relevant fragments can be broader or more concentrated. This value is also related to topic specificity and number of relevant fragments. Usually a more concentrated distribution – one or a few fragments comprise most of the probability for a certain topic – means a less specific topic. In other words, chemical topics which are only described by a few fragments, *e.g.* a phenyl and a carboxylic acid moiety, will only cover small parts of the compounds and thereby allow a more diverse composition of these topics. To calculate a single value for this, the probability determined by the most likely fragments of a topic can be used (*e.g.* the sum of the probabilities of the first five most likely fragments).

These values can help to pick the most interesting topics from a chemical topic model based on larger data sets. Figure 3.7 shows an overview of all of these values for a 100-topic model built for ChEMBL23. For example, ordering the topics by the fraction of highly probable molecules (here defined as having a probability for a certain topic greater than 0.6) in them identifies

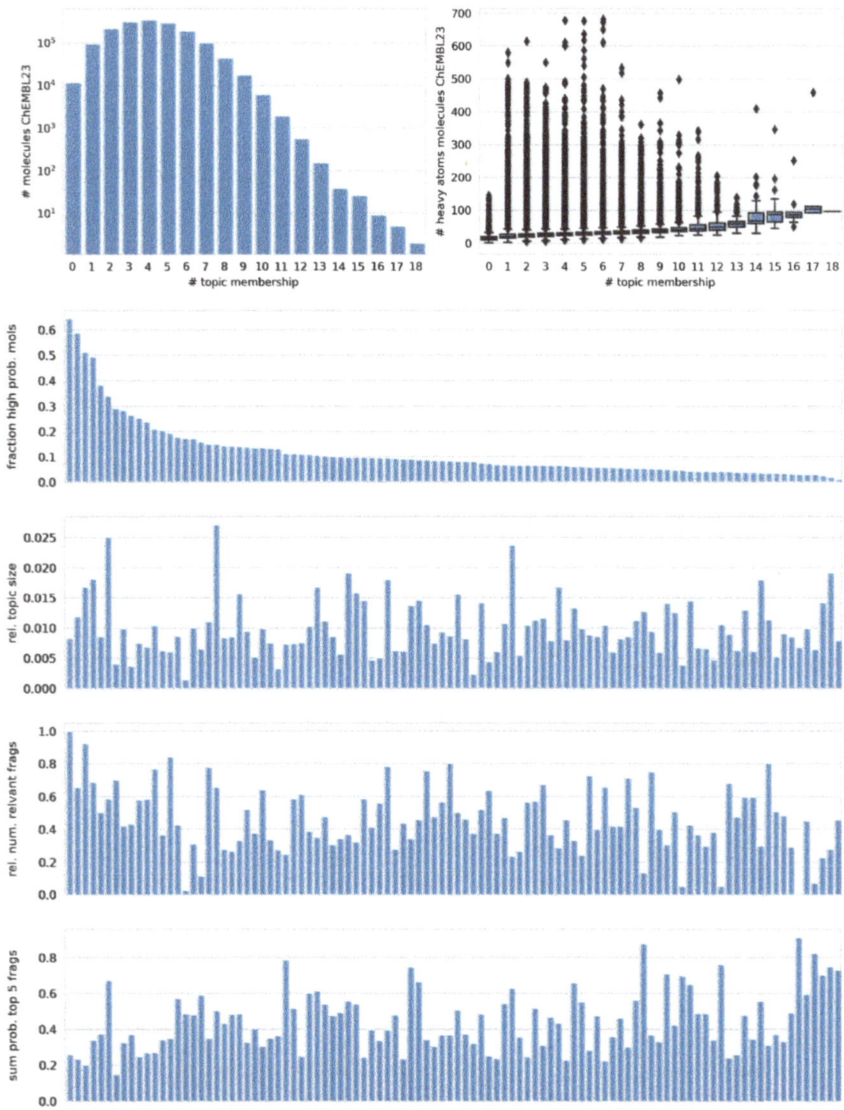

Figure 3.7 Using the probabilities of the chemical topic model to analyze a 100-topic ChEMBL23 model. The last four plots were ordered by the fraction of highly probable molecules (here defined as having a probability for a certain topic greater than 0.6) for each topic respectively.

quite interesting topics in the large heterogeneous data set of ChEMBL: the first six topics in the ranking (high to low value) correspond to well-known molecule classes/families: cyclic peptides, steroid-like natural product analogs, nucleotides, non-cyclic peptides (including proteins), glycosides and fatty-acid analogs. Figure 3.6 was created using this ranking and shows the most likely fragments for these topics. The feature that these topics have in common is that they consist of molecule classes based on unique motifs (peptide bonds or amino acids, nucleotides, unsaturated fused rings, *etc.*) which form molecules by repeating those building blocks. Therefore especially the macro-molecules like proteins or DNA obtain a high probability for those topics. The "fatty acids" topic (topic id 19 in Figures 3.6 and 3.7) builds the second biggest topic of the model. Looking at the most likely fragments it becomes obvious that this topic contains molecules that have long alkane chains (min. length of five) in general and that fatty acids create the most likely molecules in the topic because of their high repetition of this motif.

The cyclic peptides topic (topic id 83 in Figures 3.6 and 3.7) not only exhibits the largest fraction of highly probable molecules within a topic but it also contains the largest number of relevant fragments. On the other end of that scale, topic 90 has the lowest number of relevant fragments. Here the three most likely fragments make up almost 90% of the probability distribution. They describe a rather unspecific structural motif: a fluorophenyl.

Figure 3.7 top left shows that about 10K of the molecules in ChEMBL23 were not assigned to any of the topics (# topic membership = 0); these molecules have uniform probability distributions across all topics. This happens because those molecules do not contain any fragments apart from very common and very rare fragments which we filter out prior to the actual topic model building. Hence, they constitute a special set of molecules and need to be modeled separately.

3.4.1 Hierarchical Topics

An interesting way of further structuring and exploring chemical topic models is to use a hierarchical approach: a topic model is built for the whole data set which tends to be more general (*i.e.* a manageable number of topics). Then separate topic models are created for each topic of the "parent" model (*i.e.* using the molecules that are assigned to each of the topics w.r.t. the most likely topic) (see Figure 3.8).

More advanced approaches also exist which model topic hierarchies using a closed probabilistic form like hierarchical Latent Dirichlet Allocation (hLDA) or Hierarchical Dirichlet Process (HDP).[68–70] The approach described above is an approximation or simplification of these advanced models and it does not take into account the mixed-membership property of a topic model when creating the next level of a sub-topic model. Nevertheless, since chemical topic modeling has not been transferred to these more advanced concepts yet, our pragmatic approach can help extract a finer-grained structure of the data set. It can also be used, for example, to model the subset of molecules that did not retain any features after the fragment vector was filtered. The following two applications of the described hierarchical approach will demonstrate its benefits.

Modeling the ChEMBL23 data set as a topic model as shown above leaves about 0.7% of the molecules without any features/fragments (representing 11 733 molecules). Building a successive chemical topic model ($K = 100$) with only those molecules reveals that they contain special compound classes like fullerenes, special natural products (*e.g.* mycotoxins or quassinoids), cyclodextrins or cyclophosphamide. In general, these compounds contain less common sub-structural motifs like cubane, highly fluorinated alkanes, or tetrabromophenyl. Besides modeling these special compounds that could not be described by the upper level topic model, one could also investigate chemical topics that for example have a large number of relevant fragments but do not have many highly probable molecules like topic 61 for the ChEMBL23 chemical topic model (compare Figure 3.7). In this case, one can assume that these "mixed-bag" topics contain different compound classes which all contribute to the relevant fragments but which do not have a large overlap in terms of shared substructures. Deconvolution of this topic by building a secondary chemical topic model ($K=100$) reveals more homogeneous topics which cover, for example, PDE inhibitors, antibacterials (topoisomerase inhibitors) or CRF-1 receptor antagonists. Another

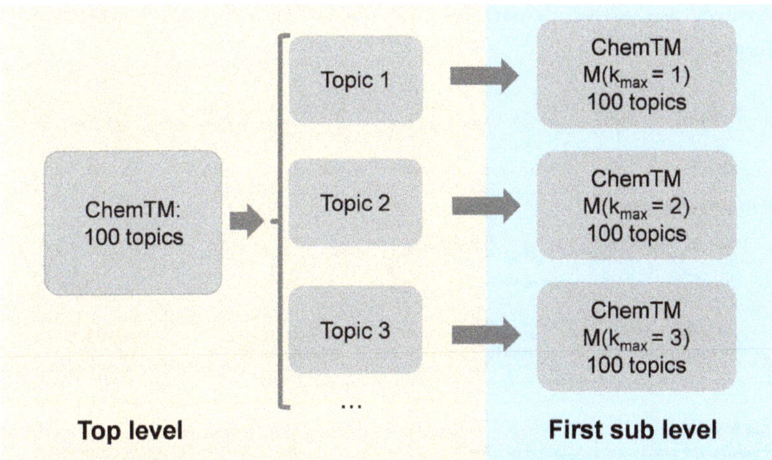

Figure 3.8 Illustration of hierarchical topic modeling. ChemTM = chemical topic model, $M(k_{max}=1)$ means all molecules which are having topic 1 as the most likely topic.

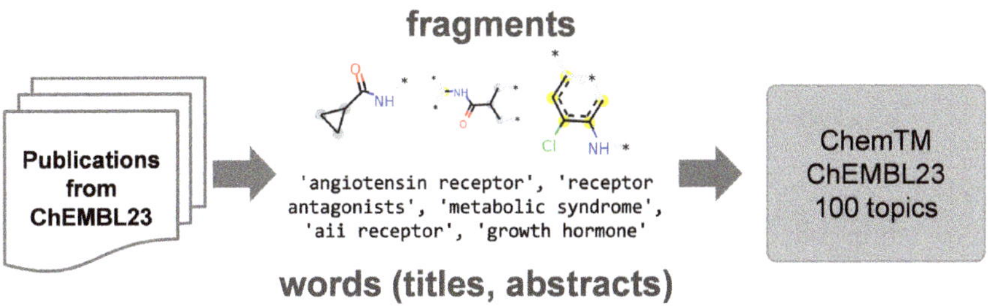

Figure 3.9 Combining text and fragments in a chemical topic model. ChemTM = chemical topic model.

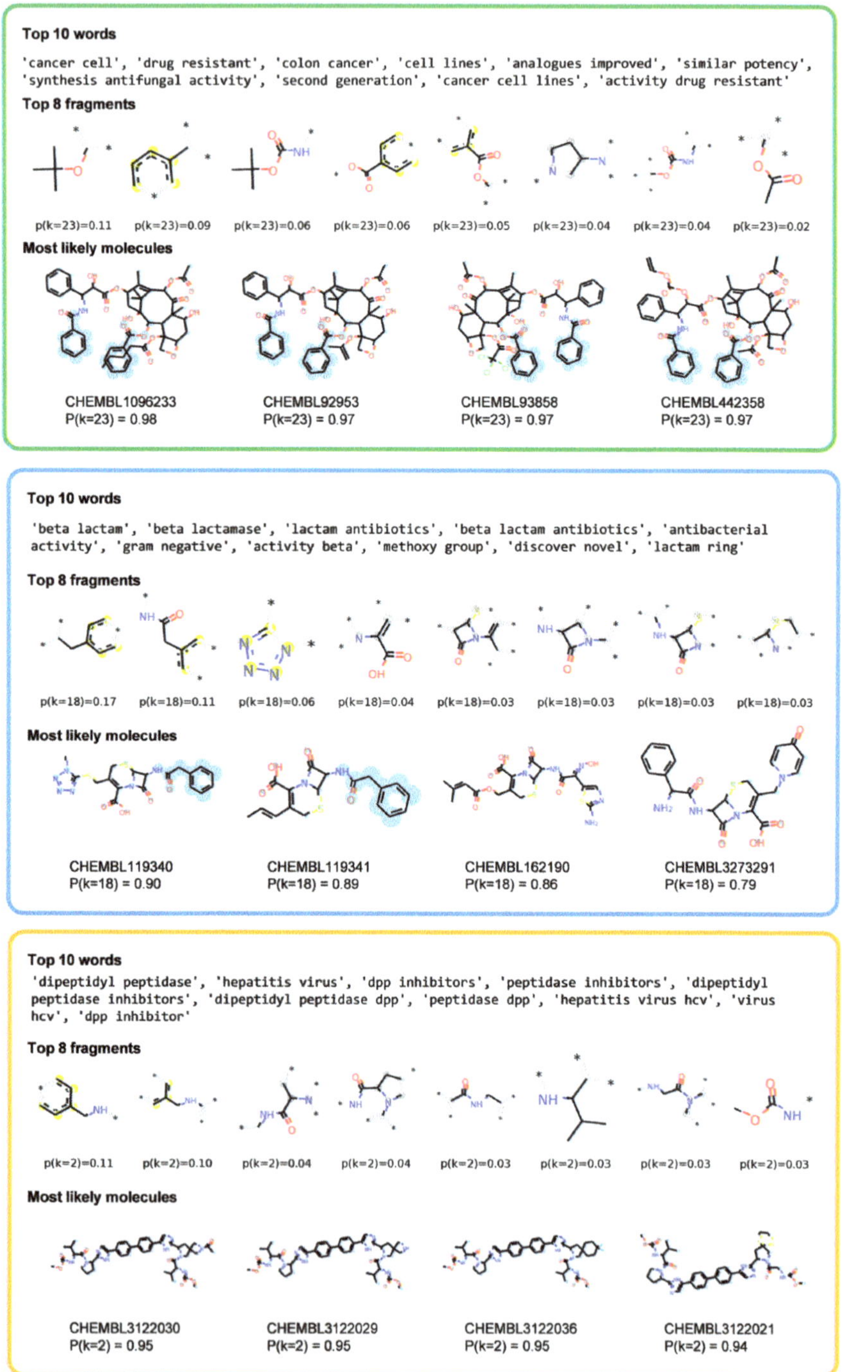

Figure 3.10 A topic model built using a combination of text and chemical features. The three panes each summarize a separate topic and are discussed in the text.

interesting aspect of this secondary chemical topic model is that it only uses about 40 topics of the 100 it could populate and that for those the average fraction of high probable molecules (*i.e.* the topic specificity) is much higher (49% ± 27% per topic) than for the upper level topic model (12% ±12% per topic). This demonstrates how the hierarchical approach allows us to iteratively refine topics within a large heterogeneous data set.

3.5 Combining Text and Chemical Information

The framework of chemical topic modeling offers the flexibility to combine different types of features with each other and use this to infer the topics from the data. Having such a feature-rich data set as the ChEMBL23 database can be useful to include features other than just the chemical structure when extracting the latent variables. The ChEMBL23 data set provides chemical data extracted from scientific publications, other data sets and patent documents. For each compound in the data set its source (paper, patent or data set) is provided and a title and abstract is available for many of the molecules obtained from scientific literature. These texts often provide information about compound class, chemical moieties, protein target, pathway, indication or disease. Analyzing those texts together with the chemical structures found in the different publications allows us to find interesting and possibly new connections in the data. In order to explore this, the 1.6 million molecules of the ChEMBL23 data set were filtered to keep only molecules obtained from publications with at most 50 compounds (to avoid including other data sets or patents). This leaves about 880K molecules from which 100K were randomly picked together with the title and the abstract of the publication they were extracted from. The molecules were tokenized as described in Section 3.3.1 and the texts from title and abstract were modeled as n-grams (bi- to pentagrams) and TF–IDF-filtered so that a maximum of 10K words was kept for the vocabulary. These two types of features – fragments from the molecules and words from the documents – were combined and given as input to a chemical/text 100-topic model (see Figure 3.9).

The resulting topic model populates 93 of the given 100 topics and it ranges from very small (one compound) to very large (~13K compounds) topics. Most of the topics have a higher fraction of relevant words than relevant fragments, however the fragments constitute the most likely features (top 20 features) for almost all topics. Overall the most likely words for each topic provide a good description of the topics. For example, one of the topics has the following most likely words (top 20 words within the relevant features): "cancer cell", "drug resistant", "colon cancer", "cell lines", "analogues improved", "similar potency", "synthesis antifungal activity", "second generation", "cancer cell lines", "activity drug resistant", "breast cancer cell", "human breast cancer cell", "superior potency", "modifications positions", "breast cancer", "slightly potent", "activity drug", "human breast cancer", "fungicidal activities", "carcinoma human".

Obviously, the topic is about cancer and chemotherapy but it also seems to have a connection to "fungicidal activities". By investigating the most likely fragments only (Figure 3.10 top) this conclusion wouldn't be possible. Looking at the most likely molecules in that topic, however, it becomes clear that they are related to the natural product taxol which is the basis of several chemotherapeutics against different kinds of cancer. An interesting observation is that the most likely words in the topic create a connection to the less well-known antifungal properties of taxol.[71] This shows how this combination of text and fragments in the chemical topic model are able to bring different aspects together and simplify the extraction of valuable information. Figure 3.10. highlights two other topics from the chemical/text topic model. The first topic is well described by the most likely words and the most likely fragments: these name or show the beta-lactam chemical moiety – a prominent motif found in many antibacterials. As expected the most likely molecules in this topic are all beta-lactam antibiotics (Figure 3.10 middle). The last example in Figure 3.10 shows a topic where the most likely fragments are peptidic bonds and the most likely words are related to "dipeptidyl peptidase inhibitors" and "hepatitis virus hcv". Inspecting the most likely molecules in that topic (Figure 3.10 bottom) reveals that those are all hepatitis C virus (HCV) replication inhibitors.[72] Apart from these, the topic also contains many dipeptidyl peptidase inhibitors with somewhat lower probabilities (0.7 or less). Both of these compound classes share a larger chemical moiety of pyrrolidines connected to amide bonds (*e.g.* proline). This shows how the combination of text and fragments may help to reveal relationships between different compounds classes and find less well-known indications of molecules.

3.6 Conclusions, Limitations and Future Work

In this chapter we described the concept of topic modeling and its application to chemical data. In general, it can be used to group and organize large and diverse sets of data. The probabilistic framework of topic modeling allows the extraction of the latent or hidden structure in the data and offers new aspects for exploring the data. The probability distributions the model infers from the chemical data can be used to retrieve the most likely fragments of a topic and to directly highlight them within the molecules assigned to this topic. This offers unique advantages over using other non-probabilistic unsupervised machine-learning methods like *K*-means clustering. It can help chemists to better explore and interpret results on large data sets. Especially for large heterogeneous data sets the iterative hierarchical application of chemical topic modeling can assist in finding a finer-grained composition of the data rather than just having a huge number of different topics. Furthermore the flexibility of the approach enables different types of data like chemical structures and text to be easily combined, as shown in Section 3.5, allowing the discovery of novel connections within the data set. Another advantage of topic modeling is that it is a mixed-membership model and thereby creates a new descriptor space – a fragment embedding – for molecules which can provide the starting point for other data science applications.

One of the challenges with chemical topic modeling, or topic modeling in general, is finding a relevant number of topics to describe a data set. Since it is an unsupervised method no correct or optimal number can be defined but it is more up to the data scientist to decide which degree of granularity would be appropriate for grouping the data set. Extensions of topic modeling exist which can help solving this challenge by inferring an optimal number of topics from the given data.[70] In general, topic modeling is still an active field of research and many other interesting extensions could be explored for application to chemical data. For example, dynamic topic models[11] could be used to derive an evolution of compound classes over a longer time frame or supervised topic models[12] to predict protein targets for new compounds.

In conclusion, chemical topic model is an interesting and interpretable alternative to clustering approaches to organizing data which offers many opportunities left to study.

References

1. J. Xie, R. Girshick and A. Farhadi, in *International Conference on Machine Learning*, 2016, p. 478.
2. C. M. Bishop, *Pattern Recognition and Machine Learning*, 1st edn, Springer, New York, 2006.
3. N. Schneider, N. Fechner, N. Stiefl and G. A. Landrum, *J. Chem. Inf. Model.*, 2017, **57**, 1816.
4. D. M. Blei, A. Y. Ng and M. I. Jordan, *J. Mach. Learn. Res.*, 2003, **3**, 993.
5. D. M. Blei, A. Y. Ng and M. I. Jordan, *Advances in Neural Information Processing Systems*, 2002, 601.
6. *Chemical topic modeling source code and notebooks*, https://github.com/rdkit/CheTo, accessed May 2019.
7. D. M. Blei, *Commun. ACM*, 2012, **55**, 77.
8. M. Hoffman, F. R. Bach and D. M. Blei, *Adv. Neural Inf. Process Syst.*, 2010, 856.
9. M. D. Hoffman, D. M. Blei, C. Wang and J. W. Paisley, *J. Mach. Learn. Res.*, 2013, **14**, 1303.
10. J. D. Lafferty and D. M. Blei, in *Advances in Neural Information Processing Systems*, 2006, p. 147.
11. D. M. Blei and J. D. Lafferty, in *Proceedings of the 23rd international conference on Machine learning*, 2006, p. 113.
12. J. D. McAuliffe and D. M. Blei, *Advances in Neural Information Processing Systems*, 2008, p. 121.
13. S. Deerwester, S. T. Dumais, G. W. Furnas, T. K. Landauer and R. Harshman, *J. Am. Soc. Inf. Sci.*, 1990, **41**, 391.
14. T. Hofmann, in *Proceedings of the Fifteenth Conference on Uncertainty in Artificial Intelligence*, 1999, p. 289.

15. N. Péladeau and E. Davoodi, in *Proceedings of the 51st Hawaii International Conference on System Sciences*, 2018.
16. G. E. Hinton and R. R. Salakhutdinov, *Science*, 2006, **313**, 504.
17. G. E. Hinton and R. R. Salakhutdinov, *Adv. Neural Inf. Process Syst.*, 2009, 1607.
18. M. Gerlach, T. P. Peixoto and E. G. Altmann, *Sci. Adv.*, 2018, **4**, eaaq1360.
19. D. Mimno, *ACM J. Comput. Cult. Herit.*, 2014, **5**(3), 1.
20. J. Boyd-Graber, Y. Hu and D. Mimno, *Found. Trends Inf. Ret.*, 2017, **11**, 143.
21. L. Liu, L. Tang, L. He, S. Yao and W. Zhou, *Biotechnol. Biotechnol. Equip.*, 2017, **31**, 630.
22. J. J. van der Hooft, J. Wandy, M. P. Barrett, K. E. Burgess and S. Rogers, *Proc. Natl. Acad. Sci. U. S. A.*, 2016, **113**, 13738.
23. J. J. van der Hooft, J. Wandy, F. Young, S. Padmanabhan, K. Gerasimidis, K. E. Burgess and S. Rogers, *Anal. Chem.*, 2017, **89**, 7569.
24. S. Rogers, C. W. Ong, J. Wandy, M. Ernst, L. Ridder and J. J. van der Hooft, *Faraday Discuss.*, 2019, DOI: 10.1039/C8FD00235E.
25. S. Rogers, M. Girolami, C. Campbell and R. Breitling, *IEEE/ACM Trans. Comput. Biol. Bioinf.*, 2005, **2**, 143.
26. X. Chen, X. Hu, X. Shen and G. Rosen, in *2010 IEEE International Conference on Bioinformatics and Biomedicine (BIBM)*, 2010, p. 149.
27. W. Zhao, W. Zou and J. J. Chen, *BMC Bioinf.*, 2014, **15**, S11.
28. L. Liu, L. Tang, W. Dong, S. Yao and W. Zhou, *SpringerPlus*, 2016, **5**, 1608.
29. J. J. Irwin, T. Sterling, M. M. Mysinger, E. S. Bolstad and R. G. Coleman, *J. Chem. Inf. Model.*, 2012, **52**, 1757.
30. S. Kim, P. A. Thiessen, E. E. Bolton, J. Chen, G. Fu, A. Gindulyte, L. Han, J. He, S. He and B. A. Shoemaker, *Nucleic Acids Res.*, 2016, **44**, D1202.
31. A. Gaulton, L. J. Bellis, A. P. Bento, J. Chambers, M. Davies, A. Hersey, Y. Light, S. McGlinchey, D. Michalovich, B. Al-Lazikani and J. P. Overington, *Nucleic Acids Res.*, 2012, **40**, D1100.
32. A. P. Bento, A. Gaulton, A. Hersey, L. J. Bellis, J. Chambers, M. Davies, F. A. Krüger, Y. Light, L. Mak and S. McGlinchey, *Nucleic Acids Res.*, 2014, **42**, D1083.
33. L. Ruddigkeit, R. Van Deursen, L. C. Blum and J. L. Reymond, *J. Chem. Inf. Model.*, 2012, **52**, 2864.
34. M. A. Clark, R. A. Acharya, C. C. Arico-Muendel, S. L. Belyanskaya, D. R. Benjamin, N. R. Carlson, P. A. Centrella, C. H. Chiu, S. P. Creaser and J. W. Cuozzo, *Nat. Chem. Biol.*, 2009, **5**, 647.
35. Enamine REAL database, https://enamine.net/library-synthesis/real-compounds/real-database, accessed May 2019.
36. J. MacQueen, in *Proceedings of the Fifth Berkeley Symposium on Mathematical Statistics and Probability*, **vol. 1**, 1967, p. 281.
37. J. H. Ward Jr, *J. Am. Stat. Ass.*, 1963, **58**, 236.
38. J. W. Raymond, C. J. Blankley and P. Willett, *J. Mol. Graphics*, 2003, **21**, 421.

39. A. Böcker, S. Derksen, E. Schmidt, A. Teckentrup and G. Schneider, *J. Chem. Inf. Model.*, 2005, **45**, 807.
40. S. Gan, D. A. Cosgrove, E. J. Gardiner and V. J. Gillet, *J. Chem. Inf. Model.*, 2014, **54**, 3302.
41. M. Steinbach, L. Ertöz and V. Kumar, in *New Directions in Statistical Physics*, Springer, Berlin, Heidelberg, 2004, pp. 273–309.
42. S. Kearnes, K. McCloskey, M. Berndl, V. Pande and P. Riley, *J. Comput.-Aided Mol. Des.*, 2016, **30**, 595.
43. W. Jin, R. Barzilay and T. Jaakkola, in *International Conference on Machine Learning*, 2018, p. 2328.
44. R. E. Carhart, D. H. Smith and R. Venkataraghavan, *J. Chem. Inf. Comput. Sci.*, 1985, **25**, 64.
45. R. Nilakantan, N. Bauman, J. S. Dixon and R. Venkataraghavan, *J. Chem. Inf. Comput. Sci.*, 1987, **27**, 82.
46. H. L. Morgan, *J. Chem. Doc.*, 1965, **5**, 107.
47. D. Rogers and M. Hahn, *J. Chem. Inf. Model.*, 2010, **50**, 742.
48. R. D. Hull, S. B. Singh, R. B. Nachbar, R. P. Sheridan, S. K. Kearsley and E. M. Fluder, *J. Med. Chem.*, 2001, **44**, 1177.
49. R. D. Hull, E. M. Fluder, S. B. Singh, R. B. Nachbar, S. K. Kearsley and R. P. Sheridan, *J. Med. Chem.*, 2001, **44**, 1185.
50. S. B. Singh, R. P. Sheridan, E. M. Fluder and R. D. Hull, *J. Med. Chem.*, 2001, **44**, 1564.
51. *Daylight Theory Manual*, http://www.daylight.com/dayhtml/doc/theory/index.pdf, accessed May 2019.
52. G. A. Landrum, *et al.*, *RDKit: Open-Source Cheminformatics Software, version 2018.09.02*, DOI: 10.5281/zenodo.2574427; http://www.rdkit.org and https://github.com/rdkit/rdkit (accessed May 2019).
53. T. Engel and J. Gasteiger, *Chemoinformatics: Basic Concepts and Methods*, John Wiley & Sons, 2018.
54. RDKit Blog, *Using the new fingerprint bit rendering code*, http://rdkit.blogspot.com/2018/10/using-new-fingerprint-bit-rendering-code.html, accessed May 2019.
55. J. Degen, C. Wegscheid-Gerlach, A. Zaliani and M. Rarey, *ChemMedChem*, 2008, **3**, 1503.
56. X. Q. Lewell, D. B. Judd, S. P. Watson and M. M. Hann, *J. Chem. Inf. Comput. Sci.*, 1998, **38**, 511.
57. H. M. Wallach, I. Murray, R. Salakhutdinov and D. Mimno, in *Proceedings of the 26th Annual International Conference on Machine Learning*, ACM , 2009, pp. 1105–1112.
58. RDKit Blog, Colliding bits III, http://rdkit.blogspot.com/2016/02/colliding-bits-iii.html, accessed May 2019.
59. A. Rajaraman and J. D. Ullman, *Mining of Massive Datasets*, Cambridge University Press, 2011.
60. J. Chang, J. L. Boyd-Graber, S. Gerrish, C. Wang and D. M. Blei, *Adv. Neural Inf. Process Syst.*, 2009, **31**, 1.
61. S. Jaeger, S. Fulle and S. Turk, *J. Chem. Inf. Model.*, 2018, **58**, 27.

62. D. Erhan, Y. Bengio, A. Courville, P. A. Manzagol, P. Vincent and S. Bengio, *J. Mach. Learn. Res.*, 2010, **11**, 625.
63. N. X. Vinh, J. Epps and J. Bailey, *J. Mach. Learn. Res.*, 2010, **11**, 2837.
64. A. Rosenberg and J. Hirschberg, in *Proceedings of the 2007 Joint Conference on Empirical Methods in Natural Language Processing and Computational Natural Language Learning (EMNLP-CoNLL)*, 2007.
65. S. Riniker, N. Fechner and G. A. Landrum, *J. Chem. Inf. Model.*, 2013, **53**, 2829.
66. L. AlSumait, D. Barbará, J. Gentle and C. Domeniconi, in *Joint European Conference on Machine Learning and Knowledge Discovery in Databases*, 2009, p. 67.
67. M. Röder, A. Both and A. Hinneburg, in *Proceedings of the Eighth ACM International Conference on Web Search and Data Mining*, 2015, p. 399.
68. T. L. Griffiths, M. I. Jordan, J. B. Tenenbaum and D. M. Blei, in *Advances in Neural Information Processing Systems*, 2004, p. 17.
69. Y. W. Teh, M. I. Jordan, M. J. Beal and D. M. Blei, in *Advances in Neural Information Processing Systems*, 2005, p. 1385.
70. J. Paisley, C. Wang, D. M. Blei and M. I. Jordan, *IEEE Transactions on Pattern Analysis and Machine Intelligence*, **37**, 2015, p. 256.
71. D. H. Young, E. L. Michelotti, C. S. Swindell and N. E. Krauss, *Experientia*, 1992, **48**, 882.
72. W. M. Kazmierski, A. Maynard, M. Duan, S. Baskaran, J. Botyanszki, R. Crosby, M. S. Dickerson, M. Tallant, R. Grimes, R. Hamatake, M. Leivers, C. D. Roberts and J. Walker, *J. Med. Chem.*, 2014, **57**, 2058.

CHAPTER 4

Deep Learning and Chemical Data

COLIN BATCHELOR,* PETER CORBETT, AILEEN DAY, JEFF WHITE AND JOHN BOYLE

Royal Society of Chemistry, Thomas Graham House, Cambridge, UK CB4 0WF
*Email: batchelorc@rsc.org

4.1 Introduction

The traditional approach to machine learning is one where the system under investigation is modelled by a set of features and trained to weight those features in a way that best reproduces the training data. The selection of features is difficult, often counter-intuitive, and is progressively revised in the light of the performance of a model. It is based on the researcher's experience and intuitions about the domain. The size of a feature set can the range from very small, for example five features in the case of Fisher's Iris dataset, to molecular fingerprints which can be thousands of bits long. One appealing aspect of deep learning is the promise of replacing feature engineering with a system that learns the features directly from the data. This has been particularly successful in image processing, where the features have historically proven difficult to specify, and in speech recognition. It has also resulted in significantly better performance in machine translation, where the feature set is reasonably well-understood but extremely large. Another attractive aspect is that the models and representations learned should be amenable to application to other tasks in the same domain. Indeed one of the earliest applications of deep learning in chemistry was in an

Drug Discovery Series No. 75
Artificial Intelligence in Drug Discovery
Edited by Nathan Brown

Published by the Royal Society of Chemistry, www.rsc.org

entry to a Kaggle competition in 2012 where the data was supplied by Merck. The winning entry used an ensemble of deep neural networks, gradient boosting and Gaussian regression. The subsequent paper describes this system, 15 large datasets, mainly in ADME, ranging in size from 2763 compounds to 318 795 compounds, that were used for training, and also a simpler system that only uses deep neural networks.[1]

Even when a trained model achieves its original purpose it can still be unsatisfactory in a number of ways, many of which derive from their consisting of thousands upon thousands of adjustable parameters. Much of a given neural network is redundant and it can be robust to significant pruning, which is the origin of the dropout method which is used to address overfitting.[2] Neural nets are difficult to interpret and are invariably thought of by their developers as "black boxes", something which is also the case for some other machine-learning methods such as tree ensembles and generalised linear models. This has legal consequences in the European Union with the advent of GDPR and the "right to explanation". A neural network trained using high-precision arithmetic can also be replaced by one that uses eight-bit or even single-bit precision with only a modest decrease in accuracy.[3]

Many of the chapters in this book focus on drug discovery and discuss systems that use descriptors that explicitly encode local or global descriptions of a chemical structure, for example SMILES or fingerprints. However, there are other forms of chemical data that are equally amenable to machine learning. This chapter is about models that take a sequence as input and output either another sequence or weighted classifications. The input sequence might be inherently numeric, as in a one-dimensional spectrum, or a text string, which can be encoded in a variety of ways. We describe a tentative deep-learning approach to elucidating NMR spectra and two significantly more mature natural-language processing (NLP) tasks: deep learning treatments of chemical named entity recognition (NER) and relationship extraction. No knowledge of natural-language processing and only a basic level of cheminformatics expertise is assumed on the part of the reader.

4.2 Background

In this section we discuss deep learning, evaluation methods and give a brief overview of natural-language processing.

4.2.1 Deep Learning

The original neural network architecture was the perceptron, a binary classifier which consisted of a single input layer connected to an output layer:

$$f(\mathbf{x}) = \begin{cases} 1 & (\mathbf{w} \cdot \mathbf{x} + b > 0), \\ 0 & \text{otherwise} \end{cases} \quad (4.1)$$

where **w** is a set of weights and b is a bias. These are only capable of learning linearly-separable problems and interest in neural networks waned until the development of multilayer neural networks.

Multilayer neural networks add an extra layer or layers in between the input and output, which are called hidden layers. Another change is that non-linearities are introduced by multiplying eqn (4.1) by an activation function. Early work, inspired by the firing rate of neurons in organisms, involved the sigmoid function $S(x) = 1/(1 + e^{-x})$, but more recent work has used various forms of linear activation functions, for example the rectified linear unit:

$$f(x) = \max(0,x), \tag{4.2}$$

which is non-linear, being 0 for $x \leq 0$. It is significantly quicker to calculate than the sigmoid function and is less likely to saturate. Recently, the most successful families of architecture of multilayer neural networks have been feed-forward networks, where each layer takes as input the output of the previous layer, and recurrent networks, where each layer takes as input the previous layer and its previous state. Feed-forward networks can be divided into fully-connected networks, where every node in one layer is connected to every node in the succeeding layer, and convolutional networks. Because the number of weights to be learned in a fully-connected network is nm, where n is the number of units in the first layer and m in the second, and because the results are not positionally invariant, convolutional neural networks (CNNs) model the weights of an $x \times x$ square, where $x \ll n$. Convolutional neural networks first came to widespread attention in 2012 when Krizhevsky *et al.* used them to classify images from ImageNet with significantly better performance than previously.[4]

A simple example of a recurrent neural network is given by Elman and applied to calculating the XOR function (a classic example of non-linearity) and predicting the next character in a text string.[5] The model is:

$$\mathbf{h}_t = \sigma_h(\mathbf{W}_h\mathbf{x}_t + \mathbf{C}_h\mathbf{h}_{t-1} + b_h) \tag{4.3}$$

$$\mathbf{y}_t = \sigma_y(\mathbf{W}_y\mathbf{h}_t + b_y), \tag{4.4}$$

which is to say that at each timestep t the hidden layer is updated based on the input vector and its state at time $t-1$, with weightings **W** and **C** applied respectively. σ_h and σ_y are activation functions (not in fact specified in the paper).

Both feed-forward and recurrent neural networks can be trained through backpropagation, which is where the loss value based on the network's performance against the training data is propagated backward through the network. Backpropagation has been invented many times. As it is an application of the chain rule, it is very similar to, for example, sensitivity analysis of differential equations. Backpropagation is one of the technical reasons why deep learning has enjoyed a surge in popularity over the past decade. Other reasons include the widespread availability of GPUs, the publicity coming from successes in playing Go and 1980s arcade games, and the low barriers to entry provided by open-source libraries with Python wrappers such as Tensorflow and PyTorch.

Another neural network architecture that has been used in the past in cheminformatics but is no longer popular is the counterpropagation neural network. They have been used in the past in cheminformatics but are more complicated to train as more than one algorithm is involved. Aires-de-Sousa *et al.* give a good description of how they work in an early application to NMR spectrum elucidation.[6] There is a single hidden layer, the Kohonen or self-organising map layer, where the units are connected to each other in a lattice. This is trained first. As an unsupervised system which can be used for clustering or dimensionality reduction it is trained for a fixed number of iterations rather than until convergence. The output, or Grossberg, layer is then trained separately using backpropagation.

The tasks we consider in this chapter are sequence processing tasks, for which feed-forward and recurrent networks can both be used but for which the most successful approaches tend to use recurrent networks. While the idea of machine learning in general is that the parameters of the model are learned from the training data, the training process itself has parameters such as learning rate (the constant multiplicative factor applied to the loss value as it is backpropagated through the network), batch size (if the training data is separated into batches) and the size of individual hidden layers. Other hyperparameters are the choice of optimization algorithm, which itself may have parameters. There are also additional sorts of layer that can be added: a pooling layer typically reduces the dimensionality of the output of the previous layer by taking the maximum value (or average, or L_2-norm) of subsets. Optimising these for deep learning is an open problem. The most straightforward approach is grid search, where the hyperparameters are systematically varied across a plausible range, but other approaches include gradient descent or even genetic algorithms.

4.2.2 Evaluation Methods

Training any supervised machine-learning algorithm requires a function to calculate a loss which can be used to adjust the weights in the system. This is an example of intrinsic evaluation, a measure of how well an algorithm performs on training or test data, as opposed to extrinsic evaluation, which is how well it performs in a context in which it is deployed.[7] In this chapter we will focus on intrinsic evaluation. For natural-language processing intrinsic evaluation is often done in terms of precision and recall based on a "gold standard" of trusted results. This is relevant to classification tasks and information retrieval. Let us assume we are looking for chemical names in text. Precision is the number of true positives divided by the sum of true positives and false positives, so an algorithm that only finds a single chemical name but gets it right will score 100%. Recall is the number of true positives divided by the sum of true positives and false negatives, so an algorithm that classifies everything as a chemical name will score 100% as well. The tradeoff between these depends on the context in which the code is deployed, but a common measure in the literature is the *F*-score, which is the harmonic mean of precision and recall.

Creating a gold standard, as the name implies, is an expensive process. It involves at least some human annotation and it is best practice to have more than one annotator. We can understand the difficulties of a given task on a given set of data by measuring to what extent annotations are stable (whether the same annotator annotates the same things in the same way over time) and reproducible (whether different annotators agree on classifications). This has traditionally in natural-language processing been measured with the κ coefficient,[8] but this has been criticised and a variety of other more sophisticated measures like α and π have been proposed as alternatives.[9] The idea of the κ coefficient and similar measures is to adjust for unbalanced datasets, which are commonplace and will flatter naive percentage agreement scores, by adjusting for chance agreement. κ values of 0.67 or above are often taken to be acceptable although the original works only consider a κ of over 0.80 to be reliable.[8]

Another important part of evaluation is to have a baseline system against which to evaluate. The better understood the task, the stronger the baseline system will be. For a new task the system might be very simple indeed, perhaps in a classification task assigning everything to the most common category, whereas for a well-established task the baseline should be the state-of-the-art.

4.2.3 Natural-language Processing

Natural-language processing has a long history going back at least to the start of the 20th century when Markov applied Markov chains to Pushkin's *Eugene Onegin*. Collobert *et al.* were responsible for one of the first applications of deep learning to general natural-language processing in 2011.[10] We say "general" but, like a great deal of NLP, the data used was all in English, and all in a single genre, newswire. This means that the resulting models will transfer imperfectly to other domains (such as scientific articles or social media) and to other languages only with great difficulty. It is worth discussing the tasks in the Collobert paper in some detail in order to put the work on chemical natural-language processing later on in this chapter in context. They developed a single multitask neural network that was jointly trained on part-of-speech (POS) tagging, chunking, named entity recognition and semantic role labelling. Many of the datasets used come from CoNLL, the annual ACL SIGNLL Conference on Computational Natural Language Learning. This conference has a different shared task each year, and the scripts they use to evaluate entries automatically are made publicly available. These datasets are derived from pre-existing corpora. The oldest one is the Brown Corpus, consisting of slightly over a million words, which was created in 1961 and revised in 1971 and 1979.[11] The Penn Treebank is somewhat larger and is based on excerpts from the *Wall Street Journal*.[12]

POS tagging is identifying whether a given word in a sentence is a noun, a verb, an adjective or some other part of speech. Practical schemes for POS tagging can be very detailed with hundreds of classes that capture different varieties of noun and adjective and classify verbs according to what kinds of arguments they take. The results can be used as input for parsing, determining the syntactic structure of a sentence, or as a filter in other tasks so that a preposition or conjunction, say, isn't identified as being a gene name. The baseline system is described in Toutanova *et al.*[13] using the Penn Treebank which uses 45 different POS tags. They take as the training set sections 0–18 (912 344 tokens) and as the test set sections 22–24 (129 654) tokens.

Chunking is a task that involves breaking up a sentence into non-overlapping grammatical chunks, mostly NPs (nouns and the adjectives and articles that precede them), VPs (usually in this case just the verb) and PPs (usually in this case just the preposition itself), but Sang and Buchholz list 11 chunk types in their specification of the CoNLL shared task.[14] The vast majority of chunks are NP, VP or PP and the eight other types are very rare. This sort of Zipfian distribution is typical of natural language processing. In addition to identifying the chunk type, another label is needed: I, O, B, E or S, indicating whether a word is Inside, Outside, Beginning, Ending a chunk, or a Singleton. There are 42 different tags. Collobert *et al.* took the data from the CoNLL 2000 shared task, based again on the WSJ data set, but with sections 15–18 (211 727 tokens) for training and section 20 (47 377 tokens) for testing.

Named entity recognition was initially developed to identify personal names, placenames (which might be geopolitical entities like Wales or purely geographical ones like the Sound of Sleat) and the names of organisations and facilities. The benchmark system here was the English language subset of the CoNLL 2003 shared task. This consisted of newswire text from Reuters: 203 621 tokens in the training set and 46 435 tokens in the test set.[15]

Semantic role labelling involves identifying the participants in the action described by a verb. These have been formalised for non-technical English in PropBank.[16] The benchmark system here was CoNLL's 2005 shared task, based again on the WSJ data set with sections 2–21 (950 028 tokens) in the training set and section 23 plus three sections from the Brown corpus as the test set (63 843 tokens).[17]

The last two of these tasks are tackled later on in this chapter: explicitly, named entity recognition where the entities are small molecules, and implicitly, a cruder form of semantic role labelling where entities that play the roles of small molecule and gene product for a narrow selection of verbs are identified.

Chemical named entity recognition involves identifying references to small molecules and other entities of chemical interest (molecular processes, named reactions, instruments, methods) in abstracts, journal articles, patents or in the free text in electronic lab notebooks. There are challenges in chemical named entity recognition which are familiar from general named entity recognition, which can be seen in the following text:[18] (previously used as an example by Batchelor *et al.*[19]):

> To realize this strategy diastereomerically and enantiomerically pure cyclic nitronates (+)-(4*S*,6*S*,7*S*,8*R*)-**5** and (−)-(4*R*,6*R*,7*R*,8*S*)-**5** (d.e. >99%, Scheme 2) were synthesized according to a previously reported procedure from commercially available nitroethane, isovanillin, cyclopentyl bromide and (+)- or (−)-*trans*-2-phenylcyclohexanols (>98% ee).

The first one is tokenisation. Some of the names of chemical species in this are single words ("nitroethane", "isovanillin"), some multiple words ("cyclopentyl bromide") and some, "(+)-(4*S*,6*S*,7*S*,8*R*)-**5**", for example, consist mainly of punctuation. In "ethyl acetate–hexane", a simple-minded tokeniser that broke on spaces but not en-rules would misidentify the correct tokens, "ethyl acetate" and "hexane" just as it would in "North Berwick–Boat of Garten", and authors do not capitalise chemical names.

The second one is autosuperordination. Some of the names, like nitroethane, cyclopentyl bromide or (+)-(4*S*,6*S*,7*S*,8*R*)-**5** can be tied to a specific molecular structure that can be written with a SMILES or InChI, while "(−)-*trans*-2-phenylcyclohexanols" is an underspecified expression that refers to a family of compounds containing a common substructure. Underspecification here is marked, as in English generally, by pluralization, just as "cheeses" refers to kinds of cheese rather than multiple portions of cheese. "cyclic nitronates" refers to a much bigger family of compounds that contain a given substructure (nitronate) that is also part of a ring.

The third one is scope. What counts as a chemical named entity within the task needs to be very carefully specified: "gold" in "gold standard" doesn't when used metaphorically but might do in the context of economics; "germane" doesn't usually but would do when it describes a hydride of germanium; and "water" sometimes counts as a chemical named entity and sometimes not, depending on the guidelines.

A fourth issue, not shown in the text, is the specification of entity boundaries. In the phrase "sodium ion" it is possible to imagine that "ion" might or might not form part of the named entity depending on the guidelines. Equally, guidelines might differ on how much of the phrase "the substituted cyclohexane" should count, and naive guidelines might call for "cyclohexane" to be annotated with the structure for cyclic C_6H_{12}, even though the referent is clearly not cyclohexane itself. Corbett *et al.*[20] have quantified the challenges above and showed that it is hard for human annotators to get agreement of above 93%. The guidelines developed in that work have been used as the basis of the gold-standard corpus[21] used for the CHEMDNER task in BioCreative.[22]

A weakness and a strength of chemistry are that systematic nomenclature is, given the correct intra-word tokenisation of a chemical entity, a weakly context-sensitive language in its own right. This is a weakness because there is an infinity of out-of-dictionary terms that chemists could use in their text, and a strength, because an unseen systematic name can be reverse engineered into a chemical structure, as demonstrated by the open-source tool OPSIN.[23]

An example of a program that does rule-based semantic role labelling for chemical text is ChemicalTagger.[24] This uses a hand-written grammar to parse sentences describing 21 different kinds of action (for example, adding, concentrating, cooling and so on) in order to extract structured information about chemical syntheses. Daniel Lowe's PhD thesis[25] describes how this can be deployed in practice to extract reaction schemes from patents.

4.3 Case Study 1: Spectroscopic Analysis

4.3.1 Background

The main tasks to which machine learning has been applied in spectrometry are the interpretation of mass spectra and NMR spectra. High-resolution mass spectrometry can identify the atoms present in a given molecule but not necessarily how they are connected to each other. Rule-based systems can help by using databases of plausible and implausible structures that fragments might belong to, and machine learning can improve on this by assigning weightings to those structures and integrating other sources of data, for example the organism or even the geographical location that a compound has been taken from. NMR can be used in chemistry for confirmatory purposes—to demonstrate that the authors have in fact synthesised the compound they claim in a given paper, and to determine the constitution and conformation of natural products. In general, structure elucidation by NMR is such a complicated problem that the aim is not a full end-to-end system but rather computer-assisted structure elucidation (CASE). A thorough review of recent work in CASE is given by Burns *et al.*[26]

Automated analysis of spectra goes back to the 1960s and the DENDRAL program, written in LISP, which analysed mass spectra within a limited chemical domain, that of mass spectrometry. The initial task they tackled was to generate all of the chemically-plausible acyclic structures consisting of C, H, N and O atoms. This was then used to find structures that matched a given mass spectrum, and these structures were then ranked using a scoring function that used lists of "known good" and "known bad" structures. Although these papers have "artificial intelligence" in the title, they probably wouldn't count as such today.[27,28] As for NMR, one of the most long-lasting contributions is that of Bremser, who developed an extremely efficient way of representing the local environments of NMR-active atoms in the 1970s.[29] Each atom in a structure is assigned a Hierarchical Organisation of Spherical Environments (HOSE) code on the basis of its neighbours in the graph and their immediate neighbours. A database can then be populated with HOSE codes and the corresponding chemical shifts. This is the approach taken by, for example, NMRShiftDB,[30] the developers of which found that a combination of HOSE codes and a database of chemical shifts slightly outperformed machine-learning methods for proton NMR[31] and by a more recent self-learning tool described by Castillo *et al.* who have made their dataset available on GitHub.[32]

Fan *et al.*[33] use deep learning to analyse Raman spectra of mixtures of compounds used in pharma: among others, solvents, amino acids and active ingredients. In order to obtain enough data for deep learning methods to work they generated 20 000 synthetic mixture spectra for each of the 167 compounds they investigated and trained on those.

An example of predicting spectra from given chemical structures is found in work by Aires-de-Sousa *et al.* in the Gasteiger group in 2002[6] who predicted proton NMR spectra based on a training set of 744 chemical shifts from 120 compounds. They used counterpropagating neural networks. Later Binev and Aires-de-Sousa showed an improvement in performance by using feed-forward neural networks.[34] Ghosh *et al.*[35] take a more intrinsically transferable approach where they generate Coulomb matrices for given structures and compare the performance of different network architectures at predicting excitation spectra. The Coulomb matrix for a given set of atoms is defined as

$$C_{ij} = Z_i Z_j ||\mathbf{R}_i - \mathbf{R}_j||^{-1} \tag{4.5}$$

where Z_i are the atomic numbers and $\mathbf{R}_i$ are the spatial positions, which can be obtained from crystal structures. Out of a multilayer perceptron, CNN and deep tensor neural network, a detailed description of which is beyond the scope of this chapter but is given by Schütt *et al.*,[36] they found the last of the three performed best. Another approach that combines quantum mechanics and machine learning is that taken by Ito *et al.*[37] who predict chemical shifts from structures using quantum mechanics and then compare 91 different machine-learning methods for calculating corrections, on a dataset of 1277 ^{1}H and 1078 ^{13}C chemical shifts.

4.3.2 Worked NMR Example

This is a deliberately simple proof-of-concept example that shows how machine learning can assist structure elucidation using NMR spectra. The task is to take a carbon-13 NMR spectrum and return the functional groups in the corresponding molecule. The data set consisted of 34 629 experimental spectra and corresponding structures obtained from NMRShiftDB[38†] and 24 900 synthetic spectra derived from literature structures in the MarinLit database[‡] using ACD/Labs predictor, which uses a large database experimental chemical shift data.[39]

The input vector was the spectrum itself over a chemical shift range of 310 to −40 ppm with a resolution of 0.1 ppm (3500 datapoints). The output was a vector of the functional groups found in the molecules. These were

---

[†] nmrshiftdb2withsignals.sd (NMRShiftDB2, http://nmrshiftdb.nmr.uni-koeln.de/)
[‡] http://pubs.rsc.org/marinlit/introduction

Table 4.1 Details of the three models used.

Model	Layer	Units	Activation
Convolution	Dense	3000	relu
	Conv1D	512	relu
	Average pooling	Pool size = 10	—
	Conv1D	512	—
	Average pooling	Pool size = 10	—
	Dense	282	linear
Dense	Dense	3000	relu
	Dense	2000	relu
	Dense	1000	relu
	Dense	2000	relu
	Dense	2000	relu
	Dense	500	relu
	Dense	282	linear
LSTM	Dense	3000	relu
	LSTM	100	—
	LSTM	200	—
	Dense	282	linear

chosen by a combination of looking at which groups are reported explicitly in MarinLit and general knowledge. The 282 different functional groups were themselves identified using RDKit[§] and a collection of 304 SMARTS patterns.

We considered three separate models which are detailed in Table 4.1 and chose the Random Forest implementation in R to serve as a baseline that was distinct from the synthetic spectra.[40] Keras[41] was used as the neural network library. The three models were a one-dimensional convolutional network model (**Convolution**), a fully-connected, or dense, network model (**Dense**) and a recurrent one (**LSTM**). Specifically the recurrent model used long short-term memory (LSTM) units, which are a form of recurrent network that contain "forget gates" to avoid the vanishing gradient problem.[42]

$$\mathbf{h}_t = \mathbf{o}_t \cdot \sigma_h(\mathbf{c}_t) \tag{4.6}$$

$$\mathbf{o}_t = \sigma_g(\mathbf{W}_o \cdot \mathbf{x}_t + \mathbf{U}_o \cdot \mathbf{h}_{t-1} + b_o) \tag{4.7}$$

$$\mathbf{c}_t = \mathbf{f}_t \cdot \mathbf{c}_{t-1} + \mathbf{i}_t \cdot \sigma_c(\mathbf{W}_c \cdot \mathbf{x}_t + \mathbf{U}_c \cdot \mathbf{h}_{t-1} + b_c) \tag{4.8}$$

$$\mathbf{f}_t = \sigma_g(\mathbf{W}_f \cdot \mathbf{x}_t + \mathbf{U}_f \cdot \mathbf{h}_{t-1} + b_f) \tag{4.9}$$

$$\mathbf{i}_t = \sigma_g(\mathbf{W}_i \cdot \mathbf{x}_t + \mathbf{U}_i \cdot \mathbf{h}_{t-1} + b_o) \tag{4.10}$$

c is the memory cell, while i and o are the input and output "gates", as opposed to the output layer. f is the "forget gate".

Table 4.2 shows micro-F scores for the three different models and the ensemble. Micro-averaging is where the F score is calculated based on all of

[§]RDKit: Open-Source Cheminformatics Software, version 2017.03.1, http://www.rdkit.org/, https://github.com/rdkit/rdkit

Table 4.2 Micro-averaged *F* scores for different deep learning models for NMR spectrum.

Model	Precision	Recall	F_1
Convolution	0.7391	0.7628	**0.7508**
Dense	0.7402	0.7340	0.7370
Ensemble	0.6973	**0.8049**	0.7472
LSTM	**0.7592**	0.6984	0.7275
Baseline: RF	0.7233	0.5633	0.6334

the predictions, whereas macro-averaging is where the *F* score for each class (here the functional group) is calculated separately and then averaged. There is a clear relationship between the number of times molecules featuring a particular group appear in the training data and the F_1 score, which is of course hardly surprising. Some functional groups show good performance.

One noisy source of further training data could be to take the experimental sections from organic chemistry papers, which often contain a chemical name and a list of peaks from ^{13}C NMR spectra, amongst others.[43] The chemical structure, and hence the functional groups, can be inferred programmatically from the chemical name.

4.4 Case Study 2: Natural Language Processing Experiments

4.4.1 Introduction

In this section, we review the application of deep learning to named entity recognition and relationship extraction with particular attention to work in this group by Corbett and Boyle.[44,45] In both cases we contrast a system that takes a feature-engineering approach with one that, as far as possible, learns the features from the input data. These were entries in two challenges in the BioCreative series, which started in 2004. Similar to the CoNLL series of conferences described earlier, the aim of the BioCreative series is to provide a set of common evaluation tasks in order to assess the state of the art for text mining applied to biological problems.[46]

4.4.2 Chemical Entity Mentions in Patents

The first case study is the BioCreative V.5 CEMP (Chemical Entity Mentions in Patents) task which involves identifying chemical named entities in patents.[47] It took 21 000 training and 9000 test abstracts from patent categories *A61K31 – Medicinal preparations containing organic active ingredients* and *A61P – Specific therapeutic activity of chemical compounds or medicinal preparations*.

The overall shape of the problem here is that the system should take the text of the patent as input and output the starting and ending character indices of the chemical entities. A naive approach to encoding the input data would be to assign each different word a different number, so-called "one-hot" encoding. A method that takes advantage of the similarities between words is that of word embeddings, for example word2vec[48] and GloVe.[49] They operate on the Firthian principle that you shall know a word by the company it keeps.

Corbett and Boyle used two systems based on LSTMs, the overall idea of which is described above. The first of the two systems was a "traditional" system that operated at the token level, combined a rich feature set initially described in Copestake and Corbett[50] with trainable token embeddings, chemical dictionaries and a single hidden recurrent layer. It generated a large model file and took hours to train. Conversely the second system, the "minimalist" one, operated at the character level, only used character embeddings, did not use any external resources, had three hidden recurrent layers, took days to train but generated a small model file.

4.4.3 Deep Learning *vs.* Feature Engineering for Relationship Extraction

The second case study is taken from the BioCreative VI challenge described in Krallinger *et al.*[51] which involves identifying chemical–protein relations. We compare the methods and results in Corbett and Boyle,[45] which focussed on transfer learning from a large corpus, with those in Peng *et al.*,[52] which, despite using deep learning, was in many ways a more traditionalist feature-engineering paper.

Relations between small molecules and gene products are key to medicinal chemistry. Because genes and gene products (proteins or RNAs) are typically referred to by the same name, the task treats them identically. There was no precisely comparable foregoing task in the literature. However, there had been a previous BioCreative task in finding chemical–disease relations,[53] for which the winning *F*-score was 57.03%, and in a recent BioCreative protein–protein interaction task the winning *F*-score was 55%, both of which scores are significantly lower than named entity recognition.[54] The gold-standard corpus was taken from PubMed and manually annotated by the organisers. Unfortunately no interannotator agreement score is reported by the organisers so we do not know what a likely upper bound is for performance on this task. Each abstract in the data set was annotated with a chemical entity, a protein and the relation between the two, which could be up-regulation/activation (CPR:3), down-regulation/inhibition (CPR:4), agonist activity (CPR:5), antagonist activity (CPR:6), the protein acting as a substrate (CPR:9), or indeed the absence of a relation. The training set consisted of 1020 abstracts which were divided into a development set of 612 and a test set of 3399.

Previous deep-learning work in this area includes that by Crichton *et al.*, who have systematically compared the performance of a multitask deep learning model against those of single task models on named entity

Table 4.3 Results of chemical–protein interaction task in Corbett and Boyle. Adapted from ref. 45, https://doi.org/10.1093/database/bay066, under the terms of a CC BY 4.0 license, https://creativecommons.org/licenses/by/4.0/.

Corpus	Precision	Recall	*F*
Development	0.5652	0.7042	0.6271
Test	0.5610	0.6784	0.6141

recognition for entities from a wide range of biomedical datasets.[55] A serious problem with multitask training for named-entity recognition tasks in the same domain is that the annotation guidelines often differ in terms of scope and in terms of boundary specifications. This has been studied quantitatively by Wang *et al.* for gene and protein annotations, who found that the *F*-score is often significantly worse for systems that have been trained on a combination of annotated corpora than on a single corpus.[56]

The key methodological advance in the Corbett and Boyle paper was to use transfer learning. This application of transfer learning is where one finds a task that a neural network can do with unannotated data, hence avoiding the discrepancies between annotation guidelines mentioned above, train it on that, and then use the output of that neural network as the input to another system that can be trained on labelled data. A previous example of transfer learning in biomedical NER is due to Giorgi and Bader who make use of silver-standard corpora.[57] A silver-standard corpus is one that has been generated by using multiple pre-existing automated taggers to tag a large collection of text and an automated process to reconcile their results. An example is the CalBC corpus described by Rebholz-Schuhmann *et al.*[58] Giorgi and Bader use a hybrid model that consists of an LSTM combined with a conditional random field (CRF) layer with separate embeddings and bi-LSTMs for tokens and characters.

The task that used unannotated data in this case was to predict the next word in a sequence, or when the bidirectional layer is running in reverse, the previous word. To build embeddings for this task they took the full texts of patents from patent categories A61K31 and A61P as above, the full texts of RSC journal papers from 2000 to the end of 2016, and 24 million paragraphs, either title or abstract, from PubMed records from 1809 to the end of 2015, resulting in a file of 6.7 billion tokens, and used GloVe to extract 300-dimensional vectors with a window size of 15 and an x_{max} (the scaling factor and cutoff for counting co-occurrences) of 100.

The output of this "pre-training" network generated embeddings that were then fed into a "recognition" network. The first two hidden layers of this were a bidirectional LSTM, followed by convolution, concatenation, another bidirectional LSTM, a pooling layer and a dense layer that produced an output vector of length six, each element being a score for a type of relation between a chemical and a protein. Table 4.3 shows the results of the system on the CHEMPROT task. The recall was markedly higher than the precision (Table 4.4).

Table 4.4 CHEMPROT results reported by Peng *et al.* Adapted from ref. 52 with permission from Oxford University Press.

System	Precision	Recall	*F*
SVM	0.6291	0.4779	0.5430
CNN	**0.6406**	0.5713	0.6023
RNN	0.6080	**0.6139**	**0.6094**
Ensemble: majority voting	0.7408	0.5517	0.6319
Ensemble: random forest	**0.7554**	**0.5524**	**0.6378**

In contrast, Peng *et al.* use an extremely rich set of features as input to a support vector machine (SVM), a convolutional neural network (CNN) and a recurrent neural network (RNN), which are then ensembled in different ways. The SVM is the most traditionalist approach, being trained on: (1) features from a window size of five tokens around the chemical and protein mentions of interest, including the lemma, POS tag and chunk type as obtained by the GENIA tagger,[59] (2) the words in between the two entity mentions in the paragraph, (3) the number of words between the two entity mentions in the sentence, (4) a hand-built list of keywords, for example "inhibit" for CPR:4, and (5) features extracted from a dependency parse of the relevant sentence.[¶] The CNN took as its input a concatenation of word, POS tag, IOB-chunk tag, whether it was a named entity (from the GENIA tagger), a dependency label and positions relative to the chemical entity and the protein entity. The RNN was a bidirectional LSTM to which the input was a concatenation of the word, distance in tokens from the relevant chemical name, distance in tokens from the relevant protein name, POS tag and IOB-chunk tag, the embeddings for all of which were updated during tagging.

The two ensembling methods were majority voting and a Random Forest classifier trained on the output of the three methods.

4.5 Conclusions and Future Work

Some points about machine learning in general and deep learning in particular are worth reiterating. In all of these tasks it is important to have a strong baseline system to compare performance against, and indeed to bear in mind the downstream task that the machine-learning system's output is

[¶]Dependency parsing is an approach to syntactic analysis which has gained ground in recent years because it is fast, requiring only a shift–reduce parser, and unlike purely context-free approaches, in principle has few requirements on word order. The notion is that each word in a sentence is connected to at least one other word by a typed dependency, for example nsubj (nominal subject of verb) or det (connecting a determiner, like "a" or "the", to the noun that succeeds it). This converts a sentence to a tree and enables each word in the sentence to be labelled with not only the types of its dependencies but its dependents and any word it depends on.

going to serve as input for. Can you think of an extrinsic as well as an intrinsic validation? An orthogonal point is that interpretability may turn out to be more important to the end user than raw performance against a given metric. To this end a current area of research is the use of surrogate models, where a simpler model is trained on the output of a more complicated one. Some researchers are using deep learning itself as a surrogate model for a more complicated, and presumably even less interpretable, system.

One problem with the field of machine learning in general is that significantly less effort goes into creating gold standards than into developing new learning algorithms. Much of the data we end up using will hence be of lower quality than we would like. A counterexample to this is the very detailed datasets collected by Merck. However, these are noisy in a different way because they are the results of bioassays and as a result show a great deal of variability from experimental conditions and simple biological variation.

A desirable feature of any machine-learning system is that a given model should be generalisable to new domains or related tasks with only a small amount of retraining. The transfer-learning model in the previous section is clearly amenable to new tasks involving chemical and biomedical texts in English. A chemistry model based on chemical structures, whether they be encoded as adjacency matrices or as SMILES, seems more likely to achieve this goal than the simplistic NMR model described in this chapter. However, we have shown in this chapter how character-based deep-learning models, which have been constructed without lengthy feature engineering, can perform at very close to human levels for a well-specified and well-understood task like chemical named-entity recognition, and this, taken together with advances described in the other chapters in this book, suggests that if you have the right class of problem then deep learning is a method in which you should invest some time.

Acknowledgements

We would like to thank Nicholas Bailey for helpful and provocative discussions.

References

1. J. Ma, R. P. Sheridan, A. Liaw, G. E. Dahl and V. Svetnik, *J. Chem. Inf. Model.*, 2015, **55**, 263–274.
2. N. Srivastava, G. Hinton, A. Krizhevsky, I. Sutskever and R. Salakhutdinov, *J. Mach. Learn. Res.*, 2014, **15**, 1929–1958.
3. M. Kim and P. Smaragdis, *CoRR*, 2016, abs/1601.06071, 06071.
4. A. Krizhevsky, I. Sutskever and G. E. Hinton, in *Advances in Neural Information Processing Systems 25*, ed. F. Pereira, C. J. C. Burges, L. Bottou and K. Q. Weinberger, Curran Associates, Inc., 2012, pp. 1097–1105.
5. J. L. Elman, *Cognit. Sci.*, 1990, **14**, 179–211.

6. J. Aires-de Sousa, M. C. Hemmer and J. Gasteiger, *Anal. Chem.*, 2002, **74**, 80–90.
7. J. Galliers and K. S. Jones, *Evaluating Natural Language Processing Systems*, University of Cambridge, Computer Laboratory Technical Report UCAM-CL-TR-291, 1993.
8. J. Carletta, A. Isard, S. Isard, J. C. Kowtko, G. Doherty-Sneddon and A. H. Anderson, *Comput. Linguist.*, 1997, **23**, 13–31.
9. R. Artstein and M. Poesio, *Comput. Linguist.*, 2008, **34**, 555–596.
10. R. Collobert, J. Weston, L. Bottou, M. Karlen, K. Kavukcuoglu and P. Kuksa, *J. Mach. Learn. Res.*, 2011, **12**, 2493–2537.
11. W. N. Francis and H. Kucera, *MANUAL OF INFORMATION to accompany A Standard Corpus of Present-day Edited American English, for use with Digital Computers*, 1979.
12. M. P. Marcus, B. Santorini and M. A. Marcinkiewicz, *Comput. Linguist.*, 1993, **19**, 313–330.
13. K. Toutanova, D. Klein, C. D. Manning and Y. Singer, *Proceedings of the 2003 Human Language Technology Conference of the North American Chapter of the Association for Computational Linguistics*, 2003, pp. 252–259.
14. E. F. Tjong Kim Sang and S. Buchholz, *Fourth Conference on Computational Natural Language Learning and the Second Learning Language in Logic Workshop*, 2000.
15. E. F. Tjong Kim Sang and F. De Meulder, *Proceedings of the Seventh Conference on Natural Language Learning at HLT-NAACL 2003*, 2003, pp. 142–147.
16. M. Palmer, D. Gildea and P. Kingsbury, *Comput. Linguist.*, 2005, **31**, 71–106.
17. X. Carreras and L. Màrquez, *Proceedings of the Ninth Conference on Computational Natural Language Learning (CoNLL-2005)*, Ann Arbor, Michigan, 2005, pp. 152–164.
18. P. A. Zhmurov, A. Y. Sukhorukov, V. I. Chupakhin, Y. V. Khomutova, S. L. Ioffe and V. A. Tartakovsky, *Org. Biomol. Chem.*, 2013, **11**, 8082–8091.
19. C. Batchelor, P. Corbett and S. Teufel, in *Handbook of Linguistic Annotation*, ed. N. Ide and J. Pustejovsky, Springer Netherlands, 2017, pp. 893–903.
20. P. Corbett, C. Batchelor and S. Teufel, *Biological, Translational, and Clinical Language Processing*, Prague, Czech Republic, 2007, pp. 57–64.
21. M. Krallinger, O. Rabal, F. Leitner, M. Vazquez, D. Salgado, Z. Lu, R. Leaman, Y. Lu, D. Ji, D. Lowe, R. Sayle, R. Batista-Navarro, R. Rak, T. Huber, T. Rocktaschel, S. Matos, D. Campos, B. Tang, H. Xu, T. Munkhdalai, K. Ryu, S. Ramanan, S. Nathan, S. Zitnik, M. Bajec, L. Weber, M. Irmer, S. Akhondi, J. Kors, S. Xu, X. An, U. Sikdar, A. Ekbal, M. Yoshioka, T. Dieb, M. Choi, K. Verspoor, M. Khabsa, C. Giles, H. Liu, K. Ravikumar, A. Lamurias, F. Couto, H.-J. Dai, R. Tsai, C. Ata, T. Can, A. Usie, R. Alves, I. Segura-Bedmar, P. Martinez, J. Oyarzabal and A. Valencia, *J. Cheminf.*, 2015, **7**, S2.

22. M. Krallinger, F. Leitner, O. Rabal, M. Vazquez, J. Oyarzabal and A. Valencia, *J. Cheminf.*, 2015, **7**, S1.
23. D. M. Lowe, P. T. Corbett, P. Murray-Rust and R. C. Glen, *J. Chem. Inf. Modell.*, 2011, **53**, 739–753.
24. L. Hawizy, D. M. Jessop, N. Adams and P. Murray-Rust, *J. Cheminf.*, 2011, **3**, 17.
25. D. Lowe, Ph.D. thesis, University of Cambridge, 2012.
26. D. C. Burns, E. P. Mazzola and W. F. Reynolds, *Nat. Prod. Rep.*, 2019, **36**(6), 919–933.
27. J. Lederberg, G. L. Sutherland, B. G. Buchanan, E. A. Feigenbaum, A. V. Robertson, A. M. Duffield and C. Djerassi, *J. Am. Chem. Soc.*, 1969, **91**, 2973–2976.
28. A. M. Duffield, A. V. Robertson, C. Djerassi, B. G. Buchanan, G. L. Sutherland, E. A. Feigenbaum and J. Lederberg, *J. Am. Chem. Soc.*, 1969, **91**, 2977–2981.
29. W. Bremser, *Anal. Chim. Acta*, 1978, **103**, 355–365.
30. C. Steinbeck and S. Kuhn, *Phytochemistry*, 2004, **65**, 2711–2717.
31. S. Kuhn, B. Egert, S. Neumann and C. Steinbeck, *BMC Bioinf.*, 2008, **9**, 400.
32. A. M. Castillo, A. Bernal, R. Dieden, L. Patiny and J. Wist, *J. Cheminf.*, 2016, **8**, 26.
33. X. Fan, W. Ming, H. Zeng, Z. Zhang and H. Lu, *Analyst*, 2019, **144**, 1789–1798.
34. Y. Binev and J. Aires-de Sousa, *J. Chem. Inf. Comput. Sci.*, 2004, **44**, 940–945.
35. K. Ghosh, A. Stuke, M. Todorovic, P. B. Jørgensen, M. N. Schmidt, A. Vehtari and P. Rinke, *Adv. Sci.*, 2017, **6**, 1801367.
36. K. T. Schütt, F. Arbabzadah, S. Chmiela, K. R. Müller and A. Tkatchenko, *Nat. Commun.*, 2017, **8**, 13890.
37. K. Ito, Y. Obuchi, E. Chikayama, Y. Date and J. Kikuchi, *Chem. Sci.*, 2018, **9**, 8213–8220.
38. C. Steinbeck, S. Krause and S. Kuhn, *J. Chem. Inf. Comput. Sci.*, 2003, **43**, 1733–1739.
39. ACD/I-Lab Version 8.0 for Microsoft Windows User's Guide, 2010.
40. A. Liaw and M. Wiener, *R. News*, 2002, **2**, 18–22.
41. F. Chollet *et al.*, *Keras*, 2015, https://github.com/fchollet/keras.
42. F. A. Gers, J. Schmidhuber and F. Cummins, *1999 Ninth International Conference on Artificial Neural Networks ICANN 99. (Conf. Publ. No. 470)*, **vol. 2**, 1999, pp. 850–855.
43. S. E. Adams, J. M. Goodman, R. J. Kidd, A. D. McNaught, P. Murray-Rust, F. R. Norton, J. A. Townsend and C. A. Waudby, *Org. Biomol. Chem.*, 2004, **2**, 3067–3070.
44. P. Corbett and J. Boyle, *J. Cheminf.*, 2018, **10**, 59.
45. J. Boyle and P. Corbett, *Database*, 2018, **2018**, bay066.
46. C. N. Arighi, Z. Lu, M. Krallinger, K. B. Cohen, W. J. Wilbur, A. Valencia, L. Hirschman and C. H. Wu, *BMC Bioinf.*, 2011, **12**, S1.

47. M. Pérez-Pérez, O. Rabal, G. Pérez-Rodriguez, M. Vazquez, F. Fdez-Riverola, J. Oyarzabal, A. Valencia, A. Lourenço and M. Krallinger, *Proceedings of the Biocreative V.5 Challenge Evaluation Workshop*, 2017, pp. 11–18.
48. T. Mikolov, K. Chen, G. Corrado and J. Dean, *arXiv preprint*, 2013, abs/1301.3781.
49. J. Pennington, R. Socher and C. D. Manning, *Empirical Methods in Natural Language Processing (EMNLP)*, 2014, 1532–1543.
50. P. Corbett and A. Copestake, *BMC Bioinf.*, 2008, **9**, S4.
51. M. Krallinger, O. Rabal, S. A. Akhondi, M. P. Pérez, J. Santamaría, G. P. Rodríguez, G. Tsatsaronis, A. Intxaurrondo, J. A. Lopez, U. Nandal, E. M. van Buel, A. P. Chandrasekhar, M. Rodenburg, A. Laegreid, M. A. Doornenbal, J. Oyarzábal, A. Lourenço and A. Valencia, *Proceedings of the BioCreative VI Workshop*, 2017.
52. Y. Peng, Z. Lu, A. Rios and R. Kavuluru, *Database*, 2018, **2018**, bay073.
53. J. Li, Y. Sun, A. P. Davis, C. J. Mattingly, D. Sciaky, R. J. Johnson, T. C. Wiegers, C.-H. Wei, R. Leaman and Z. Lu, *Database*, 2016, **2016**, baw068.
54. M. Krallinger, F. Leitner, C. Rodriguez-Penagos and A. Valencia, *Genome Biol.*, 2008, **9**, S4.
55. G. Crichton, S. Pyysalo, B. Chiu and A. Korhonen, *BMC Bioinf.*, 2017, **18**, 368.
56. Y. Wang, J.-D. Kim, R. Sætre, S. Pyysalo and J. Tsujii, *BMC Bioinf.*, 2009, **10**, 403.
57. J. M. Giorgi and G. D. Bader, *Bioinformatics*, 2018, **34**, 4087–4094.
58. D. Rebholz-Schuhmann, A. J. Jimeno-Yepes, E. M. van Mulligen, N. Kang, J. Kors, D. Milward, P. Corbett, E. Buyko, K. Tomanek, E. Beisswanger and U. Hahn, *Proceedings of the Seventh conference on International Language Resources and Evaluation (LREC'10)*, 2010.
59. Y. Tsuruoka and J. Tsujii, *Proceedings of Human Language Technology Conference and Conference on Empirical Methods in Natural Language Processing*, Vancouver, British Columbia, Canada, 2005, pp. 467–474.

Section 3: Ligand-based Predictive Modelling

CHAPTER 5

Concepts and Applications of Conformal Prediction in Computational Drug Discovery

ISIDRO CORTÉS-CIRIANO*[a] AND ANDREAS BENDER*[b,†]

[a] European Molecular Biology Laboratory, European Bioinformatics Institute, Wellcome Genome Campus, Hinxton CB10 1SD, UK; [b] Centre for Molecular Informatics, Department of Chemistry, University of Cambridge, Lensfield Road, Cambridge CB2 1EW, UK
*Emails: icortes@ebi.ac.uk; ab454@cam.ac.uk

5.1 Introduction

A major research area in machine learning is the development of algorithms to compute the reliability of individual predictions. Such reliability estimates are essential to increase the trust and application of artificial intelligence solutions to guide decision making, especially in the context of healthcare and personalized medicine, where correctly identifying, *e.g.*, which patients are likely to benefit from a particular drug treatment, has strong ethical and legal implications.[1]

Estimating the reliability of predictive models is of particular relevance in drug discovery, where both the prediction and the associated uncertainty need to be taken into account for decision making.[2] The set of molecules for which a model is expected to generate reliable predictions is termed the

[†]Current address: AstraZeneca, Clinical Pharmacology & Safety Sciences (CPSS), Data Science & AI, Cambridge, UK

Drug Discovery Series No. 75
Artificial Intelligence in Drug Discovery
Edited by Nathan Brown

Published by the Royal Society of Chemistry, www.rsc.org

applicability domain of a model.[3] Therefore, the development of algorithms to define and compute the applicability domain of predictive models has been an area of intense research in preclinical drug discovery, and due to the amount of recent research in the field we will in the following often focus on virtual screening in particular.[4–7]

The generation of a predictive model, such as a Quantitative-Structure Activity Relationship (QSAR) model, or any other property model, consists of encoding the set of molecules for which the variable of interest has been experimentally measured (*e.g.*, *in vitro* potency; the dependent or response variable) using numerical descriptors (covariates or independent variables).[8–10] Subsequently, the covariates are related to the response variable using a mathematical model, which can be simple (*e.g.*, multiple linear regression) or complex, such as Random Forests (RF), Support Vector Machines (SVM) or Deep Neural Networks, which might consist of thousands to millions of parameters.[11] These models are then used to predict the activity or property of interest for untested molecules *in silico* to prioritize for further experimental testing those with higher chances of being active. Although a plethora of algorithmically diverse approaches to encode chemical structures and to relate these to a dependent variable of interest have been developed over the last decades,[10,12] all depend on the limited amount of data available for training. This is key to drug discovery, as the amount and diversity of the training data (in this case chemical compounds) limit the set of small molecules to which a given model can be applied and give reasonable results. In practice usually more (and more homogenous) data is available in early-stage settings (such as hit discovery), and less so in later (such as *in vivo*) stages, making early stage model generation generally speaking relatively easier than late-stage models. However, what is generally applicable and a key question is how to assign the confidence, reliability, or applicability domain of a model, and while quantitatively models in different stages might have larger or smaller applicability domains, its importance is universal.

Widely-used algorithms to generate QSAR models, such as Support Vector Machines (SVM) or Random Forest (RF), output a single value (regression) or label (classification) for new instances.[13–15] In practice, the lack of uncertainty estimates for point or single-class predictions hampers the use of computational models to guide prospective screening experiments in early-stage drug discovery, given that 'desired' values of an output variable may well be associated with high uncertainty (and *vice versa*), but this will not be explicitly communicated to the user by the algorithm. However, in the same way experiments need to be repeated in order to assign both a measurement and its standard deviation of a variable, also computational predictions need to output the expectation value of the model and an associated confidence interval.

We note that there exist algorithms whose outputs are well-calibrated probability distributions rather than point predictions, *e.g.*, Gaussian Processes (GP)[16–19] or Dropout Neural Networks.[20,21] However, these are generally computationally intense (*e.g.,* GP training requires the inversion of the covariance matrix, which drastically increases their computational footprint), require the optimization of a large number of parameters, as well as prior knowledge about these.[19] Hence, most applications in the medicinal chemistry literature use alternative algorithms (*e.g.*, RF), which are faster to train (by *e.g.*, parallelizing the training phase), and require less parameter optimization.[22,23] Therefore, much effort has been invested in the community to develop uncertainty estimation methods that could be easily adapted to algorithms used in practice.[22,24–27]

To date, a plethora of diverse algorithms has been developed to define the applicability domain of virtual screening models.[3,5,7,10,12,17,18,22,24,25,28–33] Overall, existing methods harness the distance in descriptor space of new instances to those in the training set,[25,34] or intrinsic information derived from the models (*e.g.*, bagged variance, which is positively correlated with high average error in prediction[22,35]), to identify areas in chemical space underrepresented in the training data for which the models are unlikely to deliver reliable predictions. The performance of these approaches is generally quantified by testing whether the error in prediction increases as a function of the distance of the test molecules to those in the training data.[34] However, such correlations do not guarantee that the predictions are well-calibrated, *i.e.,* the fraction of instances whose true value lies within the predicted error interval is not guaranteed to be proportional to a user-defined tolerated error rate.

One method which has recently gained a lot of attention to address this problem is Conformal Prediction (CP; Table 5.1).[36–38] CP is a mathematical framework to generate confidence predictors to model the reliability of predictions in diverse tasks, from property modelling to multi-target drug design (Table 5.2). A confidence predictor is a type of predictor that guarantees that the true value will be within the predicted confidence region (in regression) or within a set of predicted labels (in the case of classification) at a given confidence level (CL). For instance, at a confidence level of 80%, the confidence intervals computed using a valid CP in a regression setting would contain the true value in at least 80% of the cases (Figure 5.1(a)). In classification tasks, the set of predicted classes for new instances will contain the true label in at least 80% of the cases. The prediction of well-calibrated (or *valid*) confidence regions by CP, which is usually not the case for other modelling methods, guarantees a lower bound for validation rates, and permits to limit the number of false positives, thus increasing the retrieval rate of active compounds in preclinical drug discovery.[39] The confidence level is commonly set at 0.80, as this confidence level represents a generally suitable trade-off between efficiency and validity (note that a CL of 80% is also reported as 0.80 in the CP literature).[40] The validity of CP holds if the randomness assumption is fulfilled, *i.e.*, the training instances are representative of those to which the models will be applied; or, in other words, that the data are independent and identically distributed (i.i.d.). In practice, CP are also valid if the slightly weaker assumption of exchangeability is

Table 5.1 Common terms used in CP.

Term	Definition
Validity	CPs are always valid provided that the randomness or exchangeability principles hold. In the case of regression, a CP is valid if the confidence level matches the fraction of instances whose true value lies within the predicted confidence region. For instance, at a confidence level of 80%, the confidence intervals would contain the true value in at least 80% of the cases. In classification tasks, CPs are valid in that the set of predicted classes for new instances will contain the true label in at least 80% of the cases.
Efficiency	In regression, the efficiency of a CP refers to the average size of the predicted confidence intervals. The tighter the intervals the more efficient a CP is. In the case of classification, efficiency refers to the fraction of single-class predictions that are correct.
Confidence level (CL)	The confidence level is defined by the modeler and refers to the minimum fraction of predictions whose true value will lie within the predicted confidence region, in the case of regression, and the fraction of instances whose true class will be among the set of predicted classes.
Error rate	The error rate refers to the fraction of instances whose true value lies outside the predicted confidence regions. If a CP is well-calibrated, the error rate should not be larger than 1-CL (see also Figure 5.1).
Non-conformity measure	Function used to evaluate the relatedness or conformity of new instances to those used for model training.

fulfilled, which assumes that the datapoints do not follow any particular order even if they are not i.i.d.[36,39] Although the exchangeability principle is generally assumed to hold when modelling preclinical data,[32,36,37] the authors have shown that this is not always the case in virtual iterative screening experiments using QSAR data sets (Figure 5.1(b)).[41] Overall, CP does not introduce more assumptions than those generally made when modelling bioactivity data.

To generate errors associated with predictions, CP evaluates the 'non-conformity' of a new instance to those used during training by applying a new *non-conformity measure*. Intuitively, the non-conformity measure quantifies how different a given instance is from those already seen during the training phase, which is quantified using a *non-conformity score* (α). Therefore, any metric quantifying the applicability domain of a model, *e.g.*, the distance of a new instance to the training set, can be used as non-conformity measure. The simplest non-conformity measure in a regression setting would be the unsigned error in prediction computed, *e.g.*, the instances or the hold-out folds in cross-validation. In classification tasks, non-conformity scores are generally computed using a class probability estimation method; for instance, the fraction of trees in a RF model voting for a particular class, the distance to the cut-off value or hyperplane that separates classes in the case of *e.g.*, SVM (Table 5.2). Note that the selection of a non-conformity measure usually exploits information already provided by the underlying

Table 5.2 Drug discovery studies listed in chronological order where CP was implemented.

Data sets (targets; number of compounds; number of datapoints)	Dependent variable or classes	Machine learning technique	Modelling type and task	Type of CP used	Prospective experimental validation?	Reference	Year	Remarks
10 publicly available data sets (Signature descriptors)	pIC_{50}	Random Forest (RF)	QSAR/regression and classification	Inductive Mondrian CP	No	Norinder *et al.*[32]	2014	Introduction of CP in QSAR modelling
12 human PARPs; 181 compounds; 2164 datapoints	Thermal shift values measured using differential scanning fluorimetry[45]	RF	PCM/Regression	InductiveCP	No	Cortés-Ciriano *et al.*[46]	2015	Proteochemometric modelling (PCM) of PARP inhibition
2 CAESAR binary classification datasets: carcinogenicity[47] (805 compounds) and mutagenicity[48] (4204)	Carcinogenic *vs.* non-carcinogenic	RF	QSAR/ Classification	Inductive Mondrian CP	No	Norinder *et al.*[39]	2015	Modelling the carcinogenicity and mutagenicity of compounds using Mondrian CP
Ames mutagenicity data set	Non-mutagenic *vs.* mutagenic	Support Vector Machines (SVM)	QSAR/ Classification	Inductive Mondrian CP	No	Ahlberg *et al.*[49]	2015	Interpretation of classification models
NCI60 data set (59 cancer cell lines; 17 142 compounds; 941 831 datapoints)	pGI_{50} (50% growth inhibition bioassay end-point)	RF	PCM/ Regression	Inductive CP	No	Cortés-Ciriano *et al.*[50]	2016	PCM of growth inhibition patterns across the NCI60 data set. Webserver[51] available at www.cclp.marseille.inserm.fr
2 datasets: Estrogen Receptor (binders/non-binders: 445/476) and Androgen	Binders *vs.* non-binders (based on IC_{50} and relative	RF	QSAR/ Classification	Aggregated Mondrian CP	No	Norinder *et al.*[54]	2016	Modelling compound binding to the estrogen and

Table 5.2 (*Continued*)

Data sets (targets; number of compounds; number of datapoints)	Dependent variable or classes	Machine learning technique	Modelling type and task	Type of CP used	Prospective experimental validation?	Reference	Year	Remarks
Receptor (binders/ non-binders: 292/ 633)[52,53]	binding affinity values)							androgen receptors
The Directory of Useful Decoys, Enhanced (DUD-E)[55]	Actives *vs.* inactives	RF	QSAR/ Classification	Aggregated Mondrian CP	No	Svensson *et al.*[56]	2017	Application of CP to guide docking experiments in an iterative fashion
18 bioactivity data sets from the ExCAPE database[8]	Actives *vs.* inactives	SVM	QSAR/ Classification	Mondrian Cross-CPs	No	Sun *et al.*[57]	2017	Large-scale analysis of the application of Mondrian CPs to model imbalance data sets with imbalance levels of 1:10 to 1:1000. Non-conformity measure based on the SVM decision function
3 datasets from PubChem (AID493091, AID2796 and AID1851) and Ames mutagenicity data set from Hansen *et al.*[58]	Actives *vs.* inactives	SVM	QSAR/ Classification	Aggregated Mondrian CP	No	Norinder *et al.*[59]	2017	Binary classification of imbalanced data using SVM and the distance to the separating hyperplane as the non-conformity measure
Data set extracted from Baba *et al.*[60] encompassing 211 compounds	Permeation rate (log Kp) through the human skin	RF and SVM	Regression	Aggregated Mondrian CP	No	Lindh *et al.*[61]	2017	Prediction of the permeation rate (log Kp) of chemical compounds through the human skin

16 cytotoxicity datasets extracted from PubChem (data set size ranging from 29 938 to 404 016 compounds)	Toxic *vs.* non-toxic	RF	QSAR/ Classification	Aggregated Mondrian CP	No	Svensson *et al.*[62]	2017	Large scale modeling of cytotoxicity using largely imbalanced dataset (0.8% average fraction of toxic compounds)
25 protein targets from ChEMBL version 23	pIC_{50}	RF	QSAR/ Regression	Cross-CP	No	Cortés-Ciriano *et al.*[41]	2018	Demonstration that the exchangeability principle does not hold when CPs trained on inactive molecules are applied generate CI for active molecules
24 protein targets from ChEMBL version 23	pIC_{50}	Fully-connected deep neural networks	QSAR/ Regression	Deep Confidence	No	Cortés-Ciriano *et al.*[63]	2018	Deep Confidence: CP strategy that used snapshot ensembles[64] to generate CPs for deep neural networks
936 primary aromatic amines (630 mutagenic and 306 non-mutagenic)	Mutagenic *vs.* non-mutagenic	RF	QSAR/ Classification	Inductive Mondrian CP	No	Norinder *et al.*[65]	2018	Prediction of aromatic amine mutagenicity
DGM/NIHS data set[66] encompassing 12 140 compounds categorized as: strong mutagenic, mutagenic and non-mutagenic.	Strong mutagenic or mutagenic *vs.* non-mutagenic	RF	QSAR/ Classification	Aggregated Mondrian CP	No	Norinder *et al.*[67]	2018	Prediction of Ames mutagenicity

Table 5.2 (*Continued*)

Data sets (targets; number of compounds; number of datapoints)	Dependent variable or classes	Machine learning technique	Modelling type and task	Type of CP used	Prospective experimental validation?	Reference	Year	Remarks
316 974 from the PubChem BioAssay database (AID540275)	Active *vs.* inactive	RF	QSAR. Classification	Aggregated Mondrian CP	No	Norinder *et al.*[68]	2018	Virtual screening of vanilloid receptor type 1 (TRPV1) agonists
15 350 data points across 31 bromodomain collected from ChEMBL 20, PubChem, ChEpiMod, GOSTAR, and proprietary AstraZeneca databases	IC_{50}, K_i, K_d, % inhibition and ΔTm	RF, SVM and Generalized Linear Models (GLM)	PCM. Classification	Mondrian Cross-CPs	Yes	Giblin *et al.*[69]	2018	Application of PCM and CP to model the activity of compounds across bromodomains. The model predictions for 1139 compounds were prospectively validated
29 property, toxicity and bioactivity data sets from ChEMBL version 19	pIC_{50}, logS, Toxicity	RF, Gradient Boosting Machines (GBM), Lasso, Ridge Regression, Bayesian Ridge Regression, Adaptive Boosting (AdaBoost), Automatic Relevance Determination (ARD), Elastic Net, and Partial Least-Squares (PLS)	QSAR and QSPR. Regression	Inductive and Aggregated CP	No	Svensson *et al.*[40]	2018	Benchmarking study of non-conformity measures using QSAR data sets

12 PubChem data sets (AIDs: 411, 868, 1030, 1460, 1721, 2314, 2326, 2451, 2551, 485 290, 485 314, 504 444)	Actives *vs.* inactives	RF	QSAR/ Classification	Aggregated Mondrian CP	No	Svensson *et al.*[70]	2018	Application of a gain-cost function including the screening cost and gain in terms of bioactivity information (*e.g.*, discovery of a new hit)
108 compounds[71]	Skin sensitizers *vs.* non-sensitizers	SVM	QSAR/ Classification	Aggregated Mondrian CP	No	Forreryd *et al.*[72]	2018	Application of CP to model the readout of the Genomic Allergen Rapid DetectionAssay used to test skin sensitizers
ExCAPE database[8]	Actives *vs.* inactives	SVM	QSAR/ Classification	Aggregated Mondrian CP	No	Lampa *et al.*[43]	2018	Prediction of off-target binding profiles across 31 targets used in early hazard assessment
8 toxicology data sets including off-target functional assays, cytotoxicity tests, mutagenicity tests, CYP450 inhibition assays, acute oral toxicity assays and transporter assays extracted from the literature and public databases	Actives *vs.* inactives	RF	QSAR/ Classification	Inductive Mondrian CP	No	Ji *et al.*[73]	2018	eMolTox: webserver for the prediction of compound toxicity
1 592 127 compounds from ChEMBL version	Compound lipophilicity (logD)	SVM	QSPR/ Regression	Cross-CP	No	Lapins *et al.*[74]	2018	Large-scale prediction of LogD. The logD

Table 5.2 (*Continued*)

Data sets (targets; number of compounds; number of datapoints)	Dependent variable or classes	Machine learning technique	Modelling type and task	Type of CP used	Prospective experimental validation?	Reference	Year	Remarks
23 used for model development. The development model was applied to 91 498 351 compounds extracted from PubChem database								prediction model is available as a REST service at https://cplogd.service.pharmb.io/
5 data sets from the UCI Machine Learning Repository[75]	Two classes	RF	Binary classification	Aggregated Mondrian Transductive CP	No	Spjuth *et al.*[76]	2018	Illustration of a novel framework (Non-Disclosed aggregated CP) to aggregate CPs from multiple sources while keeping the data on which they trained private. High impact for drug discovery research
MNIST dataset, and 3 drug discovery data sets measuring the Ames Mutagenicity, mitochondrial function, and interaction with the aryl	Two classes	SVM +	QSAR/Classification	Inductive Mondrian CP	No	Gauraha *et al.*[77]	2018	Development of inductive CPs using the Learning Under Privileged Information paradigm (LUPI) and SVM +

hydrocarbon receptor of small molecules								
550 human protein targets from ChEMBL versions 23 and 24	pChEMBL	RF	QSAR/ Classification	Inductive Mondrian CP	No	Bosc *et al.*[78]	2019	Large-scale comparison of QSAR and CP showed that, overall, CP outperforms QSAR, although for some targets the opposite was observed
56 892 compounds from 10 high-throughput screening data sets originating from PubChem[79]	Active/Inactive	Matrix Factorization (Macau)	Multi-target binary classification	Macau + Mondrian Aggregated CP	No	Norinder *et al.*[80]	2019	Multitask learning using CP and matrix factorization
24 protein targets from ChEMBL version 23	pIC_{50}	Fully-connected deep neural networks	Regression	Test-Time Dropout CP	No	Cortés-Ciriano *et al.*[81]	2019	First study integrating test-time dropout and CP

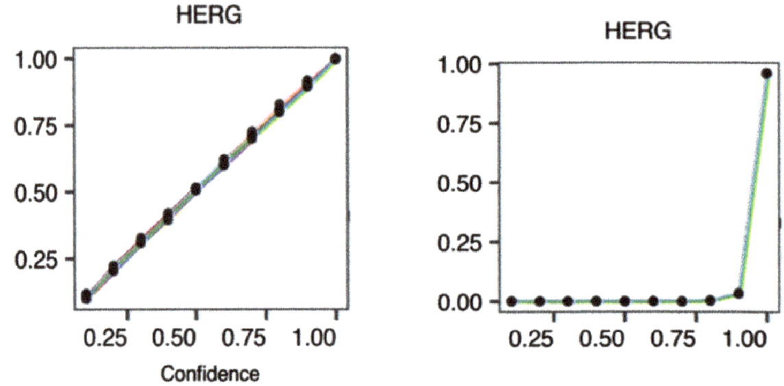

Figure 5.1 (a) Example of a calibration plot for a well-calibrated Regression Conformal Predictor (adapted from ref. 41). Calibration plots represent the fraction of confidence regions encompassing the true value (or, 1 – estimated error rate) across increasingly larger confidence levels. The *x*-axis corresponds to a user-defined confidence level, and the *y*-axis to the fraction of instances in the test set whose true value lies within the predicted confidence regions. It can be seen that the fraction of instances whose true value is within the predicted confidence interval (*y*-axis) is correlated with the confidence level, and hence this represents a 'well-calibrated' predictor in the context of CP, which can be used to assign the trust placed in future predictions of the model. (b) Example of a Regression Conformal Predictor that is not well-calibrated (adapted from ref. 41). It can be seen that the fraction of instances whose true value is within the predicted confidence interval (*y*-axis) is not correlated with the confidence level.

algorithm. For instance, in the case of RF models the fraction of trees voting for each class in classification, of the bagged variance in regression, are often used as non-conformity measures.

A second key aspect to consider is the *efficiency* of CP (Table 5.1). In regression, the efficiency of a CP is determined by the size of the predicted confidence regions. Even if a CP may be valid, it might not be useful in practice to guide decision making if the confidence regions span several bioactivity units.[40,42] In classification, efficiency is related to the fraction of single-class predictions that are correct.[38,43]

In addition to the features listed above, CP can be used in combination with any machine learning algorithm,[32,37] requires minimal computational cost beyond the training of the underlying algorithm,[32] and no parameterization is required except for the selection of a non-conformity measure.[40] Hence, CP is a technique that permits computation of well-calibrated and easy-to-interpret errors in prediction for both balanced and imbalanced data sets, as we discuss in more depth below, at minimal computational cost.[44]

The first part of this review provides an overview of the design and advantages of the CP implementations used in drug discovery settings to date. The second part revisits the molecular modelling studies where CP played a key role in either estimating the reliability of individual predictions or in guiding prospective screening experiments. Finally, we discuss the current limitations of CP and offer a perspective on future directions.

5.2 Conformal Prediction Modalities Commonly Used in Computer-aided Drug Design

In this section, we will describe the most widely used CP modalities in the drug discovery literature, and the steps required to generate them (see a general pipeline below) using a real-world bioactivity data set. We refer the reader to the work of Vovk *et al.*[32,36,37] for further details about the mathematical foundations underlying the CP framework, and to Norinder *et al.*[32] and Eklund *et al.*[44] for an introduction focused on the application of CP to virtual screening.

The CP implementations we will discuss follow the same core steps, namely:

1. choosing a non-conformity measure to evaluate the non-conformity between the training and the test instances;
2. training the machine learning model of choice, and evaluate the non-conformity values for the training examples;
3. applying the trained model to the test or external set instances;
4. for each test set instance, evaluating its non-conformity with respect to the training data using the same non-conformity measure used in step 1: the higher the conformity of the new instance, the higher the reliability of the prediction;
5. identifying reliable predictions given the user-defined significance and confidence levels; and
6. evaluating the validity and efficiency of the generated CP.

5.2.1 Inductive Conformal Prediction (ICP)

5.2.1.1 Inductive Conformal Prediction for Classification

To generate an ICP model, the data available for training is divided into (see Figure 5.2) (i) a training set (also termed proper training set[82]), consisting of *e.g.*, 70% of the data, (ii) a calibration set, encompassing *e.g.*, 15% of the data, and (iii) a test set, consisting of the remaining datapoints. Both the proper training and calibration sets are used for training, whereas the test set is used to evaluate the predictive power of the models, as well as the validity and efficiency of the CPs generated.

A machine learning model is firstly trained on the proper training set (step 1 in Figure 5.2), and subsequently applied to predict the activity label for the instances in the calibration set (step 2 in Figure 5.2; for illustration purposes, we here consider that the underlying model is always a RF model unless otherwise stated). The true and predicted class labels for the calibration set instances serve to compute non-conformity scores using the non-conformity measure of choice, for instance, the fraction of trees from a Random Forest Model voting for the class receiving most votes. The non-conformity score serves to quantify the non-conformity (or 'strangeness') of a new instance with respect to those used for training. Predictions for which most of the trees in the Random Forest predict the same label are more reliable, and hence, have a higher non-conformity score than cases where the algorithm does not delineate both classes correctly; that is, both classes are predicted to be equally likely. The model is then applied to the test set and the fraction of trees voting for the most voted class is recorded for each instance (step 3 in Figure 5.2). Finally, the *P* value for each test set instance is calculated as the fraction of instances in the calibration set with a non-conformity score equal to or smaller than the non-conformity score for the instance under consideration. Thus, the *P* value represents the ranking of the non-conformity score for a new instance with respect to the non-conformity score list generated for the calibration set. The *P* value is then compared against the significance level, ε, which is defined as 1 – CL. A prediction is considered reliable if the *P* value is higher than ε. Note that

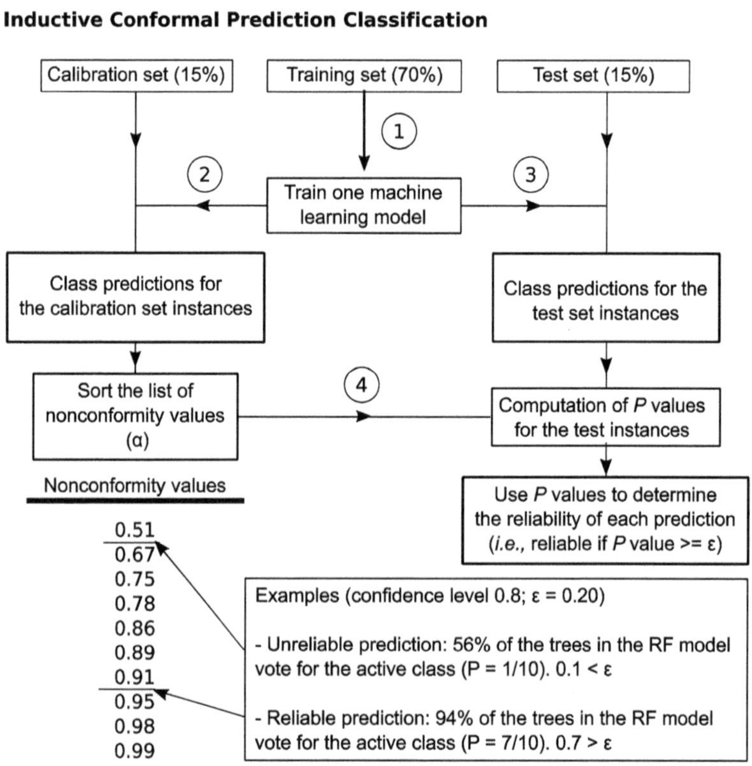

Figure 5.2 Steps for the generation of an ICP in the context of classification. As in the main text, ε denotes the significance level (Table 5.1).

the concept of *P* value used here is not equivalent to traditional *P* values used in statistics.

To illustrate this approach using real-world data, we collected pIC_{50} data for 5207 compounds against the human ether-a-go-go-related gene (hERG) potassium channel from the ChEMBL database version 23.[9] To generate a RF binary classification model, we considered as active those compounds with a pIC_{50} value ≥ 7 ($n = 332$), and assigned the remaining compounds to the inactive class ($n = 4875$). The resulting data set had an imbalance in the ratio of active to inactive compounds of ~1:15. This data set was previously used to benchmark bioactivity modelling pipelines,[83] and is publicly available at: https://github.com/isidroc/kekulescope/tree/master/datasets. Next, we calculated circular Morgan fingerprints for all compounds using RDkit (release version 2013.03.02) with a radius of 2 and a fingerprint length of 1024 bits. To generate an RF-based ICP using a confidence level of 80% we followed the steps described in Figure 5.2, with the exception that 60% of the data was used as test set in this case for illustration purposes to ensure that we had enough active compounds to compute validity estimates for the resulting CPs.

We report in Figure 5.3 the distribution of non-conformity scores (a) for the calibration and (b) test sets. Unreliable predictions on the test set (shown in red in Figure 5.3(b)) are those whose non-conformity value is smaller or equal than the 80th percentile of the list of non-conformity scores for the calibration set, indicated by the cross in Figure 5.3(a), and by the black arrow in Figure 5.3(a). As stated above, the mathematical validity of CP guarantees that at least 80% of the predictions considered reliable will be correct. However, the validity is only guaranteed globally, meaning that the error rate for the reliable predictions (*i.e.,* fraction of predictions that are incorrect) might be different across classes, as some classes are harder to predict than others. This is a major issue when modelling imbalanced data, as the error rate will be higher for, usually, the minority class.[39] In fact, in the hERG data set modeled here we observe that most active compounds are predicted as unreliable (Figure 5.3(c)). Therefore, although 80% of the predictions flagged as reliable are correct (global validity), the percentage of active compounds with a reliable prediction is only 24% (28/115; Table 5.3 and Figure 5.3), and more importantly, only 14% of the active compounds with a reliable prediction are correctly predicted as active (Table 5.3). Thus, the local validity for the active class is not guaranteed. This behavior is understandable due to the imbalance of the dataset used, which includes about 15 times more information about inactive compared to active compounds. Being aware of which predictions are more reliable than others now allows the modeler to proceed with those predictions with the desired confidence, *e.g.*, for subsequent experimental compound testing.

The lack of validity for each class (or local validity) fostered the development of *Inductive Mondrian CP*,[39,84] which permits calibration of the error rates in a class-specific manner.

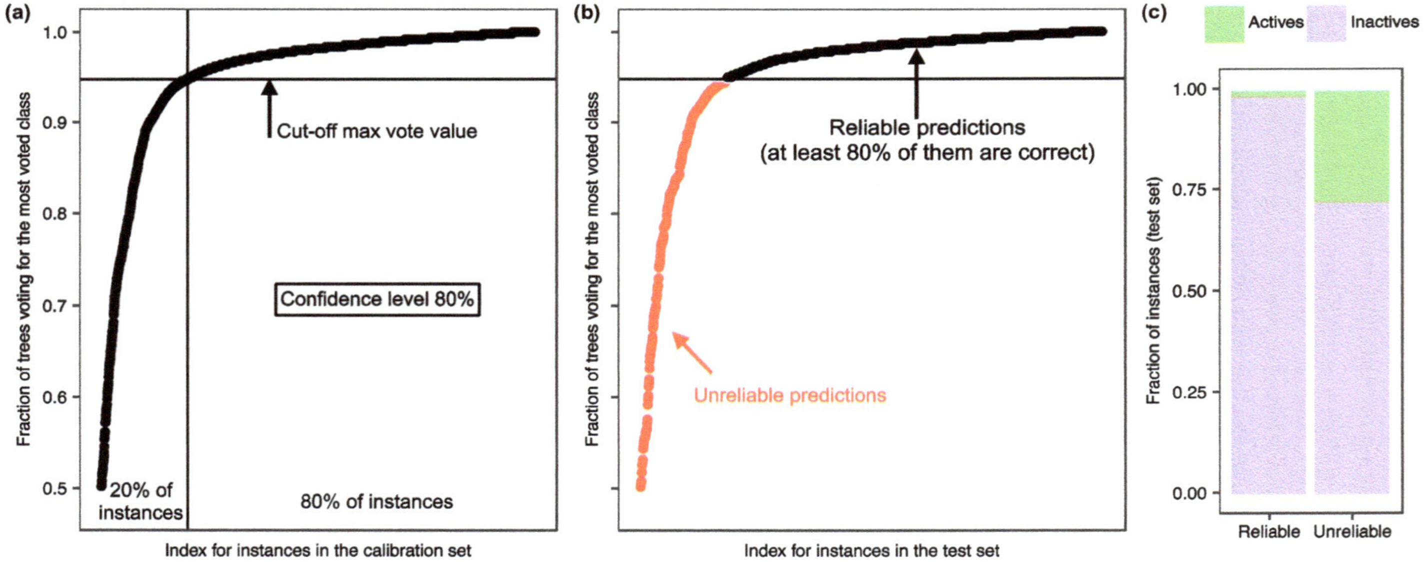

Figure 5.3 Generation of an ICP using the bioactivity data stored in ChEMBL version 23 for target hERG. The instances in the data set were assigned to the active or inactive class using a cut-off value of 7 pIC_{50} units to generate a highly-imbalanced data set (active to inactive ratio of 1:15). (a) Non-conformity values for the instances in the calibration set. In this example the non-conformity function chosen corresponds to the fraction of trees voting for the most voted class. For instance, if 85% of the trees in the RF model vote for the active class, the non-conformity score would be 0.85. The sorted list of non-conformity scores serves to calculate α_{CL}. (b) The non-conformity scores are calculated for each instance in the test set. Those predictions with α values equal or greater than α_{CL} are considered reliable. In this case, unreliable predictions are those for which roughly the same number of trees in the RF model vote for each class. In both (a) and (b) the instances in the *x*-axis have been sorted according to their non-conformity score (*y*-axis). (c) Distribution of inactive and active compounds in the set of reliable (left) and unreliable (right) predictions. The figure shows that most of the reliable predictions correspond to inactive compounds, which are easier to model given the imbalance in the data. Therefore, although at least 80% of the predictions considered reliable are correct, most of the active compounds are assigned an unreliable prediction (Table 5.3), allowing the modeler to focus on those predictions with the desired confidence in turn.

Table 5.3 Confusion matrix for the classification model trained on the hERG data set, showing that the distribution of reliable and unreliable predictions across classes is not even, and in line with the data distribution in the training data set (which is biased towards the inactive class). The results correspond to the prediction for the test set molecules, and the reliability assignment now allows for the selection of molecules with the desired confidence for subsequent steps of *e.g.*, experimental testing. See also Figure 5.3.

	Reliable predictions		Unreliable predictions	
	Active	Inactive	Active	Inactive
Predicted Active	4	0	43	44
Predicted Inactive	24	1214	44	228
Total	28	1214	87	272

5.3 Handling Imbalanced Datasets: Mondrian Conformal Prediction (MCP)

In MCP each class (*e.g.*, active and inactive) is treated separately, and the confidence in the assignment of a given instance to the classes considered is evaluated independently. That is, a list of non-conformity scores is generated for each class using the predictions for the calibration set (Figure 5.4). Thus, in a binary classification setting a compound might be classified as 'active', 'inactive', both active and inactive (class 'both'), or not assigned to either of them (class 'null' or 'empty').

Specifically, the steps to generate an inductive MCP using a RF classification model are (Figure 5.4):

- Step 1. Training an RF model on the proper training set.
- Step 2. Applying this model to classify the instances in the calibration set, and creating a list of non-conformity values (*e.g.*, fraction of trees voting for that class in the case of RF) for each category (*i.e.*, active and inactive in a binary classification setting). Next, sorting the lists of non-conformity values in increasing order, which are termed Mondrian class lists.
- Step 3. Applying the model trained in step 1 to the test set, and for each instance compute the fraction of trees voting for each class.
- Step 4. For each instance in the test set computing a *P* value for each class. The *P* value is the fraction of cases in the corresponding Mondrian class list calculated using the calibration set predictions in step 3 smaller than the vote fraction for that class. If the *P* value for a given class is above the significance level, ε, the test set instance under consideration is predicted to belong to that category. Compounds are assigned to all categories for which the *P* value is greater or equal than ε.

Hence, a given compound may be predicted as 'active', 'inactive'. However, it can also be predicted as 'both' in cases when the model does not have enough predictive power to discriminate between classes, or 'null' in cases when the instance is outside the applicability domain of the model. Thus, MCP gives an unbiased estimate of the reliability of the predictions given the training data on a class-specific manner. While this behavior might seem counterintuitive, it is actually a straightforward consequence of the types of data and evidence that might be present in a given training dataset – for example, if there are two closely related molecules to the one a prediction is made for, one of which is active and one of which is inactive, then there is evidence for *both* classes, and (at a given confidence level) neither can be chosen over the other. Likewise, if no evidence for either class is present in a data set (such as for a chemically very novel molecule), then no prediction either way can be made in practice. This aspect of modelling data is often neglected in other modelling approaches, which forces a decision onto the model, hence resulting in single labels which are often easier to deal with in practice, but which neither consider the confidence of a prediction properly,

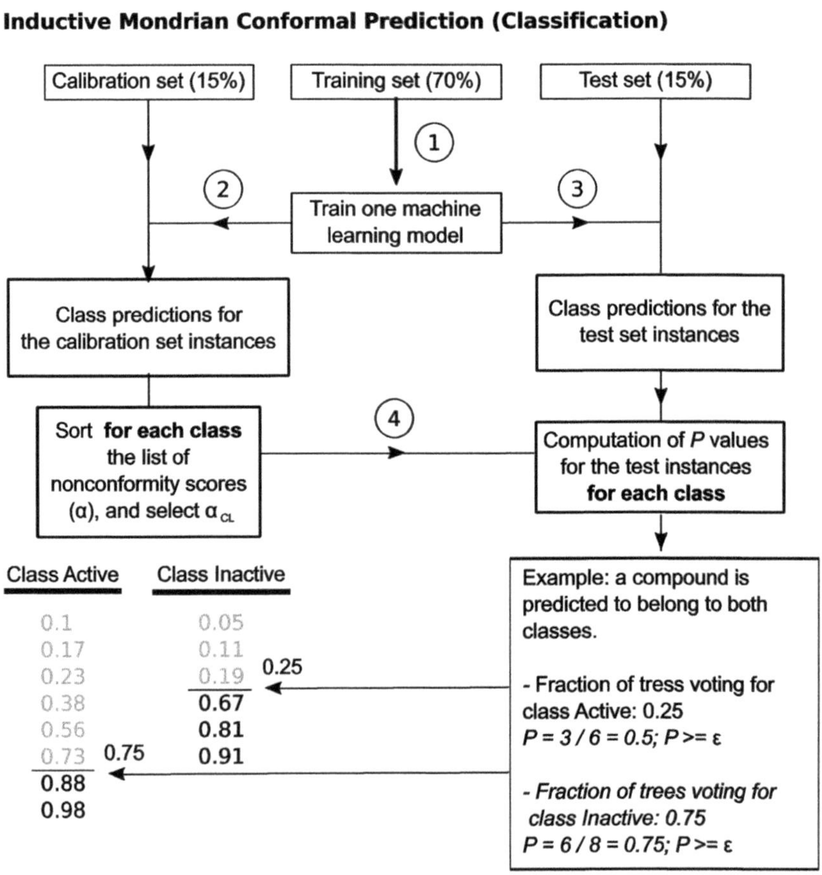

Figure 5.4 Steps required for the generation of a Mondrian CP.

nor the underlying evidence of class membership present in the available data.

The significance level, ε, indicates the maximum fraction of predictions that are incorrect. In MCP this fraction is guaranteed for each class, which means that at least $1 - \varepsilon$ of the predictions for the minority class will be correct. This is of utmost importance in drug discovery, where for example, inactive molecules usually outnumber actives by several orders of magnitude. Likewise, for models in later stages of drug discovery, being able to anticipate for which areas of chemical space no predictions can be made (and to assign a confidence to the remainder) is as important as being able to model the data in the first place.

It is important to note that increasing the confidence level generally reduces efficiency, defined as the single-label prediction rate. In fact, the number of null class predictions is *anticorrelated* with the confidence level, whereas the number of predictions predicted to belong to both categories *positively correlates* with the significance level.[32] This relationship between confidence level and one-class assignments is illustrated in Figure 5.5, where we show the results generated using a MCP trained on the hERG data set described above (see also Table 5.4 and Figure 5.3). Hence in practice the modeler needs to make choices as to which confidence level and efficiency are desired in a particular case, which involves a certain amount of subjective choice as well.

Since the introduction of MCP in the chemical-structure activity modelling community,[32] the advantages of MCP have been showcased in a number of applications, mostly to perform classification tasks using imbalanced data sets, which is the situation MCP sets out to address. Norinder *et al.*[32] applied MCP to binary classification of compounds and showed that the number of null predictions and the confidence level are inversely correlated, whereas the opposite is observed for the *both* category (Figure 5.6). Therefore, increasing the confidence might lead to assigning molecules to the both category that are already assigned only to the correct category at a lower confidence level, underlining that this parameter represents a trade-off between multiple factors, and that a higher confidence level is not always the better choice (Figure 5.6). By modelling Ames mutagenicity data using Mondrian ICP Norinder *et al.*[6] showed that the inconsistency across data sets from different sources seems to prevent the generation of valid CPs, likely in this case due to inconsistencies in categorizing moderately mutagenic compounds as mutagenic or non-mutagenic, which highlights the importance of evaluating data consistency prior to modelling.[85–87] This can also be seen as a useful feature of CP algorithms, where the model itself is able to detect how consistent the training data which it is trained with is, and in the same way it is able to assign a confidence level to the output values. Norinder *et al.*[4] also implemented Inductive MCP using Random Forest as the underlying algorithm to model the mutagenicity of 936 primary aromatic amines (630 mutagenic and 306 non-mutagenic) using Leadscope fingerprints[88] where it was found that models were valid for both classes. A recent large-scale comparison of QSAR and MCP models for ligand-target prediction using *in vitro* activity data for 550

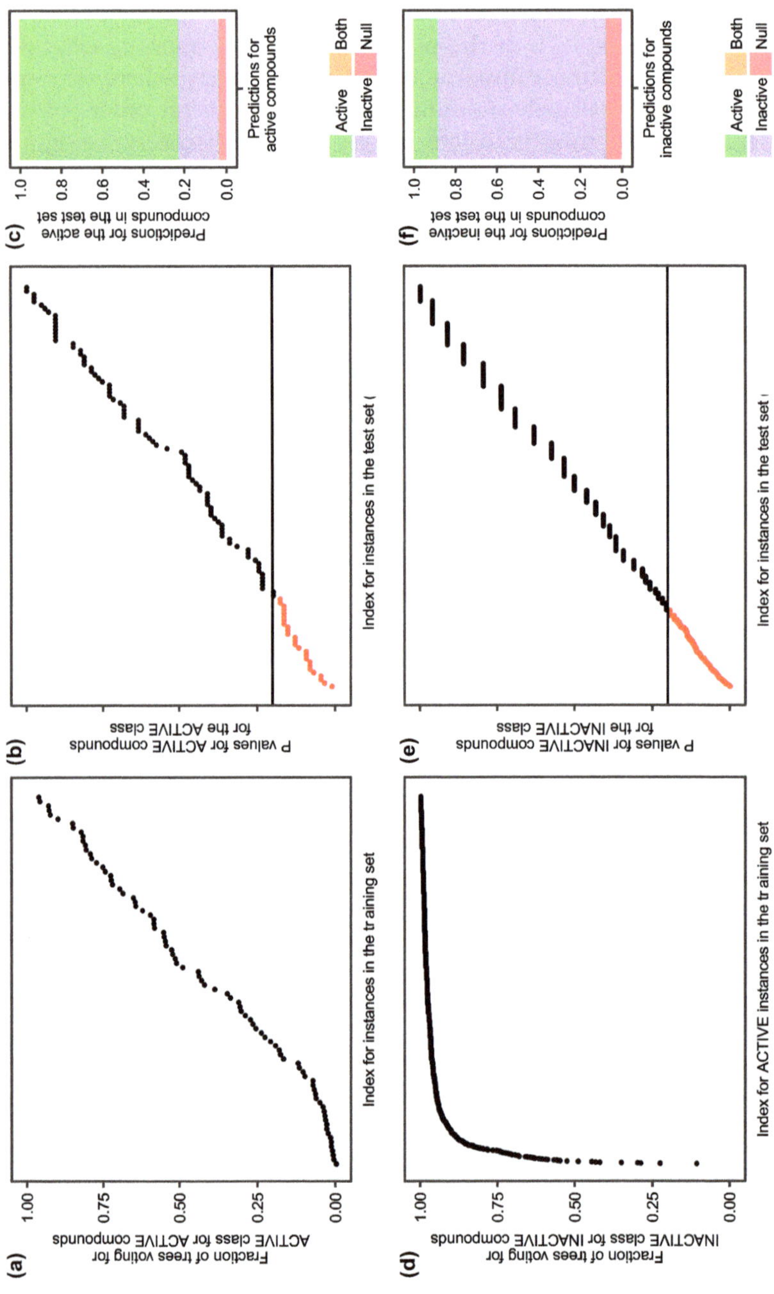
(a)
Fraction of trees voting for ACTIVE class for ACTIVE compounds
1.00
0.75
0.50
0.25
0.00
Index for instances in the tr aining set
(b)
P values for ACTIVE compounds for the ACTIVE class
Index for instances in the test set (
(c)
Predictions for the active compounds in the test set
1.0
0.8
0.6
0.4
0.2
0.0
Predictions for active compounds
Active
Both
Inactive
Null
(d)
Fraction of trees voting for INACTIVE class for INACTIVE compounds
1.00
0.75
0.50
0.25
0.00
Index for ACTIVE instances in the tr aining set
(e)
P values for INACTIVE compounds for the INACTIVE class
Index for instances in the test set
(f)
Predictions for the inactive compounds in the test set
1.0
0.8
0.6
0.4
0.2
0.0
Predictions for inactive compounds
Active
Both
Inactive
Null

human protein targets extracted from ChEMBL found that the predictive power of both approaches (in terms of correct classification rate) is overall similar for 92% of the targets considered at a confidence level of 80%.[78] The usefulness of MCP as a robust method to determine the applicability domain of predictive models for regulatory purposes has been shown by Norinder *et al.*[39] using carcinogenicity and mutagenicity data. Overall, these studies showcase the versatility of MCP to handle imbalanced data sets. However, holding out the calibration set to generate the list of non-conformity scores hampers the use of all available labelled data for model training. Therefore, several flavours of CP designed to use all available data have been developed, which we will revisit later in this review (see section 'Conformal Prediction Using All Labelled Data for Learning').

5.3.1 ICP for Regression

The underlying principles to generate Inductive CPs for regression tasks are similar to those described above for classification models. Firstly, a model is trained on the proper training set (step 1 in Figure 5.7). Subsequently, the model is applied to the calibration set (step 2 in Figure 5.7). In the case RF-based ICP models the predicted value for each instance in the validation set, $\hat{y}$, is then calculated as the average across the trees in the Random Forest, and the standard deviation of these, σ, is used as a measurement of the prediction's uncertainty.[89] Scaling the absolute error in prediction using a measure of confidence about each prediction (*e.g.*, the bagged variance[29]) serves to generate tighter predictions for those prediction that are deemed more reliable.[90] Note that without this scaling all predictions for the new molecules would be of the same size (eqn (5.1)).

The residuals and the standard deviation across the forest are used to generate a list of non-conformity scores for the calibration set as follows:

$$\alpha_i = \frac{|y_i - \hat{y}_i|}{e^{\sigma_i}} \tag{5.1}$$

Figure 5.5 Modelling the hERG data set using Mondrian Conformal Prediction. (a) Fraction of trees voting for the active class for the active compounds in the calibration set. (b) Fraction of trees voting for the active class for the active compounds in the test set. Unreliable predictions (*i.e.*, those with *P* values below the confidence level selected, 80%) are shown in red. (c) Distribution of class predictions for the active compounds in the test set. (d) Fraction of trees voting for the inactive class for the inactive compounds in the calibration set. (b) Fraction of trees voting for the inactive class for the inactive compounds in the test set. Unreliable predictions are shown in red. (c) Distribution of class predictions for the active compounds in the test set. It can be seen that the fraction of instances whose try value is among the set predicted labels is ~0.8 for both categories, which corrresponds to the selected confidence level in this case. The instances in a, b, d, and e have been sorted in increasing order according to the value of their associated non-conformity score (*y*-axis).

Table 5.4 Performance of the MCP model trained for the hERG data set. It can be seen that the predictions are globally *and* locally valid, *i.e.* across the model and also for individual classes.

Confidence level	Global validity	Validity for actives	Validity for inactives
0.1	0.14	0.14	0.14
0.5	0.5	0.41	0.51
0.8	0.81	0.77	0.81
0.9	0.92	0.9	0.92
0.99	0.99	1	0.99

where y_i is the ith instance in the validation set, and $\hat{y}_i$ and σ_i are the average and the standard deviation of the predicted activities for the ith instance across the forest, respectively. The resulting list of non-conformity scores, α, is sorted in increasing order, and the percentile corresponding to the confidence level considered is selected, *e.g.*, the 80th percentile for a confidence level of 0.80 (α_{80}). Next (step 4 in Figure 5.7), the standard deviation across the forest is used to calculate confidence regions for the data points in the test set as follows (step 4 in Figure 5.1):

$$\text{Confidence region} = \hat{y}_j \pm |y_j - \hat{y}_j| = \hat{y}_j \pm (e^{\sigma_j} * \alpha_{\text{CL}}) \tag{5.2}$$

where y_j is the jth instance in the test set, $\hat{y}_j$ and σ_j are the average and the standard deviation of the predicted activities for the jth instance across the forest, respectively, and α_{CL} is the non-conformity score for the selected confidence level.

Choosing an appropriate non-conformity measure is essential in regression to maximize efficiency, defined as the average size of the confidence regions. Svensson *et al.*[40] benchmarked the efficiency of the following six approaches to scale the errors in prediction in the non-conformity function (eqn (5.1)) using RF models and 29 public bioactivity data sets: (i) standard deviation across the trees in the RF model, (ii) the interpercentile range (10th to 90th) of the predictions across the trees in the RF model, (iii) the predicted error generated by a separate model trained on compound descriptors and using the cross-validation residuals as the dependent variable, (iv) the distance in two-component Principal Component Analysis (PCA) space computed using compound descriptors to the center of all training data, (v) the average distance to the five nearest neighbours in two-dimensional incremental PCA, and (vi) t-SNE space. Although the six non-conformity functions generated well-calibrated confidence intervals, their efficiency varied considerably, showing that the natural exponential of the RF ensemble standard deviation led to the tightest intervals, with an average prediction range of 1.65 pIC_{50} units; using the ensemble interpercentile range and a separate error model led to slightly worse performance. The exponential scaling sets the upper value for the list of non-conformity scores to be equal to the largest residual in the calibration set, as the exponential converts low σ to values close to unity.[40,63,81] Lapins *et al.*[74] included a smoothing factor to (i) make large intervals slightly smaller, thus mitigating the effect of potentially

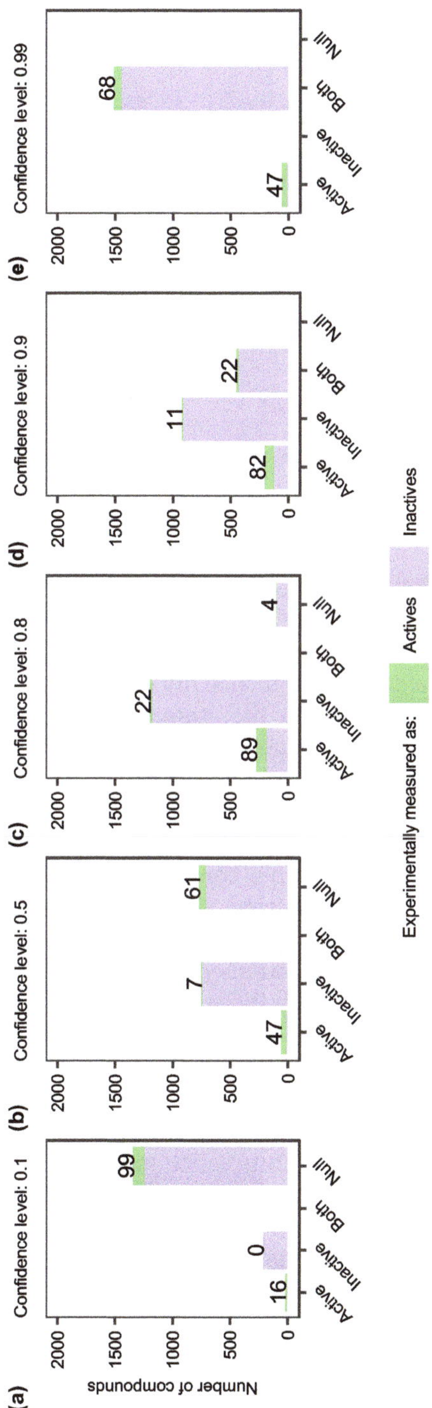

Figure 5.6 Influence of the confidence level on the efficiency of MCP classification models trained on the hERG data set and using increasingly larger confidence levels: 0.1 (a), 0.5 (b), 0.8 (c), 0.9 (d), and 0.99 (e). The numbers on top of the bars indicate the number of active compounds in each set. The *x*-axis indicates the predicted categories, whereas the colours indicate the true category for the compounds (*i.e.*, active or inactive). It can be seen that the number of instances predicted to belong to both classes (category 'both') increases with the confidence level.

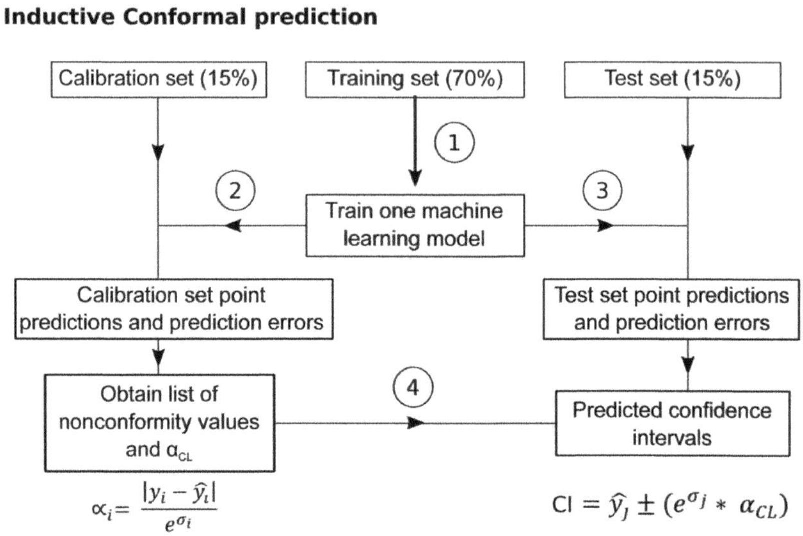

Figure 5.7 Steps for the generation of Inductive Conformal Predictors in the context of regression.

inaccurate, large σ values, and (ii) increase the size of very small intervals not reflecting the inherent uncertainty of bioactivity measurements.[85–87,91]

A number of the CPs reported in the literature applied to regression tasks use RF models as the underlying algorithm (Table 5.2), and employ the standard deviation across base learners to scale the residuals and compute non-conformity scores[40,92] (eqn (5.1)), a choice supported by comparative analysis of non-conformity measures.[40] RF models are widely used in CP because the generation of an ensemble comes at no extra computational cost,[15,32] the training can be parallelized, and the performance is stable across parameter values, thus requiring little parameter tuning. The variance across base learners (*i.e.*, the bagged variance across the trees in RF models) conveys a predictive signal to quantify the uncertainty of individual predictions, as the average RMSE on the test set increases with the variance among predictions.[22,29] However, one should note that this numerical Pearson correlation between variance and prediction error is much weaker than the inflated correlation obtained by binning the test compounds on the basis of the predicted variance,[22,29,93] which in practice means that the size of the confidence regions calculated using non-conformity measures based on the bagged variance will not be strongly correlated with the unsigned error in prediction. In other words, the average RMSE on the test set is correlated with the size of the confidence interval computed using CPs because, *on average* (but not in every individual case) the larger the variance across the ensemble, the larger the predicted confidence region will be (eqn (5.2)). Other studies have used a second machine learning model, termed the error model, to predict errors in prediction to identify which predictions are less reliable in a similar manner as the standard deviation across the forest is used in RF-based CPs (eqn (5.1)). This is achieved by training a model on the same chemical descriptors used to train a point prediction model to predict the error in prediction during cross-validation[92] or for the calibration set instances.[32] In other cases other metrics are used as covariates,[31,94] or linear methods.[95] Overall, this approach has been shown to deliver less efficient CPs as compared to those generated using the bagged variance when modelling QSAR data sets.[40]

Inductive CP has been applied to diverse regression tasks (Table 5.2), including proteochemometric modelling of PARP inhibitors[46] and the prediction of patterns of growth inhibition across cancer cell line panels.[92] However, most studies using CP for regression tasks have implemented CP modalities designed to use all available training data for learning, which we discuss in the next section.

5.3.2 Conformal Prediction Using All Labelled Data for Learning

A common disadvantage to all the CP modalities revisited so far is that not all data available for training are used for model fitting, as the calibration set needs to be kept aside to generate the list of non-conformity scores (Figures 5.2 and 5.4). However, it would be desirable to use all available data during training to increase the predictive power of the resulting models, in particular in cases where few data points are available for a given class or a particular region of the bioactivity range considered. Several CP modalities have been developed to date to solve this issue.[96,97] We here revisit the two most widely used in drug discovery in both regression and classification tasks (Table 5.2), namely Aggregated Conformal Prediction (ACP)[98,99] and Cross Conformal Prediction (CCP).[96]

- **Aggregated Conformal Prediction (ACP)**
  The ACP approach consists of generating a collection of usually 10–100[43,56,80] Inductive CPs (*e.g.*, Inductive MCP), each of them trained on the same training data but with random assignments of training set instances to the proper training and calibration sets. Thus, the instances assigned to the calibration and proper training sets are different for each model, resulting in reduced variance for the predicted confidence regions, and increased efficiency.[98] In the case of classification, each model is applied to the test (or external) set instances, and the *P* values computed with each of these models are recorded for each instance (and for each class in the case of MCP). The final *P* value is calculated as *e.g.*, the median *P* value and each instance is assigned to those classes for which the median *P* value is higher than the significance level. In the case of regression, the final confidence interval for each test set instance is also a function of the set of confidence intervals predicted by each of the set of models trained, such as half the

difference between the median of the maximum and minimum predicted values.[61]

ACP has been applied in diverse drug discovery tasks (Table 5.2), including Ames mutagenicity prediction,[67] multi-task learning using matrix factorization,[80] modelling of compound binding to the estrogen and androgen receptors,[54] and virtual screening of TRPV1 agonists.[100] Lindh *et al.*[61] generated Aggregated CPs to model the permeation rate through the human skin of 211 chemical compounds.[60] Whereas the predictive power did not increase with respect to previous models reported for the same data set, CP added the advantage of providing predictions as confidence regions, and using all available data for training, which was crucial in this particular case due to the limited size of the training data. ACP has also proved versatile to model highly imbalanced data sets. For instance, Svensson *et al.*[70] modelled 12 highly-imbalanced bioactivity data sets from PubChem using ACP and a cost-gain function taking into account screening costs and the gain of finding active hits. ACP was also shown to accurately model highly-imbalanced cytotoxicity data sets from PubChem,[62] with only 0.8% of cytotoxic compounds, showing an external validation set a sensitivity of 74% and a specificity of 65% for the single-label predictions at a confidence level of 80%. Lastly, ACP has also been applied in two iterative screening studies, where CP followed molecular docking in order to prioritize compounds for experimental testing.[56,101] Overall ACP has been found to reduce the variance and increase the efficiency of the predicted confidence intervals, while also permitting the use of all available data for training.

- **Cross-Conformal Prediction (CCP)**

  In CCP the training data set is divided into *k* non-overlapping sets, in a similar manner as performed in *k*-fold cross validation. Next, *k* ICP models are trained, each time using a different set as calibration set, thus permitting the use of all labelled data for training. For each instance in the test set *k* *P* values are generated, whose *e.g.*, mean value can be used as the output *P* value. The output *P* values are compared against the significance level to assign a class to the test instances (or classes in the case of Mondrian CCP). To date, Mondrian CCP has been mostly applied to model imbalanced bioactivity data sets (Table 5.2). Sun *et al.*[57] reported SVM-based classification Mondrian CCP models showing comparable performance to SVM models on external sets for 18 data sets from the ExCAPE database[8] with active to inactive ratios in the 1 : 10 to 1 : 1000 range. The Mondrian CCP models were well-calibrated for both the majority and the minority class. Notably, the authors reported higher efficiency for Mondrian CCP models using a non-conformity measure based on the distance to the SVM decision boundary as compared to a non-conformity measure based on the Tanimoto distance to the five nearest neighbours in the training data. Giblin *et al.*[69] conducted one of the few studies where CP guided prospective validation experiments.

Specifically, the authors built Mondrian CCP proteochemometric[102] models using bioactivity data for 6352 compounds across 31 bromodomains (15 350 data points), which showed a maximum Matthew's Correlation Coefficient of 0.83 on an external test set. Out of 1139 experimentally validated pairs, 319 compound-target pairs were confirmed, where the selection included compounds with high and low confidence values and different bioactivity profiles against sets of bromodomains. In the context of regression, CCP has been applied to model compound lipophilicity on a large scale. Lapins *et al.*[74] harnessed calculated logD values for 1 592 127 compounds extracted from ChEMBL version 23 to generate an SVM-based CCP using 10-fold cross validation. The errors in prediction were calculated as the median prediction midpoint ± the median predicted interval size across the 10 models. The final model, which achieved median confidence intervals of ± 0.39 and 0.60 log units at 80% and 90% confidence, respectively, was further applied to 91 498 351 compounds extracted from the PubChem database. These predictions were made publicly available at https://cplogd.service.pharmb.io. Overall, these studies show the versatility of CCP prediction, which permits valid and efficient CPs to be obtained while using all available labelled data for training.

5.4 Conformal Prediction Methods for Deep Learning

Deep learning is currently applied in many tasks of the drug discovery process, a trend that is only expected to rise in the coming years.[103,104] The predictions output by deep learning classification models are overall well calibrated or can be easily calibrated using *e.g.*, Platt Scaling or Isotonic Regression.[105] Therefore, there is little benefit in using CP on top of deep learning models generated for classification tasks. However, in the case of regression deep learning models simply output point predictions, thus not providing information about the reliability of individual predictions directly. In addition to variational Bayesian inference methods,[20,21] CP has been applied to deep learning models to estimate reliable confidence intervals for individual predictions. The application of CP to regression networks was initially proposed by Papadopoulos *et al.*[95] In drug discovery, the authors recently proposed two methods to obtain reliable confidence intervals for regression networks at minimal computational cost. The first approach, deep confidence,[63] harnesses the predictions generated by intermediate network states corresponding to the local minima visited during the training of a single network to build an ensemble of predictions. The variability across the ensemble can be used to scale the absolute errors in prediction in a similar way as performed using the bagged variance in the case of RF models. More recently, we proposed a second method, test-time dropout conformal prediction,[81] that consists of training a network using dropout.[106] Next, the network is applied to compute N forward passes using dropout as well. As in the case of deep confidence, the variability across these forward passes is used to scale the absolute errors in prediction. Overall, it could be shown that both deep confidence and test-time dropout conformal prediction deliver well-calibrated predicted confidence intervals that, in addition, span a narrower range of values than those computed using RF-based models.

5.5 Open-source Implementations of Conformal Prediction

The increased adoption of CP in early-stage drug discovery settings has fostered the implementation of several CP modalities in open-source software libraries in the R programming language and Python, which are widely used programming languages in medicinal chemistry applications.[107,108] The availability of predictive modelling packages in both R (*e.g.*, *caret*[109] or *camb*[110]) and Python (*e.g.*, *scikit-learn*[111] or *PyTorch*[112]) has facilitated the integration of existing computational drug discovery pipelines and CP.[56,110]

The Python *nonconformist* package (http://donlnz.github.io/nonconformist/index.html) provides functionalities to generate inductive and aggregated CPs for both regression and classification tasks. CPSign[113] (http://cpsign-docs.genettasoft.com/) is another Python implementation of CP for chemoinformatic tasks that uses SVM and Signature molecular descriptors. The R package *conformal*[46] (https://github.com/isidroc/conformal) is an object-oriented programming implementation of CP that provides *S4* classes to construct classification and regression ICP models using random forest as the underlying algorithm. Similarly, the R package *conformalClassification* (https://cran.r-project.org/web/packages/conformalClassification/index.html) permits the generation of transductive and inductive CPs based on RF models for classification tasks. The availability of well-documented and structured software will increase the reproducibility of published results, and allow for robust benchmarking studies of novel methods. Hence, we advocate for the publication of source code in future CP studies, in line with current recommendations in the modelling community.[114,115]

5.6 Current Limitations of Conformal Prediction and Future Perspectives

While CP, as described above, is able to assign confidence to predictions within a computationally efficient framework, some areas of ongoing methodological research certainly remain. A major issue in CP applied to regression is the low efficiency of most CP models, which leads to the predicted confidence regions spanning multiple *e.g.*, pIC_{50} units. Such large intervals are not informative and thus hamper the practical usefulness of CP. Substantial efforts in the community have been invested in investigating and developing non-conformity functions to reduce the size of the predicted confidence regions.[40,63,90,116] However, future research will be needed to improve current non-conformity measures in order to, ultimately, generate errors in prediction comparable to the uncertainty of the data.[85–87] In classification settings, a common problem faced is the substantial increase of instances predicted to belong to multiple categories as the confidence level is increased.[32,78] This problem is analogous to the lack of efficiency of regression models. As in the regression case, the development of improved non-conformity measures[57,80] will be needed to improve the efficiency of classification CP models. Another major shortcoming of the predicted confidence regions in the case of regression is the poor correlation between the absolute error in prediction (*i.e.*, unsigned error) and the size of the confidence interval. This is due to the fact that error models used in non-conformity measures *e.g.*, the bagged variance, do not predict accurately the error in prediction.[89,94] Thus, large confidence intervals are obtained for accurate predictions and *vice versa.* Using alternative methods to the bagged variance to compute non-conformity scores, such as quantitative metrics developed in the QSAR arena to estimate the applicability domain of the models,[26] might alleviate this issue. Similarly, algorithms other than the most widely used to date (RF, SVM and neural networks; Table 5.2) might also be considered in drug discovery applications as alternative methods to generate more efficient CPs.[80,90]

Today, Mondrian CP modalities, including Mondrian ACP and CCP, have become the standard approach to model imbalanced data sets when using CP. However, the data sets themselves which are used to model compound activity on a continuous scale are also generally biased, which leads to an uneven coverage of the chemical space across the bioactivity range considered, and hence variable errors rates across it.[94] Therefore, the development of methods to handle the uneven distribution of datapoints across the bioactivity range considered in regression models would be useful to remove biases from models, and hence guarantee that the validity and efficiency of the predicted confidence intervals are even across the bioactivity range modelled, and not only across the entire bioactivity range.

The integration of predictions generated by independent CPs is a current area of intense research, similar to ensemble approaches in other domains. Toccaceli *et al.*[117] recently introduced a method to integrate ICP models trained on different underlying algorithms. Notably, the combined models outperformed base learners (linear SVM, Gradient Boosted trees, and *k*-Nearest Neighbours) on an IDH1 bioactivity data set extracted from the ExCAPE database.[8] An alternative approach to integrate CP models, applicable to both classification and regression models, is Synergy CP,[97] which permits the aggregation of CP models trained in parallel on subsets of the training data into valid and efficient CPs. Overall, these studies represent innovative solutions that will permit not only performance improvements but also the exploitation for drug discovery of (proprietary) data dispersed across companies and institutions in distributed environments.

As stated above, the validity of CP is only guaranteed if the randomness or exchangeability assumptions hold. This assumption is, however, not usually verified in practice, and it is only assumed that the training data and the molecules to which CP models are applied are drawn from the same distribution. It is of course unlikely that the chemical space covered in the training data of virtual screening models, even if these encompass thousands of molecules, is entirely representative of the entire chemical space already comprised in academic and commercial chemical libraries, or amenable to chemical synthesis. In fact, the authors recently showed using iterative virtual screening experiments that breaching the randomness assumption leads to useless CPs.[41] This issue is of particular relevance given that many of the CP reported to date are based on few hundred datapoints (Table 5.2). Therefore, further development of methods to determine to what extent the randomness or exchangeability assumptions hold would be useful in practice to make informed decisions on the applicability of the developed CP models on the basis of the difference between the training data and those molecules to which the models are applied.

Conflicts of Interest

AB is a co-founder and shareholder of Healx Ltd. and Pharmenable Ltd.

Acknowledgements

This project has received funding from the European Union's Framework Programme For Research and Innovation Horizon 2020 (2014-2020) under the Marie Sklodowska-Curie Grant Agreement No. 703543 (I.C.C.).

References

1. E. Vayena, A. Blasimme and I. G. Cohen, Machine learning in medicine: Addressing ethical challenges, *PLOS Med.*, 2018, **15**, e1002689.
2. M. D. Segall and E. J. Champness, The challenges of making decisions using uncertain data, *J. Comput. Aided. Mol. Des.*, 2015, **29**, 809–816.
3. T. Hanser, C. Barber, J. F. Marchaland and S. Werner, Applicability domain: towards a more formal definition, *SAR QSAR Environ. Res.*, 2016, **27**, 865–881.
4. I. V Tetko, P. Bruneau, H.-W. Mewes, D. C. Rohrer and G. I. Poda, Can we estimate the accuracy of ADME-Tox predictions?, *Drug Discov. Today*, 2006, **11**, 700–707.
5. S. Weaver and M. P. Gleeson, The importance of the domain of applicability in QSAR modeling, *J. Mol. Graph. Model.*, 2008, **26**, 1315–1326.
6. M. Toplak, *et al.*, Assessment of Machine Learning Reliability Methods for Quantifying the Applicability Domain of QSAR Regression Models, *J. Chem. Inf. Model.*, 2014, **54**, 431–441.
7. T. I. Netzeva, *et al.*, Current Status of Methods for Defining the Applicability Domain of (Quantitative) Structure – Activity Relationships, *Altern. Lab. Anim.*, 2005, **33**, 155–173.

8. J. Sun, *et al.*, ExCAPE-DB: an integrated large scale dataset facilitating Big Data analysis in chemogenomics, *J. Cheminform.*, 2017, **9**, 17.
9. A. Gaulton, *et al.*, The ChEMBL database in 2017, *Nucleic Acids Res.*, 2017, **45**, D945–D954.
10. A. Cherkasov, *et al.*, QSAR modeling: where have you been? Where are you going to?, *J. Med. Chem.*, 2014, **57**, 4977–5010.
11. Y. Lecun, Y. Bengio and G. Hinton, Deep learning, *Nature*, 2015, **521**, 436–444.
12. R. Todeschini and V. Consonni, *Handbook of Molecular Descriptors*, DOI: 10.1002/9783527613106, 2008.
13. P. Rivas-Perea, *et al.*, Support Vector Machines for Regression: A Succinct Review of Large-Scale and Linear Programming Formulations, *Int. J.*, 2013, **3**, 5–14.
14. C. Cortes and V. Vapnik, Support-Vector Networks, *Mach. Learn.*, 1995, **20**, 273–297.
15. L. Breiman, Random Forests, *Mach. Learn*, 2001, **45**, 5–32.
16. O. Obrezanova, G. Csányi, J. M. R. Gola and M. D. Segall, Gaussian Processes: A Method for Automatic QSAR Modeling of ADME Properties, *J. Chem. Inf. Model.*, 2007, **47**, 1847–1857.
17. I. Cortes-Ciriano, *et al.*, Proteochemometric modeling in a Bayesian framework, *J. Cheminf.*, 2014, **6**, 35.
18. O. Obrezanova and M. D. Segall, Gaussian processes for classification: QSAR modeling of ADMET and target activity, *J. Chem. Inf. Model.*, 2010, **50**, 1053–1061.
19. C. E. Rasmussen and C. K. I. Williams, *Gaussian Processes for Machine Learning*, Mit Press, 2006.
20. Y. Zhang and A. A. Lee, *Bayesian Semi-supervised Learning for Uncertainty-calibrated Prediction of Molecular Properties and Active Learning*, 2019.
21. Y. Gal, Z. Ghahramani, Z. A. Uk and Z. Ghahramani, Dropout as a Bayesian Approximation: Representing Model Uncertainty in Deep Learning, 2015, arXiv1506.02142 arXiv.org ePrint Arch. https//arxiv.org/abs/1506.02142 (accessed Jul 10, 2018), 2015.
22. R. P. Sheridan, Using Random Forest To Model the Domain Applicability of Another Random Forest Model, *J. Chem. Inf. Model.*, 2013, **53**, 2837–2850.
23. V. Svetnik, *et al.*, Random forest: a classification and regression tool for compound classification and QSAR modeling, *J. Chem. Inf. Comp. Sci*, 2003, **43**, 1947–1958.
24. F. Berenger and Y. Yamanishi, A Distance-Based Boolean Applicability Domain for Classification of High Throughput Screening Data, *J. Chem. Inf. Model.*, 2019, **59**, 463–476.
25. R. Liu and A. Wallqvist, Molecular Similarity-Based Domain Applicability Metric Efficiently Identifies Out-of-Domain Compounds, *Altern Lab Anim.*, 2019, **59**, 181–189.
26. T. I. Netzeva, *et al.*, Current status of methods for defining the applicability domain of (quantitative) structure-activity relationships. The

report and recommendations of ECVAM Workshop 52, *Altern. Lab. Anim*, 2005, **33**, 155–173.
27. I. Sushko, *et al.*, Applicability domain for in silico models to achieve accuracy of experimental measurements, *J. Chemom.*, 2010, **24**, 202–208.
28. F. Sahigara, Defining the Applicability Domain of QSAR models: An overview, *Mol. Descriptors. Free online Resour*, 2007, 1–6.
29. D. J. Wood, L. Carlsson, M. Eklund, U. Norinder and J. Stålring, QSAR with experimental and predictive distributions: an information theoretic approach for assessing model quality, *J. Comput. Aided Mol. Des.*, 2013, **27**, 203–219.
30. T. S. Schroeter, *et al.*, Estimating the domain of applicability for machine learning QSAR models: a study on aqueous solubility of drug discovery molecules, *J. Comput. Mol. Des.*, 2007, **21**, 485–498.
31. R. P. Sheridan, The Relative Importance of Domain Applicability Metrics for Estimating Prediction Errors in QSAR Varies with Training Set Diversity, *J. Chem. Inf. Model.*, 2015, **55**, 1098–1107.
32. U. Norinder, *et al.*, Introducing Conformal Prediction in Predictive Modeling. A Transparent and Flexible Alternative To Applicability Domain Determination, *J. Chem. Inf. Model.*, 2014, **54**, 1596–1603.
33. A. Schwaighofer, *et al.*, Accurate solubility prediction with error bars for electrolytes: a machine learning approach, *J. Chem. Inf. Model.*, 2007, **47**, 407–424.
34. R. Liu, K. P. Glover, M. G. Feasel and A. Wallqvist, General Approach to Estimate Error Bars for Quantitative Structure–Activity Relationship Predictions of Molecular Activity, *J. Chem. Inf. Model.*, 2018, **58**, 1561–1575.
35. I. Cortes-Ciriano, D. S. Murrell, G. J. P. van Westen, A. Bender and T. Malliavin, Prediction of the Potency of Mammalian Cyclooxygenase Inhibitors with Ensemble Proteochemometric Modeling, *J. Cheminf.*, 2014, **7**, 1.
36. V. Vovk, A. Gammerman and G. Shafer, *Algorithmic Learning in a Random World*, Springer, 2005.
37. G. Shafer and V. Vovk, A Tutorial on Conformal Prediction, *J. Mach. Learn. Res.*, 2008, **9**, 371–421.
38. V. Vovk, V. Fedorova, I. Nouretdinov and A. Gammerman, Criteria of efficiency for conformal prediction, *Lecture Notes in Computer Science (including subseries Lecture Notes in Artificial Intelligence and Lecture Notes in Bioinformatics*, 2016, **vol. 9653**, 23–39.
39. U. Norinder, L. Carlsson, S. Boyer and M. Eklund, Introducing conformal prediction in predictive modeling for regulatory purposes. A transparent and flexible alternative to applicability domain determination, *Regul. Toxicol. Pharmacol.*, 2015, **71**, 279–284.
40. F. Svensson, *et al.*, Conformal Regression for Quantitative Structure–Activity Relationship Modeling—Quantifying Prediction Uncertainty, *J. Chem. Inf. Model.*, 2018, **58**, 1132–1140.

41. I. Cortes-Ciriano, N. C. Firth, A. Bender and O. Watson, Discovering highly potent molecules from an initial set of inactives using iterative screening, *J. Chem. Inf. Model.*, 2018, **58**, 2000–2014.
42. U. Johansson, H. Linusson, T. Löfström and H. Boström, Interpretable regression trees using conformal prediction, *Expert Syst. Appl.*, 2018, **97**, 394–404.
43. S. Lampa, *et al.*, Predicting Off-Target Binding Profiles With Confidence Using Conformal Prediction, *Front. Pharmacol*, 2018, **9**, 1256.
44. M. Eklund, U. Norinder, S. Boyer and L. Carlsson, The application of conformal prediction to the drug discovery process, *Ann. Math. Artif. Intell.*, 2015, **74**, 117–132.
45. E. Wahlberg, *et al.*, Family-wide chemical profiling and structural analysis of PARP and tankyrase inhibitors, *Nat. Biotechnol.*, 2012, **30**, 283–288.
46. I. Cortés-Ciriano, A. Bender and T. Malliavin, Prediction of PARP Inhibition with Proteochemometric Modelling and Conformal Prediction, *Mol. Inform*, 2015, **34**, 357–366.
47. N. Fjodorova, M. Vračko, M. Novič, A. Roncaglioni and E. Benfenati, New public QSAR model for carcinogenicity, *Chem. Cent. J.*, 2010, **4**, S3.
48. T. Ferrari and G. Gini, An open source multistep model to predict mutagenicity from statistical analysis and relevant structural alerts, *Chem. Cent. J.*, 2010, **4**, S2.
49. E. Ahlberg, O. Spjuth, C. Hasselgren and L. Carlsson, in *Interpretation of Conformal Prediction Classification Models*, Springer, Cham, 2015, pp. 323–334. DOI: 10.1007/978-3-319-17091-6_27.
50. I. Cortés-Ciriano, *et al.*, Improved large-scale prediction of growth inhibition patterns using the NCI60 cancer cell line panel, *Bioinformatics*, 2016, **32**, 85–95.
51. I. Cortes-Ciriano, *et al.*, Cancer Cell Line Profiler (CCLP): a webserver for the prediction of compound activity across the NCI60 panel, *bioRxiv*, 2017, 105478.
52. G. G. J. M. Kuiper, *et al.*, Comparison of the Ligand Binding Specificity and Transcript Tissue Distribution of Estrogen Receptors α and β, *Endocrinology*, 1997, **138**, 863–870.
53. M. O. Taha, M. Tarairah, H. Zalloum and G. Abu-Sheikha, Pharmacophore and QSAR modeling of estrogen receptor β ligands and subsequent validation and in silico search for new hits, *J. Mol. Graph. Model.*, 2010, **28**, 383–400.
54. U. Norinder, A. Rybacka and P. L. Andersson, Conformal prediction to define applicability domain – A case study on predicting ER and AR binding, *SAR QSAR Environ. Res.*, 2016, **27**, 303–316.
55. M. M. Mysinger, M. Carchia, J. J. Irwin and B. K. Shoichet, Directory of useful decoys, enhanced (DUD-E): better ligands and decoys for better benchmarking, *J. Med. Chem.*, 2012, **55**, 6582–6594.
56. F. Svensson, U. Norinder and A. Bender, Improving Screening Efficiency through Iterative Screening Using Docking and Conformal Prediction, *J. Chem. Inf. Model.*, 2017, **57**, 439–444.

57. J. Sun, *et al.*, Applying Mondrian Cross-Conformal Prediction To Estimate Prediction Confidence on Large Imbalanced Bioactivity Data Sets, *J. Chem. Inf. Model.*, 2017, **57**, 1591–1598.
58. K. Hansen, *et al.*, Benchmark Data Set for in Silico Prediction of Ames Mutagenicity, *J. Chem. Inf. Model.*, 2009, **49**, 2077–2081.
59. U. Norinder and S. Boyer, Binary classification of imbalanced datasets using conformal prediction, *J. Mol. Graph. Model.*, 2017, **72**, 256–265.
60. H. Baba, J. Takahara and H. Mamitsuka, In Silico Predictions of Human Skin Permeability using Nonlinear Quantitative Structure–Property Relationship Models, *Pharm. Res.*, 2015, **32**, 2360–2371.
61. M. Lindh, A. Karlén and U. Norinder, Predicting the Rate of Skin Penetration Using an Aggregated Conformal Prediction Framework, *Mol. Pharm.*, 2017, **14**, 1571–1576.
62. F. Svensson, U. Norinder and A. Bender, Modelling compound cytotoxicity using conformal prediction and PubChem HTS data, *Toxicol. Res.*, 2017, **6**, 73–80.
63. I. Cortés-Ciriano and A. Bender, Deep Confidence: A Computationally Efficient Framework for Calculating Reliable Prediction Errors for Deep Neural Networks, *J. Chem. Inf. Model.*, 2019, **59**, 1269–1281.
64. G. Huang *et al.*, Snapshot Ensembles: Train 1, get M for free, 2017, *arXiv1704.00109 arXiv.org ePrint Arch.* https//arxiv.org/abs/1704.00109 (accessed Jul 10, 2018).
65. U. Norinder, *et al.*, Predicting Aromatic Amine Mutagenicity with Confidence: A Case Study Using Conformal Prediction, *Biomolecules*, 2018, **8**, 85.
66. M. Honma, *et al.*, Improvement of quantitative structure–activity relationship (QSAR) tools for predicting Ames mutagenicity: outcomes of the Ames/QSAR International Challenge Project, *Mutagenesis*, 2019, **34**, 3–16.
67. U. Norinder, E. Ahlberg and L. Carlsson, Predicting Ames Mutagenicity Using Conformal Prediction in the Ames/QSAR International Challenge Project, *Mutagenesis*, 2018, DOI: 10.1093/mutage/gey038.
68. U. Norinder, D. Mucs, T. Pipping and A. Forsby, Creating an efficient screening model for TRPV1 agonists using conformal prediction, *Comput. Toxicol*, 2018, **6**, 9–15.
69. K. A. Giblin, S. J. Hughes, H. Boyd, P. Hansson and A. Bender, Prospectively Validated Proteochemometric Models for the Prediction of Small-Molecule Binding to Bromodomain Proteins, *J. Chem. Inf. Model.*, 2018, **58**, 1870–1888.
70. F. Svensson, A. M. Afzal, U. Norinder and A. Bender, Maximizing gain in high-throughput screening using conformal prediction, *J. Chemin-form.*, 2018, **10**, 7.
71. H. Johansson, M. Lindstedt, A.-S. Albrekt and C. A. Borrebaeck, A genomic biomarker signature can predict skin sensitizers using a cell-based in vitro alternative to animal tests, *BMC Genomics*, 2011, **12**, 399.

72. A. Forreryd, U. Norinder, T. Lindberg and M. Lindstedt, Predicting skin sensitizers with confidence — Using conformal prediction to determine applicability domain of GARD, *Toxicol. Vitr.*, 2018, **48**, 179–187.
73. C. Ji, F. Svensson, A. Zoufir and A. Bender, eMolTox: prediction of molecular toxicity with confidence, *Bioinformatics*, 2018, **34**, 2508–2509.
74. M. Lapins, *et al.*, A confidence predictor for logD using conformal regression and a support-vector machine, *J. Cheminform.*, 2018, **10**, 17.
75. D. Dua and C. Graff, *UCI Machine Learning Repository, University of California*, School of Information and Computer Sciences, Irvine. Available at: http://archive.ics.uci.edu/ml. (Accessed: 1st July 2019), 2017.
76. O. Spjuth, L. Carlsson and N. Gauraha, Aggregating Predictions on Multiple Non-disclosed Datasets using Conformal Prediction, *ArXiv*, 2018, **abs/1806.04000**, https://arxiv.org/abs/1806.04000.
77. N. Gauraha, L. Carlsson and O. Spjuth, Proceedings of the Seventh Workshop on Conformal and Probabilistic Prediction and Applications *Conformal Prediction in Learning Under Privileged Information Paradigm with Applications in Drug Discovery*, 2018 http://proceedings.mlr.press/v91/gauraha18a.html.
78. N. Bosc, *et al.*, Large scale comparison of QSAR and conformal prediction methods and their applications in drug discovery, *J. Cheminform.*, 2019, **11**, 4.
79. A. de la Vega de León, B. Chen and V. J. Gillet, Effect of missing data on multitask prediction methods, *J. Cheminform.*, 2018, **10**, 26.
80. U. Norinder and F. Svensson, Multitask Modeling with Confidence Using Matrix Factorization and Conformal Prediction, *J. Chem. Inf. Model.*, 2019, acs.jcim.9b00027.
81. I. Cortés-Ciriano and A. Bender, Reliable Prediction Errors for Deep Neural Networks Using Test-Time Dropout, *J. Chem. Inf. Model.*, 2019, **59**, 3330–3339.
82. M. Eklund, U. Norinder, S. Boyer and L. Carlsson, Benchmarking Variable Selection in QSAR, *Mol. Inf.*, 2012, **31**, 173–179.
83. I. Cortés-Ciriano and A. Bender, KekuleScope: prediction of cancer cell line sensitivity and compound potency using convolutional neural networks trained on compound images, *J. Cheminform.*, 2019, **11**, 41.
84. V. Vovk, S. C. H. Hoi and W. Buntine, Conditional validity of inductive conformal predictors, *Mach. Learn.*, 2013, **92**, 349–376.
85. C. Kramer, T. Kalliokoski, P. Gedeck and A. Vulpetti, The experimental uncertainty of heterogeneous public K(i) data, *J. Med. Chem.*, 2012, **55**, 5165–5173.
86. I. Cortés-Ciriano and A. Bender, How consistent are publicly reported cytotoxicity data? Large-scale statistical analysis of the concordance of public independent cytotoxicity measurements, *ChemMedChem*, 2015, **11**, 57–71.
87. T. Kalliokoski, C. Kramer, A. Vulpetti and P. Gedeck, Comparability of mixed IC_{50} data - a statistical analysis, *PLoS One*, 2013, **8**, e61007.

88. G. Roberts, G. J. Myatt, W. P. Johnson, K. P. Cross and P. E. Blower, LeadScope: Software for Exploring Large Sets of Screening Data, *J. Chem. Inf. Comput. Sci.*, 2000, **40**, 1302–1314.
89. B. Beck, A. Breindl and T. Clark, QM/NN QSPR Models with Error Estimation: Vapor Pressure and LogP, *J. Chem. Inf. Comput. Sci.*, 2000, **40**, 1046–1051.
90. H. Papadopoulos, V. Vovk and A. Gammerman, Regression conformal prediction with nearest neighbours, *J. Artif. Intell. Res.*, 2011, **40**, 815–840.
91. T. Kalliokoski, C. Kramer and A. Vulpetti, Quality Issues with Public Domain Chemogenomics Data, *Mol. Inform*, 2013, **32**, 898–905.
92. I. Cortés-Ciriano, *et al.*, Improved large-scale prediction of growth inhibition patterns on the NCI60 cancer cell-line panel, *Bioinformatics*, 2016, **32**, 85–95.
93. H. Wainer, M. Gessaroli and M. Verdi, Visual Revelations, *CHANCE*, 2006, **19**, 49–52.
94. R. P. Sheridan, Three useful dimensions for domain applicability in QSAR models using random forest, *J. Chem. Inf. Model.*, 2012, **52**, 814–823.
95. H. Papadopoulos and H. Haralambous, Reliable prediction intervals with regression neural networks, *Neural Networks*, 2011, **24**, 842–851.
96. V. Vovk, Cross-conformal predictors, *Ann. Math. Artif. Intell.*, 2015, **74**, 9–28.
97. N. Gauraha and O. Spjuth, *Synergy Conformal Prediction*, 2018 https://www.semanticscholar.org/paper/Synergy-Conformal-Prediction-Gauraha-Spjuth/32046d4cc060fe8d57afef821787f6c8c2936f80.
98. L. Carlsson, M. Eklund and U. Norinder, in *Aggregated Conformal Prediction*, Springer, Berlin, Heidelberg, 2014, pp. 231–240. DOI: 10.1007/978-3-662-44722-2_25.
99. H. Linusson, *et al.*, On the Calibration of Aggregated Conformal Predictors, *Proc. Mach. Learn. Res.*, 2017, **60**, 1–20.
100. T.-H. Ha, *et al.*, TRPV1 antagonist with high analgesic efficacy: 2-Thio pyridine C-region analogues of 2-(3-fluoro-4-methylsulfonylaminophenyl)propanamides, *Bioorg. Med. Chem.*, 2013, **21**, 6657–6664.
101. L. Ahmed, *et al.*, Efficient iterative virtual screening with Apache Spark and conformal prediction, *J. Cheminform.*, 2018, **10**, 8.
102. I. Cortes-Ciriano, *et al.*, Polypharmacology Modelling Using Proteochemometrics: Recent Developments and Future Prospects, *Med. Chem. Commun.*, 2015, **6**, 24.
103. K. A. Carpenter, D. S. Cohen, J. T. Jarrell and X. Huang, Deep learning and virtual drug screening, *Future Med. Chem.*, 2018, **10**, 2557–2567.
104. H. Chen, O. Engkvist, Y. Wang, M. Olivecrona and T. Blaschke, The rise of deep learning in drug discovery, *Drug Discov. Today*, 2018, **23**, 1241–1250.
105. A. Niculescu-Mizil and R. Caruana, Predicting good probabilities with supervised learning, in *Proceedings of the 22nd International Conference*

on Machine Learning – ICML '05, ACM Press, 2005, pp. 625–632, DOI: 10.1145/1102351.1102430.
106. N. Srivastava, G. Hinton, A. Krizhevsky and R. Salakhutdinov, Dropout: A Simple Way to Prevent Neural Networks from Overfitting, *J. Mach. Learn. Res.*, 2014, **15**, 1929–1958.
107. Nowotka, Papadatos, Davies, Dedman and Hersey Want Drugs? Use Python, 2016, arXiv1607.00378 arXiv.org ePrint Arch. https//arxiv.org/abs/1607.00378 (accessed Jul 10, 2018).
108. S. Mente and M. Kuhn, The use of the R language for medicinal chemistry applications, *Curr. Top. Med. Chem.*, 2012, **12**, 1957–1964.
109. M. Kuhn, Building Predictive Models in R Using the caret Package, *J. Stat. Softw.*, 2008, **28**, 1–26.
110. D. S. Murrell, *et al.*, Chemically Aware Model Builder (camb): an R package for property and bioactivity modelling of small molecules, *J. Cheminform.*, 2015, **7**, 45.
111. F. Pedregosa, *et al.*, Scikit-learn: Machine Learning in Python, *J. Mach. Learn. Res.*, 2011, **12**, 2825–2830.
112. A. Paszke, *et al.*, Automatic differentiation in PyTorch, *Advances in Neural Information Processing Systems*, 2017, **30**, 1–4.
113. S. Arvidsson, *CPSign Documentation — CPSign 0.7.8 Documentation*, 2016.
114. W. P. Walters, Modeling, informatics, and the quest for reproducibility, *J. Chem. Inf. Model.*, 2013, **53**, 1529–1530.
115. G. A. Landrum and N. Stiefl, Is that a scientific publication or an advertisement? Reproducibility, source code and data in the computational chemistry literature, *Future Med. Chem.*, 2012, **4**, 1885–1887.
116. H. Papadopoulos, A. Gammerman and V. Vovk, Normalized Nonconformity Measures for Regression Conformal Prediction, in *Proceedings of the IASTED International Conference on Artificial Intelligence and Applications, AIA*, ACTA Press, 2008.
117. P. Toccaceli and A. Gammerman, Combination of inductive mondrian conformal predictors, *Mach. Learn.*, 2019, **108**, 489–510.

CHAPTER 6

Non-applicability Domain. The Benefits of Defining "I Don't Know" in Artificial Intelligence

DAMJAN KRSTAJIC

Research Centre for Cheminformatics, Jasenova 7/17, 11030 Beograd, Serbia
Email: Damjan.Krstajic@rcc.org.rs

6.1 Introduction

The saying "I know that I know nothing" is attributed to the Greek philosopher Socrates,[1] and one might wonder what the actual point of it is. As a young man Socrates visited and conversed with various scholars in ancient Athens, and he realised that they all talked about how much they knew, but really did not. Therefore, by admitting how much he does not know he becomes wiser than them. In simple terms the main lesson from Socrates is that by accepting the limitations of our knowledge we would avoid ignorance and wrongdoing. We think that the same applies to artificial intelligence (AI).

Substantial investments and many hopes are placed in AI research today. We are overwhelmed with news about how AI will transform future science and our everyday life. However, is AI capable of answering "I don't know"? How are the limitations of AI being defined? Who is responsible for doing that? What are the benefits of knowing them? As far as we are aware, there is not a great deal of interest in answering these questions.

Drug Discovery Series No. 75
Artificial Intelligence in Drug Discovery
Edited by Nathan Brown

Published by the Royal Society of Chemistry, www.rsc.org

The drug discovery process is a journey into the unknown and is an expensive one. Lab technology has advanced so much that one may process large amounts of chemical compounds in a relatively short time. This creates a situation where a single human being, or even a team, is not capable of managing the selection of such a large number of chemical compounds for each iteration. Inevitably this provides an opportunity to apply AI in drug discovery. However, we think that there are two important questions related to the use of AI in drug discovery:

(a) When to trust AI in drug discovery?
(b) How to develop and extend AI continuously as a dynamic system capable of accepting changes as lab results emerge?

In our opinion, knowing the limitations of AI is crucial to answering both questions, and it is the responsibility of the creators of AI systems to inform users about them. However, defining the limitation of AI is not an easy task at all. The creators of AI systems are probably aware that their AI has limitations, but are they able to define them? And how?

We consider that asking AI creators to design AI with an "I don't know" answer will force them to come up with a solution regarding the limitations of the AI systems they develop. For us humans, the answer "I don't know" comes naturally as the result of our introspection, and we rarely hear anyone asking themselves: "How do I know that I don't know?". However, the question of how AI would know that it does not know is probably ignored because there is no obvious answer to it, as AI does not have the luxury of introspection.

There is another aspect of the "I don't know" answer that may be easily overlooked. "The more I learn, the more I realise how much I don't know." is a quote by Albert Einstein, and it reflects the relationship between ignorance and knowledge. From our point of view, understanding our ignorance is a fundamental part of our knowledge and is therefore important in teaching, as well as in science. If AI is to drive research, or is to be a useful tool in science, then acquiring the knowledge of what AI does not know is, in our view, the way forward. We have named the research field which is devoted to "I don't know" answers in AI the *non-applicability domain*, and in the following text we provide some arguments for its importance, as well as benefits to be gained from its development.

As Rene Descartes suggested in the *Discourse of the Method,*[2] we should commence with the simplest problems and those easiest to resolve, hoping that step by step we shall be able to resolve more complex issues. Therefore, our aim is first to introduce the "I don't know" answer in the simplest cases. One of AI's many applications is to be a forecasting tool, and binary classification is one of the simplest prediction problems. Therefore, we first define a procedure which implements the introduction of the "I don't know" answer for binary classifiers. Later we illustrate it with a simulated example and show how it may be useful in the drug discovery process. We end with the criticism and questions we have encountered so far regarding the non-applicability domain and our proposed approach, followed by our answers.

6.2 Predictive Models

Let's look at today's software expert systems that provide predictions in various parts of our lives. They contain some useful knowledge. For example, weather forecasts ease our daily routines, fraud detection systems fight crime, and so on. Their value is unquestionable. Most of them provide a predicted number or category, but as far as we are aware none of them produce an "I don't know" answer. You can obtain a prediction for the weather at a specific location for 30 days in advance, but experts would tell you that it is only reliable for the next few days. Unfortunately, predictions by software systems in other fields of expertise do not come with such an easy understanding of trustworthiness.

We are interested in scenarios where the goal is to create a predictive model $F()$ which predicts a variable Y using values of variables $X_1,\ldots,X_m$. It can be viewed as the relationship $Y=F(X_1,\ldots,X_m)$. If our predictive model is a statistical model then it is created using previously known values $(Y_i,X_{i1},\ldots,X_{im})$ $i=1..N$, which we refer to as the *learning data*. The quality of the predictive model $F()$ is usually measured by how well it predicts a previously unseen set of samples, which we refer to as the *validation data*. In the validation dataset we know Y, as well as $X_1,\ldots,X_m$, and we measure how much $F(X_1,\ldots,X_m)$ differs from Y. We expect that predictive models that perform well on the validation data may be used in practice.

The whole purpose of creating a predictive model $F()$ is to apply it in a situation where we do not know Y, but we have information regarding $X_1,\ldots,X_m$. We are then faced with the problem of the uncertainty of our predictions and the consequences of them being true or false. We argue that there are cases where it is better to admit that we cannot predict variable Y, and that this may provide practical benefits. Due to their simplicity, we use binary classification models to present our arguments.

Binary classification models are predictive models where Y has only two values, which we shall refer to as positive and negative. Their common measure of quality is the misclassification error, *i.e.* the proportion of wrong predictions. Furthermore, in binary classification settings it is common for a predictive model $F()$ not only to predict whether something is positive or negative but also to estimate its probability with values in the interval [0,1]. The probability equal to 0 means that a sample is negative without any uncertainty, while if it is 1 then the sample is positive, again without any uncertainty. However, what does it mean if the probability is equal to 0.5? It could mean that the sample is so different from the learning dataset that there are no arguments for the model to classify it as belonging to either of the two categories. Or, it could be that the model recognises an equally large number of arguments in favour of and against classifying it into any of the two categories. In the former instance a human reply would be equal to

"I don't know", while the latter would be "50:50". Regardless of this subtle and important difference , binary classification models have so far had to produce either positive or negative predictions. Therefore, in the case of probability 0.5 a rule might be created that it is, for example, always predicted positive, or always predicted negative, or randomly selected between positive or negative.

Krstajic *et al.*[3] recently proposed an introduction of a third prediction category in binary classifications, which could mean either "I don't know" or "50:50". They refer to it is as the Uncertain category, but we now think that the term NotAvailable is more appropriate. We shall later discuss the issues regarding the naming of such a category and why we chose NotAvailable. We shall use the findings of Krstajic *et al.*[3] as the basis to extend their research. When referencing their results, in order not to confuse the reader, we shall use the term NotAvailable even though they have used the term Uncertain throughout their paper.

6.3 Defining NotAvailable Predictions

Obviously a sample with the predicted probability equal to 0.5 is a situation where we cannot decide whether it is positive or negative, and one may argue that it should be treated as NotAvailable. However, such samples are very rare in reality. But what about a sample with probability equal to 0.49999? Why wouldn't its predicted category also be NotAvailable as well? If so, where could we draw the line between positive and NotAvailable? And between negative and NotAvailable?

Krstajic *et al.*[3] first tried to define NotAvailable predictions with intervals of probability predictions. What would happen if, for example, all samples with predicted probabilities in the range [0.49, 0.51] were to be categorised as NotAvailable? Or if the range is [0.48, 0.52]? Krstajic *et al.*[3] refer to the intervals of probabilities that define NotAvailable samples as intervals of uncertainty. They presented cases which show the following pattern. As the interval of uncertainty widens ([0.49, 0.51], [0.48, 0.52], ...) the misclassification error is reduced and the proportion of NotAvailable predictions increases.

We shall here present their findings in the case of mutagenicity predictions. They used a publicly available dataset of 4335 chemical compounds, 2400 categorised as "mutagen" (positive) and the remaining 1935 compounds as "nonmutagen" (negative).[4] Chemical descriptors were calculated for each compound, and these were inputs $(X_1,...,X_m)$ for creating predictive models.[5]

They repeated the following process 100 times. The dataset was split into equally sized halves ensuring that each half contained the same proportion of "mutagen" and "nonmutagen" samples. One half was used as the learning dataset to create a predictive model and the other half as the validation dataset. For each validation sample the model predicted a category ("mutagen" or "nonmutagen") as well as the probability of it being "mutagen". They examined what would happen if all samples with probabilities in the range [0.49, 0.51] were to be categorised as NotAvailable, and then for the range [0.48, 0.52], and so on until [0.3, 0.7]. They applied the following two model building techniques: ridge logistic regression[6] and random forest.[7] In Figures 6.1 and 6.2 the relationship between the intervals of uncertainty and average misclassification errors (from 100 repeats) as well as the average proportion of NotAvailable is shown for each model building technique.

In both cases the introduction of NotAvailable predictions caused other predictions to be on average more accurate.

Here we have gains in certainty and losses in productivity. When the costs of NotAvailable outweigh the improvements in accuracy, then obviously one would not introduce the NotAvailable category. Similarly, when the benefits of having more reliable predictions outweigh the fact that there will be some samples we cannot predict, then the benefits of their introduction are self-evident. In that case one may judge the pros and cons between the improvements and costs and decide which interval of uncertainty will be used for defining the NotAvailable predictions.

We shall presume that we cannot measure costs and benefits, or that we are dealing with a zero sum game, that is to say that the costs of introducing NotAvailable predictions are the same as its benefits. Either way we are faced

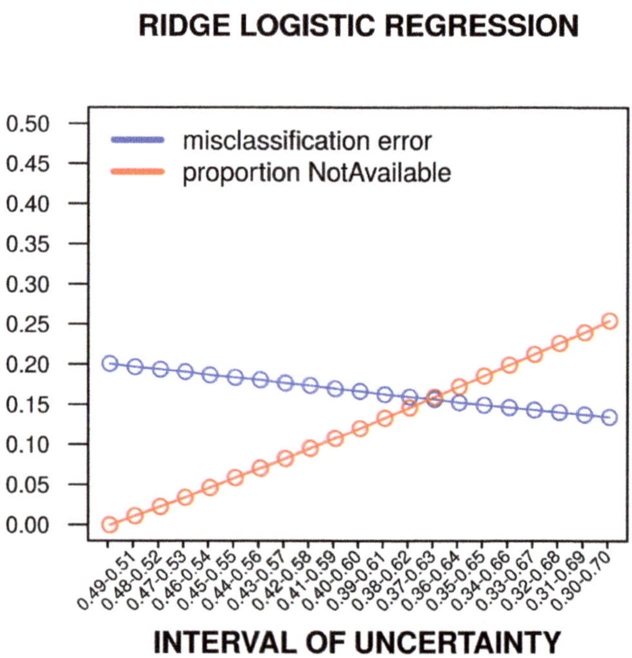

Figure 6.1 Relationship between the intervals of uncertainty and average misclassification errors (from 100 repeats) as well as the average proportion of NotAvailable for ridge logistic regression.

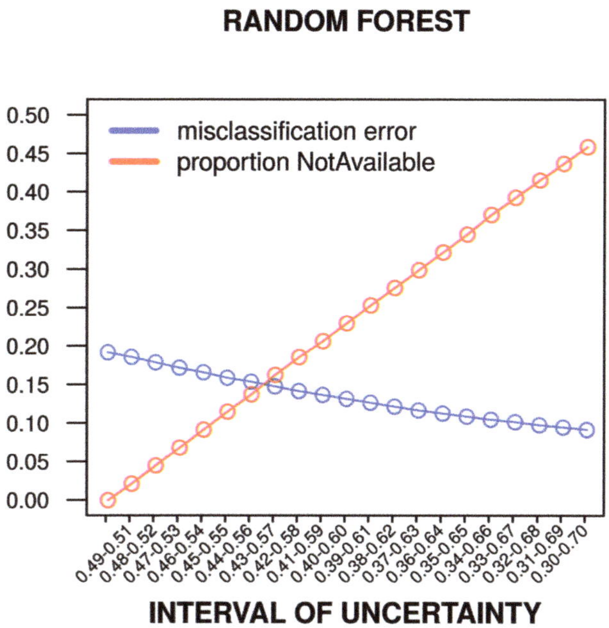

Figure 6.2 Relationship between the intervals of uncertainty and average misclassification errors (from 100 repeats) as well as the average proportion of NotAvailable for random forest.

with a situation where we do not know if, and which interval of uncertainty to pick. Nevertheless, we would definitely like to know which predicted samples are not reliable and not to be taken into consideration. Remember the analogy between a weather forecast for the next few days and the one for the month ahead. Is there a way we could locate a "blind spot" in our predictive models so to speak? Krstajic *et al.*[3] proposed a solution for statistical models, and it is related to how they are created.

6.4 All Leave One Out Models

What would happen if we were to remove a single sample from the learning dataset, re-create a predictive model on the remaining $(N-1)$ samples and use that instead? Let us say that we have removed the ith sample and the newly predictive model is $Fi()$. Can we say that in the cases when $F()$, the original model, and $Fi()$ predict a different category we should accept the prediction by the original model? In our opinion, the answer is negative. Based on such premises Krstajic *et al.*[3] suggested the following rule for defining NotAvailable predictions. If a learning dataset consists of N samples then one would create N binary classification models on $(N-1)$ samples in exactly the same way as the original model $F()$ was created. This would mean that when predicting an unseen sample we would have N predicted categories from the N models. The sample would be categorised as NotAvailable if and only if it has opposite predicted categories among the N predictions. Otherwise all N models would predict unanimously, either all positive or all negative, and that would be the predicted category. In other words, if we create a model *Fi*( ) in exactly the same way as the original model *F*( ), but without the i-th learning sample, and the model *Fi*( ) predicts differently from the original model *F*( ) then we say that the model's answer ought to be "I don't know" and the prediction is NotAvailable. This would mean that instead of applying a single binary classification model in practice one would need to execute a collection of N binary classification models which we refer to as *All Leave One Out Models* (ALOOM). Note that when the ALOOM is applied, then there is no need to execute the original model *F*( ). It is presumed that the situation where the ALOOM predicts one category unanimously and the original model *F*( ) predicts the opposite is not realistic. However, in order to compare the original model *F*( ) and the ALOOM, we shall execute them both.

Krstajic *et al.*[3] applied the ALOOM on the same mutagenicity dataset using the same 100 split into halves and applying the same two types of model: ridge logistic regression and random forest. In Table 6.1 we show the mean misclassification error and mean percentage of NotAvailable predictions generated with the ALOOM for each model type as well as the results of the original model *F*( ). It can be seen from Table 6.1, and Krstajic *et al.*[3] have demonstrated it on other datasets as well, that the ALOOM approach generates a different percentage of NotAvailable predictions, depending on the type of model.

Their initial idea was to use the ALOOM to find an optimal interval of uncertainty. Unfortunately, they have found that certain validation samples may have close probabilities, but may not be equally stable. For example, the original model may predict sample A to be negative with probability 0.48 and another sample B to be negative with lower probability 0.46. However, the ALOOM may predict A to be negative and B to be NotAvailable, even though the originally predicted probability of A is closer to 0.5. It is possible that ALOOMs predict A to be negative (with probability below 0.5), while one or more of them may predict B to be positive (with probability over 0.5). This shows that the prediction of B is unstable, and that it should not be taken into account, and therefore deserves to be categorised as NotAvailable. In Table 6.2 we show examples of pairs of compounds where a compound with

Table 6.1 Mean misclassification error and mean percentage of NotAvailable predictions generated with the ALOOM for each model type as well as the results of the original model *F*( ).

Model type	Aloom misclassification error	% Not available	Original model's misclassification error
Random forest	0.112	19.73	0.192
Ridge logistic regression	0.174	6.59	0.201

Table 6.2 Examples of pairs of compounds where a compound with originally predicted probability closer to 0.5 was not categorised as NotAvailable, while another compound with original predicted probability further away from 0.5 was categorised as such. Minimum and maximum ALOOM's predicted probabilities are also shown.

Model type	Compound name	Original model's predicted probability	Aloom prediction	Aloom minimum probability	Aloom maximum probability
Random forest	64	0.572	Positive	0.510	0.664
Random forest	1	0.690	Not available	0.494	0.706
Ridge logistic regression	214670	0.454	Negative	0.398	0.498
Ridge logistic regression	214959	0.383	Not available	0.249	0.559

original predicted probability closer to 0.5 was not categorised as NotAvailable, while another compound with original predicted probability further away from 0.5 was categorised as such.

Even though one may not find an optimal interval of uncertainty with the ALOOM procedure, it was shown that the ALOOM as such may be the procedure for defining NotAvailable predictions.

6.5 Benefits of Defining NotAvailable Predictions

In our view the main benefit accrued from introducing the NotAvailable predictions is that we can identify some predictions that should not be taken into consideration. This does not, however, mean that all predictions which are not NotAvailable are reliable, although we can expect them to be on average more accurate.

There are real-life situations where models need to predict binary values, and decisions depend on their predictions. For example, in medical diagnostics we have tools that predict whether a person has an aggressive cancer or not. In toxicology we have models which predict whether a compound is toxic or not. Those are situations where we have true or false predictions. However, in situations where the resulting actions are not strictly binary (to pull a trigger or not), where we have various options to execute, then having NotAvailable predictions may be very beneficial. In case of toxicology, all NotAvailable compounds would automatically be sent to the laboratory for *in vitro* screening. In the case of cancer predictions, a patient with a Not Available prediction regarding the existence of an aggressive cancer may provide the basis for an oncologist to seek additional tests. Furthermore, even in situations where we have strictly binary actions, the information that the prediction is NotAvailable ought to be useful to a decision maker, so that he/she would not take it into consideration.

In our opinion, identifying some predictions which should not be taken into consideration is the main benefit of defining NotAvailable predictions, but not the only one. Far from it. We think that it may be used, for example, to improve current predictive models. In theory we have a learning dataset, we create a predictive model and we assess it on a validation dataset. However, in practice we might know that our predictive model is not good enough prior to assessing it on the validation dataset. We would then like to improve it by enlarging the learning dataset. Sometimes we do not have any control over how the additional new samples for the learning dataset are selected. However, when we do, we suggest randomly selecting samples which are predicted as NotAvailable by the current model.

Almost a century ago Sir Ronald Fisher suggested randomisation as a way to avoid confounding factors in the data.[8] Therefore, we would randomly select new samples for the learning dataset. However, it seems obvious to us that it is, figuratively speaking, better to learn something we do not know than something we do. Our thesis is that in the long run the predictive model would be improved with less cost by gaining knowledge of something it cannot predict and combined, of course, with randomisation. Therefore, when the goal is to improve predictive models, we consider that enlarging the learning dataset with randomly selected samples with NotAvailable predictions is better than merely randomly selecting new samples.

We think that drug discovery may provide a good testbed for this approach. It is an iterative process of designing and synthesising new molecules, followed by assessing their chemical properties with *in vitro* assays. After each iteration we learn from past molecules and try to design new ones that would bring us closer to finding the desired drug molecule. Predictive models are used to guide us in selecting molecules for the next iteration. Ideally, after each iteration, and prior to selecting molecules for the next iteration, we would use information regarding molecules from previous *in vitro* assays as a learning dataset to create new predictive models. In other words, after each iteration we update our learning datasets with new information and recreate predictive models. If the goal is to improve predictive models after each iteration, then our thesis is that randomly selecting molecules that are predicted as NotAvailable would improve the model in fewer iterations than if we just randomly select new molecules.

Currently we are not aware that anybody in the industry is using molecules that models are struggling to predict to drive the process of drug discovery. Therefore, we are not able to provide a real-life example of its use. However, we shall use simulation to show the potential benefits of such an approach.

6.6 Simulation Study

The idea is to mimic the iterative process of building predictive models and to show that by updating the learning dataset with randomly selected molecules that were predicted as NotAvailable, we would, in the long run, generate better models than by updating it with just randomly selected molecules. The iterative process of applying predictive models in drug discovery may, with some simplifications, be described as follows:

(A) build a predictive model using chemical compounds with known property values
(B) predict chemical compounds with unknown property values
(C) select *X* compounds from the set of predicted chemical compounds, and send them to a lab to be calculated
(D) go to (A).

Therefore, in order to simulate an iterative process we would need to have a large dataset of known experimental values, which we would need to divide into two subsets. The first subset consists of chemical structures and the experimental values, and we shall refer to it as the training set. As regards the second subset, we simulate that their experimental values are not known to us, and we shall refer to it as the test set. So to simulate the iterative process we would need to execute the following actions:

(A) build a predictive model on the training settings
(B) predict the test set
(C) select *X* compounds from the test set
(D) the training set is updated with selected *X* compounds, while the test set does not contain them
(E) go to (A).

Here we would like to simulate and compare two different approaches for selecting compounds:

a. randomly select *X* chemical compounds from the test set
b. randomly select *X* chemical compounds with NotAvailable prediction from the test set.

The major problem with simulating the above described iterative process is to have enough compounds at our disposal. With each iteration the training set will become larger, and we can expect to see improvements. Furthermore, the test set would get smaller, which means that we shall have fewer NotAvailable predictions in the test set relatively quickly. In reality, this does not happen because with virtual screening we have a vast amount of possible test compounds. Furthermore, the choice of *X*, the number of selected compounds from the test set, makes the simulation more complicated.

6.6.1 Design of the Experiment

We consider that we can mimic an iterative approach and assess the usefulness of selecting NotAvailable predictions by repeating the following single iteration many times with different pairs of training and test sets.

(a) select an optimal predictive model on the training set
(b) build the optimal predictive model on the training set using the ALOOM approach
(c) predict the test set
(d) let K be the number of NotAvailable predictions in the test set
(e1) (scenario 1) select *K* NotAvailable test compounds
(e2) (scenario 2) randomly select *K* test compounds
(f) update the training set with the selected *K* compounds and remove them from the test set
(g) select an optimal model on the new training set
(h) build the optimal predictive model on the new training set
(i) predict the new test set.

The aim is to assess which scenario produces better predictions on the new test set. In order to do so we ought to repeat scenario 2 many times so that we can estimate its average performance. Note that here we are not randomly selecting some compounds from the test set which were predicted as NotAvailable, but all of them. This is done to reduce the number of repeats in the simulation and to simplify the process of comparing two scenarios.

6.6.2 Results of the Experiment

In our simulation we used the same publicly available mutagenicity dataset of 4335 chemical compounds. We chose the training set to have 10% of the whole dataset, while the remaining 90% will make up the test set. The split is stratified, *i.e.* both sets contain the same proportion of "mutagen" and "nonmutagen" samples. Our aim is to simulate an initial training set of modest size and to have a sufficiently large test set. For each execution of scenario 1 we repeated scenario 2 100 times. We consider that it is a sufficient number of repeats for comparison. Furthermore, the whole process was executed for 100 pairs of training and test sets each created in the ratio of sizes 1 : 9.

The predictive models were created with two methods: ridge logistic regression and random forest. The optimal predictive model for ridge logistic regression was selected with 10 times repeated 10-fold cross-validations,[9] while for random forest there was no selection of optimal models as default parameters were used and the number of trees was 1000.

In Table 6.3 the results of the experiment are shown. The mean classification errors are shown for both scenarios and for each model type. The average difference in misclassification error between scenario 2 and scenario 1 is 1.92% for ridge logistic regression and 2.06% for random forest, all in favour of scenario 1. Random forest had on average more NotAvailable predictions than ridge logistic regression (759.7 *vs.* 517.4), and therefore a larger new training set. We suspect that this explains why random forest performed better than ridge logistic regression in the simulation.

Table 6.3 Results of the simulation. Mean percentage of NotAvailable predictions, sizes of new training and new test sets are shown for each model type, as well as mean classification errors for each scenario.

Model type	% Not available predictions	Size of new training dataset	Size of new test dataset	Scenario 1 misclassification error (95% CI)	Scenario 2 misclassification error (95% CI)
Random forest	19.47% (759.7/ 3901.5)	1193.2	3141.8	0.201 (0.187–0.213)	0.222 (0.202–0.244)
Ridge logistic regression	13.26% (517.4/ 3901.5)	960.9	3374.1	0.214 (0.204–0.226)	0.233 (0.221–0.247)

6.6.3 Discussion

Our intention was to check for the potential benefits of selecting chemical compounds predicted as NotAvailable in order to enhance the model. We are of the opinion that the simulation confirms the potential benefits, but it is no way near a proper confirmation. The simulation is performed on a single dataset with training size 433 or 434 and test size 3901 or 3902, and therefore we cannot conclude anything from it. Nevertheless, it shows a possible concrete benefit. Even a minute improvement in a single iteration of the drug discovery process, when accumulated across many projects and many iterations, will reduce the costs.

6.7 Questions and Criticism

Below are some questions and criticisms we have heard so far regarding the introduction of the NotAvailable predictions, and they are followed by our answers.

6.7.1 Question 1

Why is the category named NotAvailable? What was wrong with Uncertain? What about other names like Unknown, Unreliable, Unpredictable?

The main drawback from having the category called Uncertain is that it implies that other predictions are somehow certain, which is not true. We expect them on average to be more accurate, but individually they are not more certain. As we see it, how we name the predictions that should not be taken into consideration may have implications on how the user perceives and understands the remaining predictions.

We have contemplated various other names such as Unknown, Unpredictable, Unreliable, but we have realised that we are giving names to samples that they do not deserve. It is the decision from the predictive model's perspective that defines them. If the model has difficulties in predicting a sample's category, it is the model that is uncertain, and not the sample *per se*. In that case the model is colloquially saying "I don't know" and the predicted category is not available. By saying NotAvailable we are not implying anything regarding the quality of predictions that are not NotAvailable, except that they are – available.

6.7.2 Question 2

A user of a binary model is currently informed about the probabilities of its predictions, and it is up to the user to decide how to act upon that information. Is not the process being unnecessarily complicated by introducing NotAvailable predictions?

In our opinion, it is the responsibility of the predictive model's creator to make sure that the user is informed as to when it should not be applied. If we remove one sample from a learning dataset and that changes the predicted category, should we allow the user not be informed about it? We think that there are situations where the creator of a predictive model understands better than the user when the model's predictions should not be taken into consideration. It is true, however, that this makes the process more complicated. We think that there are situations where the price of the complication is worth it. As far as we are concerned, an obvious example would be the use of medical diagnostics tools that predict whether a person has an aggressive cancer or not.

6.7.3 Question 3

How would one compare two models with NotAvailable predictions? Is this not opening a door for the use of poor predictive models and allowing possible misuse?

As we have shown there are models (such as random forest) which generate more NotAvailable predictions than others (such as linear models). We currently do not have an answer to how to compare them, and that is a task for our future research. We do, however, think that the NotAvailable predictions ought to be introduced after the original model is selected, and that the performance of the original model on the validation set ought to be reported.

6.7.4 Question 4

Why would one use the ALOOM approach if one could define the NotAvailable predictions by selecting an interval of uncertainty?

We do not think that there is anything wrong with defining NotAvailable predictions using an interval of uncertainty. However, one ought to make sure that the predicted probabilities are well calibrated.[10] Generally speaking, well-calibrated probabilities mean that for 100 samples with predicted probability W, the proportion of correct predictions amongst them is expected to be W. In our experience it is not easy to generate a well-calibrated model, and we are not aware of any model selection process which optimises the goal of achieving well-calibrated models.

On the other hand, the ALOOM is a non-parametric approach where the model, so to speak, defines its own "blind spot". It is our proposal as to how to create a binary classifier with the "I don't know" answer. Of course, it is not a bullet-proof approach, and there might be better ways to select predictions that should not be taken into consideration.

6.7.5 Question 5

What is the difference between the non-applicability domain and the applicability domain? Why is there need for another research field?

Over the past two decades there has been active research within the QSAR community to estimate confidence of predictions. It was obvious to earlier researchers that certain models may only predict well on a subspace of chemical structures that are similar to molecules in a learning data. So the application domain represents the chemical space from which a model is derived and where a prediction is considered to be reliable.[11] We agree with QSAR researchers that there is a need and practical benefit in separating reliable from unreliable predictions, but our approach is different.

In the applicability domain the goal is to define a subspace where the model may be used, while in the non-applicability domain the aim is to define samples whose predictions should not be taken into consideration. Figuratively speaking, there may be questions regarding our confidence in what we think we know (application domain), but if we do not know something, we can be very confident that we do not know it (non-applicability domain). In other words, we think that there are more chances to make a mistake when defining the application domain than the non-application domain.

Furthermore, one might argue that, by defining the non-applicability domain, one is indirectly specifying the applicability domain. Even though that has not been the intention we cannot find any arguments against the statement. However by defining the applicability domain one is not necessarily specifying the non-applicability domain. In other words, if we define the "I don't know" answer then the model is applicable on samples that are different from "I don't know". However, if we define samples where the models may be used, it does not mean that other samples are equal to "I don't know".

In our view the non-applicability domain is relevant not only in predictive systems but in AI research as a whole. How can AI systems be created which would know when they do not know? We are not aware of any research that is trying to answer this important question. The ALOOM approach, which we suggested, is a simple example of how to create a predictive model that, loosely speaking, knows when it does not know.

As far as we are concerned, the non-applicability domain is a research field in AI which requires much more focus than it currently has.

6.7.6 Question 6

There is an active research field called conformal predictions which is trying to resolve the issues discussed here. How does it fit into the non-applicability domain? What is the difference between the ALOOM approach and conformal predictions?

There is a chapter in this book devoted only to conformal predictions (CP) and we advise readers to study it. We will briefly describe the relationship between CP and the ALOOM approach, as well as the non-applicability domain.

In CP one needs to define a confidence level which is usually expressed as a percentage, as well as the non-conformity measure, a measure of how different a new sample is from the learning samples. The ALOOM, on the other hand, is a nonparametric approach, and there is no need to define any parameters.

In CP, binary classifiers may return the following four predictions: {positive}, {negative}, {positive and negative} and {null}. So instead of NotAvailable, there are two possibilities: "both" and "empty". However, the percentages of "both" and "empty" predictions depend on the CP's confidence level. The higher the CP's confidence level the more "both" and fewer "empty" predictions will appear, and *vice versa*. As far as we are concerned, in the context of the non-applicability domain one may treat "both" and "empty" predictions as NotAvailable. However, we would like to point out that we are not aware of any theoretical or empirical work that would suggest that the CP's "both" and "empty" predictions are similar to the ALOOM's NotAvailable.

6.7.7 Question 7

How can the ALOOM approach be applied on a large learning dataset?

For the datasets containing a large number of samples the ALOOM approach is not a practical solution. How to generate a meaningful "I don't know" answer for such learning datasets is one of the goals of our future research. However, if there were indications that the CP's "both" and "empty" predictions are similar to the ALOOM's NotAvailable, then one might suggest using CP on large learning datasets as a replacement for the ALOOM. Likewise, one might then suggest applying the ALOOM approach on small learning datasets instead of CP.

6.7.8 Question 8

Are there any suggestions of how to define NotAvailable in other settings, like regression or multi-class?

At the moment we suppose that the ALOOM approach may be applicable in a multi-class setting in the same way as in a binary setting, while for regression it seems to us that the definition of the "I don't know" answer may depend on the context.

6.8 Final Remarks

Consider this question. Person A knows more than person B. However, person A cannot admit "I don't know", while person B is capable of doing so. Who will you trust more?

In our view, being able to say "I don't know" is vital for trust whether we are talking about people or software expert systems however you call them. But trust is not the only benefit. Information regarding our ignorance, or that of future AI, ought to help us in defining the boundaries of our knowledge as well as potential directions for our future research. We have presented a procedure to create binary statistical models with, in our view, a meaningful "I don't know" answer. Furthermore, we have shown how the knowledge of a model's ignorance thus defined may be used to further enhance the model's capabilities. It is an example of how a well defined ignorance may improve a learning process and research.

In the past we have, together with our colleagues, worked on automating decision-making in drug discovery.[12,13] Even though having an "I don't know" answer may at first seem as a complication, it partially solves the problem of decision-making under uncertainty, and therefore ought to improve its automation.

From our point of view, the research progress in AI depends on creating tools which would, figuratively speaking, know when they don't know. We believe that it is not only possible but also that it is the responsibility of AI creators to provide design solutions for their non-applicability domains. As well as investing in extending the capabilities of AI, it would be wise to assess their limitations. Otherwise, we might end up blindly believing in AI without giving it a proper chance to succeed.

Abbreviations

AI	artificial intelligence
ALOOM	all leave one out models
CP	conformal predictions

Acknowledgements

First and foremost, the author would like to thank Dr Nathan Brown for giving him an opportunity to write about a subject in which, at one point, nobody seemed interested. Some thoughts and findings in the chapter are the direct result of his encouragement. The author is grateful to professor David E Leahy and Vladan Djordjevic for their time in discussing the subject and reading earlier drafts of the chapter. Last but not the least, the author would like to thank his mother, Linda Louise Woodall Krstajic, for correcting English typos and language improvements in the text.

References

1. T. G. West and G. S. West, ed. *Four Texts on Socrates: Plato's Euthyphro, Apology, and Crito, and Aristophanes' Clouds*. Cornell University Press, 1998.
2. R. Descartes, *Discourse on the Method*, Perennial Press, 2018.
3. D. Krstajic, L. Buturovic, S. Thomas and D. E. Leahy, *Binary classification models with "Uncertain" predictions*, arXiv preprint arXiv:1711.09677, 2017, Dec 4.
4. J. Kazius, R. McGuire and R. Bursi, Derivation and validation of toxicophores for mutagenicity prediction, *J. Med. Chem.*, 2005, **48**(1), 312–320.
5. QSARdata. https://cran.r-project.org/package=QSARdata.
6. A. E. Hoerl and R. W. Kennard, Ridge regression iterative estimation of the biasing parameter, *Commun. Stat. Theory Methods*, 1976, **5**(1), 77–88.
7. L. Breiman, Random forests, *Mach. Learn.*, 2001, **45**(1), 5–32.
8. R. A. Fisher, *The Design of Experiments*, Oliver And Boyd, Edinburgh, London, 1937.
9. D. Krstajic, L. J. Buturovic, D. E. Leahy and S. Thomas, Cross-validation pitfalls when selecting and assessing regression and classification models, *J. Cheminf.*, 2014, **6**(1), 10.
10. A. P. Dawid, Calibration-based empirical probability, *Ann. Stat.*, 1985, **13**(4), 1251–1274.
11. OECD (Organisation for Economic Co-operation and Development). *Guidance Document on the Validation of (Quantitative) Structure Activity Relationship [(Q) SAR] Models.*
12. J. Cartmell, S. Enoch, D. Krstajic and D. E. Leahy, Automated QSPR through competitive workflow, *J. Comput.-Aided Mol. Des.*, 2005, **19**(11), 821–833.
13. J. Cartmell, D. Krstajic and D. E. Leahy, Competitive Workflow: novel software architecture for automating drug design, *Curr. Opin. Drug Discovery Dev.*, 2007, **10**(3), 347–352.

Section 4: Structure-based Predictive Modelling

CHAPTER 7

Predicting Protein-ligand Binding Affinities

JOSÉ JIMÉNEZ-LUNA[a] AND GIANNI DE FABRITIIS*[a,b]

[a] Computational Science Laboratory, Universitat Pompeu Fabra, PRBB, C Dr Aiguader 88, 08003, Barcelona, Spain; [b] ICREA, Passeig Lluis Companys 23, 08010 Barcelona, Spain
*Email: gianni.defabritiis@upf.edu

7.1 Introduction

Drug discovery is inherently a multiobjective optimization problem[1,2] as several variables need to be taken into account, *e.g.* solubility,[3] toxicity,[4,5] selectivity[6,7] or kinetics.[8,9] Among these, perhaps the most important is potency, which measures how strongly a small molecule binds to its protein target to produce a desired effect or inhibition. Chemists typically study this variable through experimental affinity measurements (*e.g.* K_i, K_d, IC_{50}) in different types of assays in the laboratory, such as phenotypic or cell-based ones.

Experimentally determining binding affinities is a costly process, and therefore computational approaches to predict these quantities *in silico* were consequently developed in order to prioritize testing of compounds. In fact, quantitative structure–activity relationship (QSAR) approaches, based on fitted linear or empirical models of molecules have been common among those in a computational chemist's toolbox for the last 30 years. With the advent of increasing available affinity data coming from compound databases such as ChEMBL[10] and protein–ligand ones, such as PDBbind[11] and cheaper computational resources,[12] opportunities to explore more

Drug Discovery Series No. 75
Artificial Intelligence in Drug Discovery
Edited by Nathan Brown

Published by the Royal Society of Chemistry, www.rsc.org

data-hungry machine learning approaches have become prevalent in the last decade. In the case of structure-based approaches, these typically allowed more flexibility by not requiring an explicit mathematical relationship of the protein–ligand complex[13] and their affinity, which in practice resulted in greatly improved performances compared to classical QSAR approaches.

In this chapter we provide readers with an introduction to the field of structure-based potency prediction *via* machine learning. Though the work here does not intend to be an exhaustive review of proposed approaches, which at the time of writing continues to grow at a fast pace, we believe a disciplined introduction on the basics regarding this area, namely classification and scope of models, descriptor generation and evaluation standards to be beneficial for the community. Finally an overview of the most important techniques in the last few years, to the best of our knowledge, is provided.

7.2 A Brief Background on Classical Methodologies

A scoring function $f: \mathbb{X} \rightarrow \mathbb{R}$ maps a ligand, or a protein–ligand complex $x \in \mathbb{X}$, to a quantity which is either its binding affinity or a proportional proxy. Over the years, many types have been proposed, most claiming advantage of some form over already existing approaches. They can arguably be classified into three different categories depending on the nature of their modelling,[14] potential-, simulation- and data-based. In this section we provide a small background on previous work focusing on the subject of binding affinity prediction, based on such classification, before describing the more recent machine-learning approaches.

7.2.1 Potential-based

Methods in this category model binding affinities as the sum of statistical potentials between protein and ligand atoms:

$$\Delta G = \sum_{i \in A_l} \sum_{j \in A_p} \omega_{ij}(r), \tag{7.1}$$

where A_ℓ and A_p are the sets of atoms in the ligand and protein respectively and

$$\omega_{ij}(r) = -k_B T \log\left(\frac{\rho_{ij}(r)}{\rho^*_{ij}}\right), \tag{7.2}$$

where $\rho_{ij}(r)$ is the number density of atom pair ij at distance r, ρ^*_{ij} is the same quantity at a reference state with no interatomic interactions, k_B is the Boltzmann constant and T a temperature. The reasoning behind this approach is probabilistic: if a certain interatomic contact appears more frequently than expected of its reference state, it is energetically favorable and *vice versa*. Potential-based approaches have been widely used mainly thanks to the simplicity in their construction. Some popular implementations of these approaches are SMoG,[15] Muegge's PMF,[16] DrugScore,[17] IT-Score[18] and KECSA,[19] among many others.[20–24]

7.2.2 Simulation-based

Molecular mechanics force fields such as AMBER[25] or CHARMM[26] are regularly used for predicting protein–ligand interactions.[27–33] Often, approximate solutions such as the Poisson–Boltzmann or Poisson Generalized Born models are used, where van der Waals, electrostatic, and desolvation terms are taken into account:

$$\Delta G = \Delta E_{\text{vdw}} + \Delta E_{\text{electrostatic}} + \Delta E_{\text{H-Bond}} + \Delta G_{\text{desolvation}}. \tag{7.3}$$

Full simulation based approaches such as free energy perturbation (FEP),[34–40] or thermodynamic integration (TI)[41,42] have shown to provide excellent performance despite being computationally demanding. These methods can also take advantage of the advances of modern force fields, quantum mechanic methods and solvation models. However, recent evaluations[43] have shown that their performance is very sensitive to starting parameters such as force-field selection or treatment of waters, limiting their applicability in prospective scenarios. Other related approaches, such as linear interaction energy (LIE),[44] linear response approximation (LRA)[45] and MM-PBSA/GBSA[46] methods have shown alternate successes and failures.[47] These are typically named end-point approximation methods as they only consider both protein and ligand in their bound and unbound states.

7.2.3 Data-based

In this category we find classical scoring functions that use statistical methods such as linear regression or partial least squares (PLS)[48] to adjust the contribution of several physico-chemical terms (or other descriptors) towards an affinity prediction. Therefore a set of known protein–ligand complexes with affinity data is needed to find the aforementioned coefficients for an optimal fit. For instance, X-Score[49] adopts the following functional form:

$$\begin{aligned}\Delta G_{\text{bind}} = {} & \beta_0 + \beta_1 \Delta G_{\text{vdw}} + \beta_2 \Delta G_{\text{H-bond}} \\ & + \beta_3 \Delta G_{\text{deformation}} + \beta_4 \Delta G_{\text{hydrophobic}} + \varepsilon, \quad \varepsilon \sim (0, \sigma^2)\end{aligned} \tag{7.4}$$

where ΔG_{vdw} accounts for the van der Waals interaction between the protein and ligand, $\Delta G_{\text{H-bond}}$ for hydrogen bonding, $\Delta G_{\text{deformation}}$ for the deformation effect and $\Delta G_{\text{hydrophobic}}$ for hydrophobic interactions. An estimate of these coefficients $\hat{\boldsymbol{\beta}}$ is determined through a least squares fit. Many other empirical scoring functions have been developed over the years, such as cyScore,[50] ChemScore,[51] GlideScore-XP,[52] LudiScore,[53] among many others.[54,55]

At first glance, one may be right to think that modern machine-learning scoring functions belong to this category, as they use a training set and several fit parameters to perform predictions. The main difference is that classical empirical scoring functions assume a fixed, pre-defined mathematical relationship among handcrafted features to model the target affinity. Such is not the case for modern machine-learning scoring functions, which automatically extract the most important features from a closer representation from the real data, in practice allowing them to be considerably more flexible and consequently provide better performance. Recent advances in modern machine-learning scoring functions are summarized in the subsequent section.

7.3 Modern Machine-learning Scoring Functions

Research on machine-learning based scoring functions is currently a very active field, with dozens of proposed approaches just in the last ten years. In this section, we first describe the domain applicability of the proposed models, depending on the nature of their training data. We then make a thorough summary on the types of features (or predictors) used for the training of models in the literature, to later discuss recent developments on specific structure-based machine-learning algorithms. Finally we describe several approaches towards model interpretability and discuss about the availability of the proposed algorithms at the time of writing.

7.3.1 Domain Applicability

It is common in the development of scoring functions to distinguish between two scenarios, depending on the nature of the data at hand. If the goal itself is to predict binding affinity or a proportional proxy, one needs continuous binding constants, and typically tackles the problem as a regression task. Other scenarios may feature binary data (*i.e.* active/inactive), which is better suited for a more classical virtual screening scenario, where the goal is to select, from a very big database, as many active molecules (also called binders) against a target as possible. Since only two classes are considered in this scenario, binary classification models are commonly used to tackle this problem instead. Note that any regression-based scoring function can be used as a binary classification one given an appropriate threshold.

It is also worth mentioning that, in this chapter, we focus on the structure-based case, that is, we are interested in a model that generalizes not only in chemical space, but also across different targets, which is common at earlier phases of drug discovery, such as target identification. If the study at hand only takes into account a single protein target, ligand-based models are significantly more popular, the archetypal application in this case being lead optimization.

7.3.2 Descriptors

In this section we describe different sets of descriptor studies used in the development of machine-learning models for the prediction of both continuous and binary binding affinity estimates. It was the norm until recently for researchers not only to design their own set of features, but to perform descriptor selection as well. This process of handcrafted feature creation and filtering was subsequently abandoned, when it was shown that modern deep learning architectures could perform automatic feature extraction from a closer representation of the original data. This allowed models to reach a more diverse, non-linear, latent set of features that in practice obtains better performance when enough data is available.

We, therefore, explicitely distinguish in the next sections between handcrafted and automatic feature extraction.

7.3.2.1 Handcrafted Features

Since the inception of the (arguably) first structure-based machine-learning-based scoring functions in 2004, researchers have used an increasing number of feature sets in order to discover the better performing ones. Although we do not provide an exhaustive list, among the most popular that we have found are:

- occurrences of each protein–ligand atom pair at different distance thresholds[56–60]
- electronegativities of ligand and protein atom types[61]
- atom and group interactions such as van der Waals, electrostatics, hydrogen-bonding, π-system, aromatic and metals[62,63]
- energy terms representing desolvation and entropic losses[62]
- geometrical description of the binding, such as shape and surface property matching[62]
- property-encoded shape distributions[64]
- protein–ligand interaction fingerprints[65–67]
- intermolecular interaction terms extracted from AutoDock Vina[68] and BINANA[69,70]
- distance-based fuzzy membership functions accounting for attraction and repulsion terms between atoms[71]
- knowledge-based potentials combining SYBYL atom types[72]
- other structural features regarding β-contacts, crystallographic-normalized B factors and polar and hydrophobic contact surfaces[73]
- multiscale weighted labeled algebraic subgraphs.[74]

Interestingly, it was found that a more detailed description of the protein–ligand interaction landscape did not necessarily result in better performance in the standard PDBbind benchmark,[75] despite some studies considering more than 100 feature subsets.

7.3.2.2 Automatic Feature Extraction

One of the reasons for the success of deep learning[76] architectures in fields like computer vision[77] and natural language processing[78,79] was the fact that modern neural networks perform automatic feature extraction. Among the first structure-based approaches that adopt this strategy we acknowledge the one taken by DeepVS.[80] In particular, in this approach, the local context of an atom in a protein–ligand complex is embedded into several learned fixed sized vectors using several basic features such as atom types, partial charges and distances to the closest ligand neighbors and amino acids. The definition of the atom context, however, is user dependent, since a number of neighboring ligand and protein atoms need to be pre-specified. For each type of basic feature, a column lookup operation is performed in a pre-defined learnable embedding matrix W_t for each possible discretized value of a feature. The embedding of an atom z_i is then constructed by the concatenation of the column vectors:

$$z_i = \{z_{\text{atom}}, z_{\text{distance}}, z_{\text{change}}, z_{\text{aminoacid}}\} \tag{7.5}$$

Since the number of atoms varies with the particular system, this representation needs then to be summarized in a single latent vector v, representing an embedding for the entire complex, which we describe in the next section. This set of descriptors suffer mainly from two problems: the first being the pre-specification of context, and the second the need to discretize continuous atom features into bins to extract its corresponding learned latent space from W_t.

Other approaches have more closely followed computer vision architectures. Using the example of image classification, a convolutional neural network (CNN) learns which picture patches are the most informative in order to arrive to a correct classification, only using pixel information. A similar argument can be made for proteins and small chemical compounds: they are structures in three-dimensional space with different physiochemical values as properties. Identically to a two-dimensional image, which can be represented using three different two-dimensional arrays, or channels, representing its colors (*i.e.* RGB color space), a protein or a ligand could potentially also be represented with k three-dimensional arrays, and modern computer vision architectures would be readily applicable. In fact, this is a promising direction that several researchers have taken in the last few years: the same way that a pixel in an image holds three values for its colors, a voxel in a three-dimensional image (in our case, a protein or ligand) can feature different values representing different molecular properties. Directly translating atomic positions into volumetric space can result in a very sparse representation, and therefore most works use a distance-based interpolation over pre-defined atom types. For instance, Ragoza *et al.*[81] use the following functional:

$$A_1(d,r) = \begin{cases} \exp\left(-\dfrac{2d^2}{r^2}\right) & \text{if } 0 \le d < r \\ \dfrac{4d^2}{e^2r^2} - \dfrac{12d}{e^2r} + \dfrac{9}{e^2} & \text{if } r \le d < \dfrac{3r}{2}, \\ 0 & \text{if } d \ge \dfrac{3r}{2} \end{cases} \tag{7.6}$$

where d is the distance of each atom to a particular voxel, and r is the atom's van der Waals radius. Jiménez *et al.*,[82] on the other hand, use the following functional in the K_{DEEP} protein–ligand affinity predictor:

$$A_2(d,r) = 1 - \exp\left(-\left(\frac{r}{d}\right)^{12}\right) \tag{7.7}$$

In terms of channel selection, the vast majority of approaches use some pre-defined notion of atom typing available in other applications, such as the ones defined in smina[83] or in the AutoDock PDBQT format (hydrophobic, aromatic, hydrogen-bond acceptor/donor, positive/negative ionizable and metals).[84] A general occupancy channel is also typically included to include explicit geometrical information of the molecular object (Figure 7.1). This particular representation of biomolecular complexes has not only been used for protein–ligand binding affinity prediction, but for protein binding site prediction,[85] pharmacophore elucidation[86] and *de-novo* generation of molecules,[87] with varying success.

These sets of descriptors represent the three-dimensional structure of proteins in space, but they also suffer from several issues. In particular, three-dimensional arrays take significantly more space in memory than images in computer vision applications. Furthermore, voxelizing the binding site entails choosing a window, since the shape of the arrays needs to be fixed for their use in CNN architectures, while protein pockets and ligands can significantly differ in size. Finally, the representation is neither rotationally nor translationally invariant, properties which are desirable when modelling biomolecular complexes with the aforementioned models.

Given the success of graph convolution architectures in ligand-based approaches,[88] some attention has been given to adapting the same framework

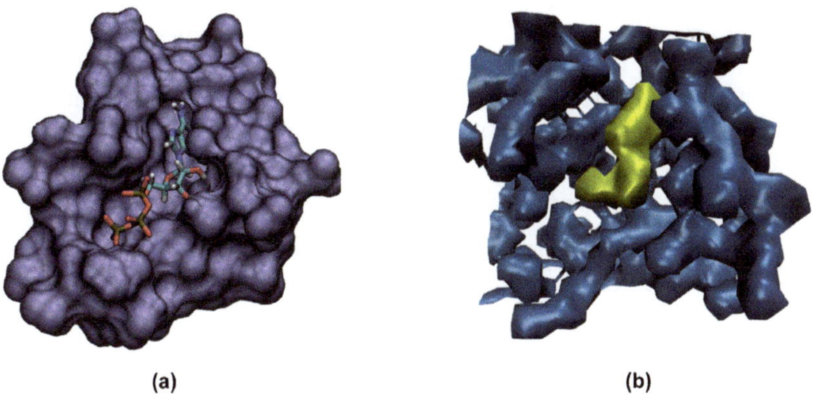

Figure 7.1 (a) PDB Id 2HMU pocket and bound ligand ATP. (b) Voxel representation of the hydrophobic channel for both protein (blue) and ligand (yellow). Reproduced from ref. 82 with permission from American Chemical Society, Copyright 2018.

for structure-based approaches. In particular, PotentialNet[89] uses the concept of adjacency from an atomic distance matrix $R \in \mathbb{R}^{N \times N}$, where N is the number of atoms in a pre-defined environment of the binding site of the system. While ligand-based graph-convolution architectures use the notion of bonds to represent adjacency, in a co-crystal it can encompass a wider range of chemical interactions among neighbors, such as π–π stacking, hydrogen bonds or hydrophobic contact. In fact, a simple distance-based threshold may serve to construct an adjacency matrix $A^{N \times N \times c}$, where c represents the number of edge types. Ordering the rows of the adjacency matrix by the membership of each atom to their corresponding protein or ligand complex, A can be seen as a block matrix, where the diagonal blocks are interactions inside the same complex, while the off-block elements represent interactions between the protein and ligand atoms:

$$A = \begin{bmatrix} A_{11} & A_{12} & \cdots & A_{1N} \\ A_{21} & A_{22} & \cdots & A_{2N} \\ \vdots & \vdots & \ddots & \vdots \\ A_{N1} & A_{N2} & \cdots & A_{NN} \end{bmatrix} = \begin{bmatrix} A_{L:L} & A_{L:P} \\ A_{P:L} & A_{P:P} \end{bmatrix}, \tag{7.8}$$

where $A_{ij} = 1$ for an occurring interaction and 0 in any other case. We will explain how this input is used in a graph-based convolutional neural network in the next subsection.

7.3.3 Models

In this section we will discuss recent advances in the modelling of protein–ligand binding affinities with modern structure-based machine-learning architectures. Mostly we focus here on those models that perform automatic feature extraction (see previous section) out of a representation closer to its source. In particular, we briefly describe custom embedding approaches, discuss three-dimensional convolutional neural networks whose corresponding input is the atom-type voxelization, and finally graph-convolution neural networks, which in turn use a distance-based representation between atoms and their features in a system.

7.3.3.1 Custom Embedding Approaches

Here we describe the approach of DeepVS.[80] Once atom feature vectors z_i have been computed, they are mapped to:

$$u_i = \tanh(W^e z_i + b^e), \tag{7.9}$$

where W^e, b^e are a set of learnable weights and biases, respectively, which are shared for all embeddings z_i. A maximum-over-columns operation is then applied to obtain a fixed-size vector representation for the entire complex, which is used by subsequent fully connected layers to obtain the desired scalar output.

7.3.3.2 3D-convolutional Neural Networks

The output ϕ of a neuron in a regular fully connected layer is obtained by multiplying an input $\boldsymbol{x}$ with some learnable weights $\boldsymbol{w}$, adding a bias b and applying a non-linearity f:

$$\varphi = f\left(\sum_i w_i x_i + b\right). \tag{7.10}$$

Convolutional neural networks are specifically designed to work with spatial inputs, such as images (or voxels, in our case), trying to emulate the response of an individual neuron to visual stimuli. This fact allows us to encode certain properties into the architecture that cannot be assumed in fully connected ones. Concretely, the layers of a three-dimensional neural network are arranged in four dimensions: height, width, depth and number of channels, with each neuron only locally connected to a localized region of the preceding layer, since it is impractical to connect to all of the previous neurons.[90] In the case of voxels, the output of a neuron is called a feature map ϕ, which is a three-dimensional tensor, obtained through discrete convolution of a filter W_i over an input feature map $z_i(x, y)$:

$$\varphi = f\left(\sum_i W_{i*} z_i(x,y,z) + b\right), \tag{7.11}$$

where $*$ represents a three-dimensional discrete convolution operation (*i.e.*, $w*f(x,y,z) = \sum_{s=-a}^{a} \sum_{t=-b}^{b} \sum_{l=-c}^{c} w(s,t,l) f(x-s, y-t, z-l)$).[91] The connectivity of a neuron is controlled by a parameter named a kernel or filter size, and its locality is only defined across the spatial dimensions, while full connectivity is applied to all feature channels. Parameters in a convolutional neural network are also typically shared in the channel dimension: if one feature is useful for a particular position in the image, it should also be so for a different one. This simplification results in a significant reduction of learnable parameters, which in turn makes current implementations more computationally approachable at a scale.

Apart from convolution, other layers are commonly used in the development of this type of neural network. For instance, pooling layers apply a non-learnable transformation to an input, typically reducing its size in order to simplify further calculations and reduce the number of parameters in the network, operating independently on each channel. Normalization layers, such as batch normalization,[92] ensure that the input distribution remains similar across batches, while dropout layers randomly drop neuron connections with the hope of avoiding overfitting. Once the size of the three-dimensional filters has been reduced enough, these are typically flattened out to a vector, so that regular fully connected layers can be applied afterwards to ensure a one-dimensional output corresponding to the predicted binding affinity.

Approaches based on three-dimensional CNNs, such as K_{DEEP}[82] have been shown to work particularly well in practice, scoring first in several targets of the D3R Grand Challenge 4.[93] Other approaches to tackle lead optimization of congeneric series, such as DeltaDelta,[94] have been recently developed.

7.3.3.3 Graph-based Models

In a regular convolutional neural network layer, the output of each layer is composed of the convolution of the previous one by the use of linear kernels and non-linearities, effectively gathering information from neighboring pixels. Similarly, a graph has an inherent structure that can be efficiently exploited: each node (or atom) v_i can have a vector of features $\mathbf{x}_i$, and a set of neighbors based on an adjacency matrix A, as described in the previous section. Each node also features a latent representation h_i, which is iteratively updated by several functions (Figure 7.2):

$$h_i^{(t+1)} = U\left(h_i^{(t)}, \sum_{v_j \in N(v_i)} m^{(t)}\left(h_j^{(t)}\right)\right), \tag{7.12}$$

where:

- U is a differentiable *update* function that updates the latent representation of a node depending on the one from its neighbors
- m is a differentiable *message* function sending a transformation of the hidden states from nodes v_j to v_i
- $N(v_i)$ is the set of neighbors of node v_i.

In order to obtain a single representation for the entire graph, a node-order invariant *readout* function R (also known as graph gather) is typically

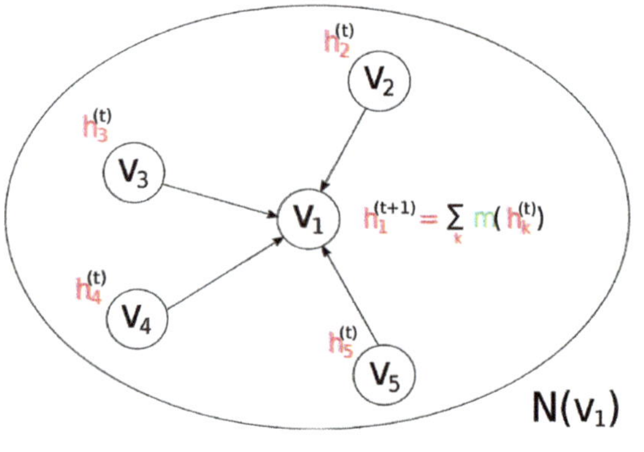

Figure 7.2 In graph convolution, the hidden state h of each node v is iteratively updated with its neighboring ones *via* a differentiable message function m.

applied. In general, update, message and readout functions can be fully parameterized by neural networks, and such is the case in most applications.[95–101] While this is the general scheme[88] for most architectures, we here focus on gated graph neural networks (GGNNs),[102] which use a gated recurrent unit (GRU)[103] module as the update function and independent linear message functions for each edge type:

$$h_i^{(t+1)} = \mathrm{GRU}\left(h_i^{(t)}, \sum_e W^{(e)} A^{(e)}\left(h^{(t)}\right)\right), \tag{7.13}$$

where $A^{(e)}$ and $W^{(e)}$ are the adjacency and learnable weight matrices for edge type e, respectively. A simple readout function which sums over the final node embeddings is applied to obtain the desired output size:

$$h^{(0)} = \sum_{r=1}^{N} \left(\sigma(i(h^{(K)}, x) \odot j(h^{(K)}))\right)_r, \tag{7.14}$$

where $\sigma(x) = 1/(1+\exp(-x))$ is the sigmoid function, i, j are arbitrary learnable functions and $\odot$ is the element-wise multiplication. Once a single vector of the desired size is obtained per graph, standard fully connected layers can be used for classification or regression tasks. A structure-based generalization of this approach is the one proposed by PotentialNet, which introduces non-linearity in the message function:

$$h_i^{(t)} = \mathrm{GRU}\left(h_i^{(t-1)}, \sum_e^{N_{\mathrm{et}}} \sum_{j \in N^{(e)}(v_i)} \mathrm{NN}^{(e)}\left(h_j^{(t-1)}\right)\right), \tag{7.15}$$

where $\mathrm{NN}^{(e)}$ represents a standard feed-forward neural network for edge type e and $N^{(e)}(v_i)$ are the neighbors for node i with edge type e. Two different *stages* are defined in the PotentialNet architecture, depending on where message functions are applied. In stage 1, named covalent propagation, only graph convolutions over ligand bonds are applied, similarly to the ligand-based counterpart. In stage 2 both bond-based and spatial-based graph convolutions are implemented, effectively propagating information between ligand and protein atoms. This stage is known as dual non-covalent and covalent propagation. Finally in stage 3, a ligand-based readout function is applied in order to obtain a fixed-size feature vector (Figure 7.3).

7.3.4 Interpretability

The ability to interpret predictions is one of the most important features when it comes to convince computational and medicinal chemists of the usefulness of a particular model. When we speak about interpretability we refer to that of its parameters, that is, the influence of each input towards its predicted affinity value. When simpler data-based scoring functions are

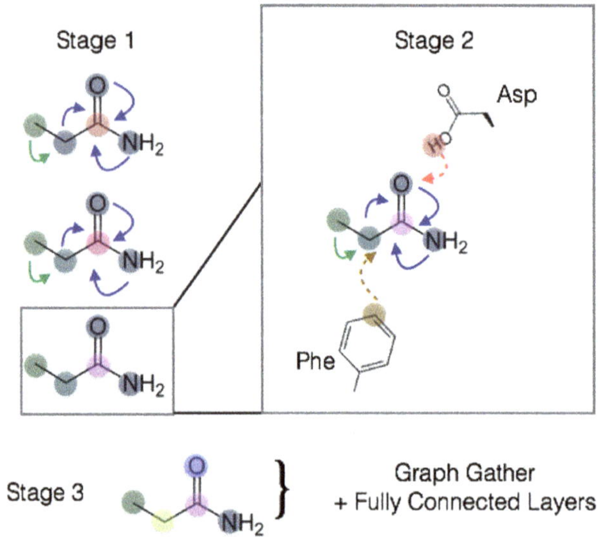

Figure 7.3 Depiction of the several stages defined by the PotentialNet approach. Stage 1 only takes into account updates over ligand bonds, while stage 2 also considers updates over neighboring protein atoms, based on a distance threshold.
Reproduced from ref. 89, https://doi.org/10.1021/acscentsci.8b00507, with permission from American Chemical Society, Copyright 2018.

considered, such as those using a linear model, interpretation is straightforward since each corresponding coefficient β_i is interpreted as the individual contribution of each of the input variables towards the prediction.

When it comes to modelling data, two different and opposing approaches arose in the statistics literature.[104] The first one assumes that the relationship between a target variable and a certain set of predictors follows a particular, but known, functional form f, which typically depends on certain parameters θ fit using available data. On the other hand, modern machine-learning approaches do not assume a particular functional form f, and instead choose non-linear algorithmic approaches that approximate the target variable as well as possible. In practice, the latter approaches have been shown to provide superior predictive performance. Earlier data-driven approaches are direct descendants of the first category, while more modern scoring functions are closer in nature to the second. A negative consequence of this is the fact that the latter do not satisfactorily address model interpretability. In this sense, there has been some recent effort in the community to satisfy the need for interpretable machine-learning models.[105,106] In this section we summarize several presented approaches towards that end.

7.3.4.1 Masking

Masking is a popular yet simple approach applied in computer vision to find out which parts of the input a trained convolutional neural network finds important in its corresponding prediction. In masking, parts of the input, in this case the ligand or the protein are sequentially removed and re-scored, the difference between the non-altered image and the altered one representing the importance of that particular part. For the ligand, one can remove either individual atoms, or choose to remove entire molecular subgraphs (fragments). Similarly, for proteins, individual residues can be removed to find their individual contribution. Masking, however, is computationally demanding due to the number of evaluations needed, as they grow polynomially with ligand subgraph generation.

7.3.4.2 *Atomic Gradient*

When training 3D-based convolutional neural network-based models, one typically minimizes a loss function by optimizing a set of learnable parameters. This process requires numerically computing the gradients of the loss with respect to them, which can then be naturally extended to computing the corresponding gradient with respect to the input representation. The negative of the aforementioned gradient can be interpreted as the directions in three-dimensional space which maximize the network's predicted binding affinity. The functions mapping a particular atom with a type to a voxel density are differentiable with respect to distance d and the gradient of the neural network scoring function f with respect to atom coordinates $\boldsymbol{a}$ is found *via* the chain rule and aggregating all grid points $\boldsymbol{G}_a$ overlapping each atom with a particular type:

$$\frac{\partial f}{\partial \boldsymbol{a}} = \sum_{A \in \boldsymbol{G}_a} \frac{\partial f}{\partial A} \frac{\partial A}{\partial d} \frac{\partial d}{\partial \boldsymbol{a}} \tag{7.16}$$

7.3.4.3 *Layer-wise Relevance Propagation*

Layer-wise relevance propagation[107] defines a measure R_d over the voxels x_d of a volumetric input which decomposes the output of a neural network binary classifier into a sum of relevances:

$$f(x) \approx \sum_{d=1}^{V} R_d, \tag{7.17}$$

with the qualitative interpretation that $R_d < 0$ contributes negative evidence for a classification, and R_d contributes positively. The goal is to find a separate relevance per layer of the classifier, with the constraint that their sums are as close as possible:

$$f(x) = \cdots = \sum_{d \in l+1} R_d^{(l+1)} = \sum_{d \in l} R_d^{(l)} = \cdots = \sum_{d} R_d^{(l)}. \tag{7.18}$$

Iterating eqn (1.18) from the last classification layer to the first one yields the desired heatmap over voxels defined by eqn (1.17). However, a decomposition satisfying said constraint is not unique, and one popular alternative is to propagate relevances according to their corresponding neuron activations $z_{ij}=x_i w_{ij}$. The relevance of node i at layer l is defined as the sum of the ones from deits following nodes j, weighted by z_{ij}:

$$R_i^{(l)} = \sum_j \frac{z_{ij}}{\sum_{ij} z_{ij}} R_j^{(l+1)}. \tag{7.19}$$

Layer-wise relevance propagation distributes the output value of the network as an explanation for the reason a particular input generated it. Similarly to the atomic gradient approach, this method only requires a single backwards pass through the network to obtain the desired results. However, there are issues with this proposed methodology, especially when propagating through nodes whose activation is zero. Several solutions to this problematic have been proposed in the literature, such as the alpha-beta decomposition or the conserved layer-wise relevance propagation.[108] Examples of the previously mentioned interpretability techniques can be checked in Figure 7.4.

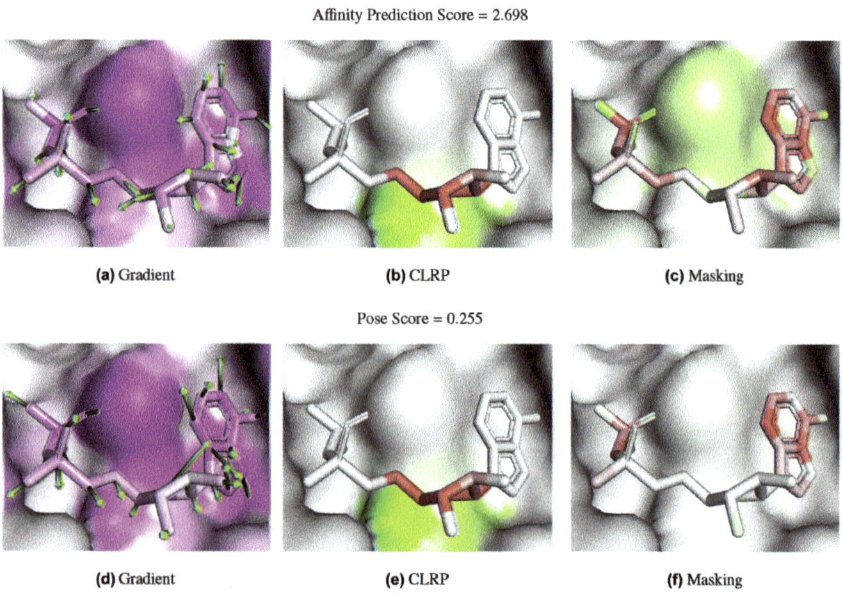

Figure 7.4 Masking, atomic gradient and layer-wise relevance propagation interpretability techniques for PDB. Id. 1o0h. For both masking and layer-wise relevance propagation approaches green and red represent a positive and negative contribution to binding, respectively, while for the atomic gradient technique its norm is represented in a purple scale.
Reproduced from ref. 108 with permission from Elsevier, Copyright 2018.

7.3.4.4 Class Activation Maps

Class activation maps[109] use the fact that the filters of convolutional neural networks behave as object detectors without explicit supervision in classification tasks (such as in virtual screening). A technique named global average pooling[110] is used to output the spatial average of the feature map of each unit at the last convolutional layer, whose weighted sum is used to generate the final output. In a similar way, a weighted sum of the feature maps of the last convolutional layer is used to obtain class activation maps. Formally, let $f_k(x, y, z)$ represent the activation of unit k in the last convolutional layer at location (x, y, z). Then the result of applying global average pooling for unit k is $F_k = \sum_{x,y,z} f_k(x,y,z)$, and therefore for a given class c, the input of the softmax classification layer is $S_c = \sum_k w_k^c F_k = \sum_{x,y,z} \sum_k w_k^c f_k(x,y,z)$, where w_k^c is the weight that corresponds to class c for unit k. The class activation map M_c at location (x, y, z) is then defined by:

$$M_c(x,y,z) = \sum_k w_k^c f_k(x,y,z) \tag{7.20}$$

Therefore $S_c = \sum_{x,y,z} M_c(x,y,z)$, and thus M_c indicates activation importance at (x, y, z). Upsampling is performed in order to map these features to the original input size, motivated by the fact that each unit is activated by some visual pattern in its receptive field.[110]

More modern techniques, such as gradient class activation maps[112] remove the restriction of a particular model architecture for producing activation maps by letting the gradient information flow into the last available convolutional layer. That is, in order to obtain the localization map $L^c_{\text{Grad-CAM}} \in \mathbb{R}^{u \times v \times o}$ of width u, height v and depth o, we compute the gradient of the score for class c, y^c with respect to A^k, the feature maps of the convolutional layer, (*i.e.* $\frac{\partial y^c}{\partial A^k}$). These then are average-pooled in order to obtain neuron importance weights a_k^c:

$$a_k^c \propto \sum_{i,j,l} \frac{\partial y^c}{\partial A^k_{ijl}}, \tag{7.21}$$

where a_k^c represents the contribution of feature map k towards c. We then apply these in a weighted sum of activation maps and use a non-linearity:

$$L^c_{\text{Grad-CAM}} = g\left(\sum_k a_k^c A^k\right). \tag{7.22}$$

Typically g is a ReLU activation function since we are interested only in those features with a positive influence towards a particular classification.

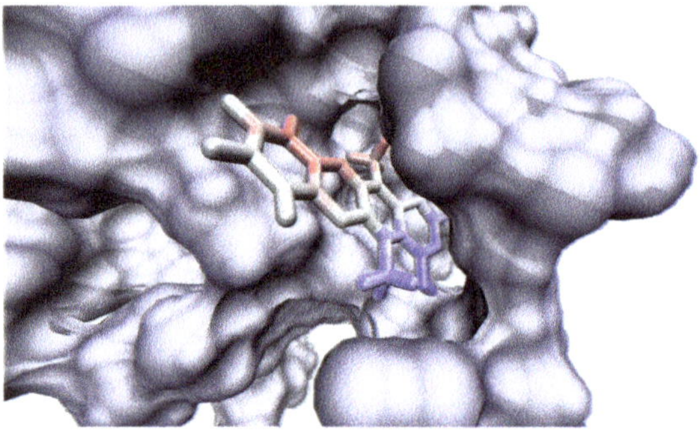

Figure 7.5 Class activation map representation taken from a public example submission to Bindscope,[111] available through the PlayMolecule.org repository of applications. Influence over a positive prediction is depicted on a blue to red scale.

A combination of up-sampling with bi-linear interpolation and guided backpropagation[113] is then used to reshape filters to the original input size.

7.3.5 Implementation and Availability

Reproducibility is a known issue in the machine learning community,[114,115] although most of the models cited here provide either source code or a web-based service. In particular, in this chapter we have mainly focused on three approaches: DeepVS[80] provides code implemented in the R programming language for training and application in virtual screening,[†] where no license of use is specified. Another software that follows the same approach is gnina,[‡] [81] with Python code for performing molecular docking and virtual screening available under an Apache license. K_{DEEP}[82] and Deltadelta[94] are available through the PlayMolecule.org repository of applications (Figure 7.5), where users can freely submit their own protein–ligand complexes, although backend code is unavailable. Finally, PotentialNet[89] does not provide an online service nor available code.

7.4 Available Data and Evaluation

In this section we focus mainly on providing basic guidance towards the training and evaluation of a new machine-learning-based scoring function. In particular we first detail the most popular databases with available binding affinity data, then later discuss the most common evaluation procedures when comparing it to other approaches, both in terms of metrics and cross-validation procedures.

[†]https://github.com/Rietaros/deepvs
[‡]https://github.com/gnina/gnina

7.4.1 Scope and Databases

In terms of available data to train structure-based models, there are several databases that can be accessed, depending on the nature of the drug discovery project at hand. In the earlier phases of a drug discovery project one may be interested in targeting a particular protein or phenotypic assay and therefore exploring a wide variety of ligands from a library of compounds. In this sense, one is concerned with training models that explore as wide a chemical space as possible, so as to know which scaffolds interact more strongly with the target of interest. To train these models several databases of diverse compounds and targets have been developed over the years. Such an example is PDBbind, which extracts and curates ligand-binding affinities from the literature for most types of biomolecular complexes deposited in the Protein Data Bank (PDB). It releases yearly, with the latest (as of the time of writing) being the 2018 release, featuring 19 588 manually curated protein–ligand complexes and their corresponding affinities. A *refined* set is selected out of all the available compounds, following filters regarding the quality of the data, excluding complexes with a resolution higher than 2.5 Å, an R-factor higher than 0.25, ligands bound through covalent bonds, ternary complexes or steric clashes, affinity not reported either in K_d or K_i, falling out of a desired range ($K_d < 1$ pM) among other criteria. Finally, a high-quality *core set* is extracted out of the refined one, with the intent of validating scoring functions and therefore providing a standard benchmark. It is common in the development of scoring functions to train models on the difference between the refined and core sets and testing only on the latter.[116–118]

The number of protein–ligand complexes available in the PDBbind database has substantially grown since its inception in 2002 (Figure 7.6), although the number of compounds in each set follow different trends. In particular, the general set size has increased more than tenfold in this period, while the refined set, having stricter inclusion requirements has grown only fivefold since its birth. The core set, on the other hand, has remained relatively stable for benchmarking, with sizes ranging from 195 to 290 compounds. Overall, the PDBbind database is one of the, if not the most, extensive collections of quality protein–ligand complexes and affinity data available today, making it the *de facto* choice for developing novel structure-based scoring functions. BindingMOAD is another well known structure-based affinity database.[119,120]

When it comes to binary affinity data, resources like the Database of Useful Decoys DUD[121] and its enhanced version DUD-E, are commonly used. The latest release, as of the time of writing, features 22 886 active compounds drawn from ChEMBL and their binary activity label against 102 protein targets, an average of 224 compounds each, as well as 50 decoys drawn from ZINC[122] per binder with similar physico-chemical properties but different two-dimensional topologies.

Once a hit, or posteriorly a lead, has been identified we focus on *lead optimization*. In this phase of drug discovery, the chemical structure of such

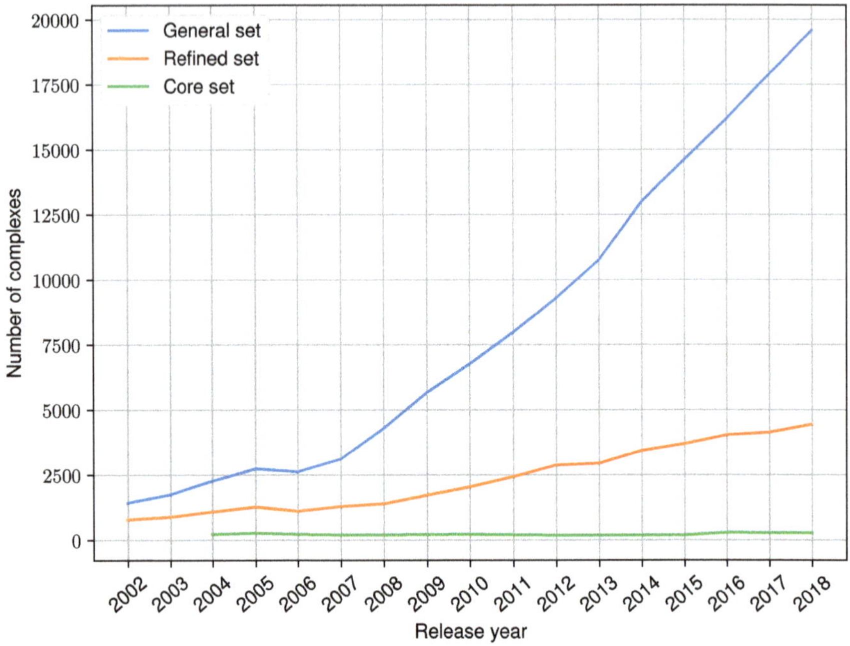

Figure 7.6 Evolution of the number of protein–ligand complexes available in the PDBbind database from 2002 to 2018.

a lead is typically modified by a team of medicinal chemists with the intent of improving its potency, selectivity, and many other pharmacokinetic and toxicological parameters. These modifications result in a congeneric series, a set of ligands with few atom changes between them, usually around a unique or small number of different scaffolds for which there are experimental structures of the complex with the target protein. The scope of this scenario is completely different from the previous one, since a diverse set does most likely not contain enough information so as to distinguish similar molecules whose potency against a target in practice can differ in less than 1 Kcal mol^{-1}. In this sense, several simulation-based approaches (which we introduced in the first section) have been developed to estimate the relative binding affinity between a pair of analogues, with relative success. Despite this, these methods suffer from several issues, such as system preparation, treatment of waters, force-field selection, protein flexibility and computational cost, making their prospective application difficult in practice. Due to this, machine learning approaches based on convolutional neural networks have recently been developed for this task,[94] showing promising results.

In terms of available databases for congeneric series, the BindingDB[123] protein–ligand validation sets provide (as of the time of writing) 645 congeneric series, which can serve as a base for prototyping models. However, it is likely that the congeneric series of the project of interest has no relation with freely available ones, so that new models have to be built from scratch to take into account its particularities. In this sense, simulation-based approaches have an advantage over data-driven alternatives, such as machine learning models, as they do not require prior knowledge of affinity data.[124] However, the latter can always be incrementally trained, increasing its accuracy when more data is available and taking into account particularities of the congeneric series of interest without relying on a physical model.

7.4.2 Evaluation

We divide this subsection into two brief parts. In the first we discuss several commonly used metrics in the evaluation of scoring functions, while in the second we focus on different data splitting procedures, which can significantly condition results.

7.4.2.1 Metrics

Scoring functions are evaluated *via* several metrics, depending on the setup of the study and its goals. In the standard regression setup, eaxamples are the root mean squared error (RMSE), Pearson's correlation coefficient (R):

$$\mathrm{RMSE}(\mathbf{y},\hat{\mathbf{y}}) = \sqrt{\frac{1}{n}\sum_{i=1}^{n}(y_i - \hat{y}_i)^2}, \tag{7.23}$$

$$R(\mathbf{y},\hat{\mathbf{y}}) = \frac{\sum_{i=1}^{n}(y_i - \bar{y})(\hat{y}_i - \bar{\hat{y}})}{\sqrt{\sum_{i=1}^{n}(y_i - \bar{y})^2}\sqrt{\sum_{i=1}^{n}(\hat{y}_i - \bar{\hat{y}})^2}}, \tag{7.24}$$

where y_i and $\hat{y}_i$ represent experimental and predicted affinity i. Spearman's ρ is also a commonly used metric, computed as the Pearson correlation coefficient of the ranked experimental and predicted variables. In a retrospective lead optimization scenario, when one has the chronological experimental order of the series, a simulation-based approach is also appropriate for data-driven scoring functions. This evaluation represents a paradigm where the model chooses the ligands to synthesize next, in each iteration increasing its training pool. The evaluation ends when the model picks the ligand with the highest affinity in the series, and its order is then compared with the experimental one achieved by medicinal chemists. If the model was able to retrieve the mentioned compound faster, it is taken as an indication that such models are able to accelerate the lead optimization process.

In terms of binary classification (active/inactive) scoring functions, one is instead interested in retrieving as many active compounds as possible from a given library, therefore metrics such as the enrichment factor (EF) are commonly used:

$$\mathrm{EF}_{k\%} = \frac{\frac{N_a(k\%)}{N(k\%)}}{\frac{N_a}{N}}, \tag{7.25}$$

where $N_a(k\%)$ and $N(k\%)$ are the number of active and total molecules in the top $k\%$ ranked molecules in the library according to the scoring function, and N_a and N are the number of actives and total number of molecules in the entire library. More classical classification metrics are also commonly used, such as the area under the receiver operating curve (AUC) or BEDROC.[125]

7.4.2.2 Splitting Procedures

When evaluating scoring functions, if a standard benchmark to compare against is not set beforehand it is important to check model performance under different scenarios. In particular, the most common type of split is some variation of *k*-fold cross-validation, where we evaluate our model several times in *k* different splits, using the remaining ones as training data. How the splits are performed can significantly vary the results: the most common strategy being randomly assigning each protein–ligand pair. It is well known that this type of split can produce over-optimistic results, particularly in cases when the test set is far (in a statistical distribution sense) from the data we have previously used to train our model (*e.g.* different chemical spaces). Several authors have then proposed several alternatives to provide more realistic performance measurements and minimize the possibility of bias, such as scaffold-based splits.[126] In situations where a time-stamp is available, such as in lead discovery, a temporal-based split can also be appropriate.[127] Finally, when developing structure-based approaches, it is also common to consider the performance of the model both in a intra-target and inter-target sense, that is, within and between proteins,[128] the latter kind of split typically performed *via* means of sequence information.

7.5 Discussion

In this chapter our intention was to provide readers with an overview of modern machine-learning structure-based approaches for the prediction of protein–ligand affinities. The success and attention of deep learning methods in this task are unquestionable, with modern models focusing on a representation of the system that is closer to reality, changing the classical feature handcrafting paradigm. In this sense, popular approach focuses on a voxelized representation of the protein, to later use readily available computer vision techniques such as three-dimensional CNNs. On the other hand, family of approaches is a direct descendent of recent ligand-based models, extending the concept of graph convolution to include the protein structure.

While deep learning approaches are attractive for their automatic feature extraction capabilities, among other things, their use in structure-based approaches is necessarily limited. It is a well known fact that deep learning approaches are very data-hungry, and significantly more so than other simpler models if satisfactory performance is to be achieved. For instance, modern CNN architectures for computer vision are trained on the ImageNet database of images,[129] which contains more than 14 m labeled instances. In contrast, the most extensive structure-based affinity database, namely PDBbind, only features around 20 k protein–ligand pairs. This scarcity necessarily forces researchers to design more data-efficient models than the current state of the art.

Interpretability of deep-learning-based models is also one of the topics that has grabbed significant attention in recent years, as we explained in the corresponding subsection of this chapter. While these tend to be significantly more accurate than other alternatives, they are commonly treated as black boxes, and this comes with two costs: first they are harder to debug, meaning it would be difficult to identify whether a model is learning an inherent bias in the data or the real signal. Secondly, all proposed models come with a domain of applicability, in most cases being the prioritization of compound testing. After all, if the decision a model makes is hard to understand for a team of medicinal or computational chemists, justifying its usefulness becomes as difficult.

References

1. I. Kola and J. Landis, Can the Pharmaceutical Industry Reduce Attrition Rates?, *Nat. Rev. Drug Discovery*, 2004, **3**(8), 711.
2. C. A. Nicolaou and N. Brown, Multi-objective Optimization Methods in Drug Design, *Drug Discovery Today: Technol.*, 2013, **10**(3), e427–e435.
3. L. Di, P. V. Fish and T. Mano, Bridging Solubility Between Drug Discovery and Development, *Drug Discovery Today*, 2012, **17**(9–10), 486–495.
4. J. Lin, D. C. Sahakian, S. M. De Morais, J. J. Xu, R. J. Polzer and S. M. Winter, The Role of Absorption, Distribution, Metabolism, Excretion and Toxicity in Drug Discovery, *Curr. Top. Med. Chem.*, 2003, **3**(10), 1125–1154.
5. A. Mayr, G. Klambauer, T. Unterthiner and S. Hochreiter, DeepTox: Toxicity Prediction Using Deep Learning, *Front. Environ. Sci.*, 2016, **3**, 80.
6. D. J. Huggins, W. Sherman and B. Tidor, Rational Approaches to Improving Selectivity in Drug Design, *J. Med. Chem.*, 2012, **55**(4), 1424–1444.
7. J. Jiménez, D. Sabbadin, A. Cuzzolin, G. Martnez-Rosell, J. Gora, J. Manchester, J. Duca and G. D. Fabritiis, PathwayMap: Molecular Pathway Association with Self-Normalizing Neural Networks, *J. Chem. Inf. Model.*, 2019, **59**(3), 1172–1181.
8. C. David, Swinney. The Role of Binding Kinetics in Therapeutically Useful Drug Action, *Curr. Opin. Drug Discovery Dev.*, 2009, **12**(1), 31–39.
9. A. Mardt, L. Pasquali, H. Wu and F. Noé, VAMPnets for Deep Learning of Molecular Kinetics, *Nat. Commun.*, 2018, **9**(1), 5.
10. A. Gaulton, L. J. Bellis, A. P. Bento, J. Chambers, M. Davies, A. Hersey, Y. Light, S. McGlinchey, D. Michalovich and B. Al-Lazikani, *et al.*,

ChEMBL: a large-scale bioactivity database for drug discovery, *Nucleic Acids Res.*, 2011, **40**(D1), D1100–D1107.
11. R. Wang, X. Fang, Y. Lu and S. Wang, The PDBbind Database: Collection of Binding Affinities for Protein-ligand Complexes with Known Three-Dimensional Structures, *J. Med. Chem.*, 2004, **47**(12), 2977–2980.
12. J. W. Scannell, A. Blanckley, H. Boldon and B. Warrington, Diagnosing the Decline in Pharmaceutical R&D efficiency, *Nat. Rev. Drug Discovery*, 2012, **11**(3), 191.
13. Q. U. Ain, A. Aleksandrova, F. D. Roessler and P. J. Ballester, Machine-learning scoring functions to improve structure-based binding affinity prediction and virtual screening, *Wiley Interdiscip. Rev.: Comput. Mol. Sci.*, 2015, **5**(6), 405–424.
14. J. Liu and R. Wang, Classification of Current Scoring Functions, *J. Chem. Inf. Model.*, 2015, **55**(3), 475–482.
15. R. S. DeWitte and E. I. Shakhnovich, SMoG: De Novo Design Method Based on Simple, Fast, and Accurate Free Energy Estimates. 1. Methodology and Supporting Evidence, *J. Am. Chem. Soc.*, 1996, **118**(47), 11733–11744.
16. I. Muegge and Y. C. Martin, A General and Fast Scoring Function for Protein-ligand Interactions: a Simplified Potential Approach, *J. Med. Chem.*, 1999, **42**(5), 791–804.
17. H. Gohlke, Manfred Hendlich, and Gerhard Klebe. Knowledge-based Scoring Function to Predict Protein-ligand Interactions, *J. Mol. Biol.*, 2000, **295**(2), 337–356.
18. S.-Y. Huang and X. Zou, An Iterative Knowledge-based Scoring Function to Predict Protein-ligand Interactions: I. Derivation of Interaction Potentials, *J. Comput. Chem.*, 2006, **27**(15), 1866–1875.
19. Z. Zheng and K. M. Merz Jr., Development of the Knowledge-based and Empirical Combined Scoring Algorithm (KECSA) to Score Protein-ligand Interactions, *J. Chem. Inf. Model.*, 2013, **53**(5), 1073–1083.
20. M. Hendlich, P. Lackner, S. Weitckus, H. Floeckner, R. Froschauer, K. Gottsbacher, G. Casari and M. J. Sippl, Identification of Native Protein Folds Amongst a Large Number of Incorrect Models: The Calculation of Low Energy Conformations from Potentials of Mean Force, *J. Mol. Biol.*, 1990, **216**(1), 167–180.
21. M. J. Sippl, Calculation of Conformational Ensembles from Potentials of Mean Force: An Approach to the Knowledge-based Prediction of Local Structures in Globular Proteins, *J. Mol. Biol.*, 1990, **213**(4), 859–883.
22. P. D. Thomas and K. A. Dill, An Iterative Method for Extracting Energy-like Quantities from Protein Structures, *Proc. Natl. Acad. Sci. U. S. A.*, 1996, **93**(21), 11628–11633.
23. P. D. Thomas and K. A. Dill, Statistical Potentials Extracted from Protein Structures: How Accurate are They?, *J. Mol. Biol.*, 1996, **257**(2), 457–469.

24. H. Lu and J. Skolnick, A Distance-dependent Atomic Knowledge-based Potential for Improved Protein Structure Selection, *Proteins: Struct., Funct., Bioinf.*, 2001, **44**(3), 223–232.
25. J. Wang, R. M. Wolf, J. W. Caldwell, P. A. Kollman and D. A. Case, Development and Testing of a General AMBER Force Field, *J. Comput. Chem.*, 2004, **25**(9), 1157–1174.
26. E. Kenno Vanommeslaeghe, C. Hatcher, S. Acharya, S. Kundu, J. Zhong, E. Shim, Darian, P. Olgun Guvench, I. Lopes and Vorobyov, *et al.*, CHARMM General Force Field: A Force Field for Drug-like Molecules Compatible with the CHARMM All-atom Additive Biological Force Fields, *J. Comput. Chem.*, 2010, **31**(4), 671–690.
27. S. Doerr and G. De Fabritiis, On-the-fly Learning and Sampling of Ligand Binding by High-throughput Molecular Simulations, *J. Chem. Theory Comput.*, 2014, **10**(5), 2064–2069.
28. G. Martinez-Rosell, M. J. Harvey and G. D. Fabritiis, Molecular-simulation-driven Fragment Screening for the Discovery of New CXCL12 Inhibitors, *J. Chem. Inf. Model.*, 2018, **58**(3), 683–691.
29. I. Buch, T. Giorgino and G. D. Fabritiis, Complete Reconstruction of an Enzyme-inhibitor Binding Process by Molecular Dynamics Simulations, *Proc. Natl. Acad. Sci. U. S. A.*, 2011, **108**(25), 10184–10189.
30. N. Ferruz, S. Doerr, M. A. Vanase-Frawley, Y. Zou, X. Chen, E. S. Marr, R. T. Nelson, B. L. Kormos, T. T. Wager, X. Hou and A. Villalobos, Dopamine D3 Receptor Antagonist Reveals a Cryptic Pocket in Aminergic GPCRs, *Sci. Rep.*, 2018, **8**(1), 897.
31. N. Ferruz, G. Tresadern, A. Pineda-Lucena and G. D. Fabritiis, Multibody Cofactor and Substrate Molecular Recognition in the Myo-inositol Monophosphatase Enzyme, *Sci. Rep.*, 2016, **6**, 30275.
32. V. Limongelli, M. Bonomi and M. Parrinello, Funnel Metadynamics as Accurate Binding Free-energy Method, *Proc. Natl. Acad. Sci. U. S. A.*, 2013, **110**(16), 6358–6363.
33. J. S. Patel, A. Berteotti, S. Ronsisvalle, W. Rocchia and A. Cavalli, Steered Molecular Dynamics Simulations for Studying Protein-ligand Interaction in Cyclin-dependent Kinase 5, *J. Chem. Inf. Model.*, 2014, **54**(2), 470–480.
34. L. Wang, B. J. Berne and R. A. Friesner, On Achieving High Accuracy and Reliability in the Calculation of Relative Protein-Ligand Binding Affinities, *Proc. Natl. Acad. Sci. U. S. A.*, 2012, **109**(6), 1937–1942.
35. E. B. Lenselink, J. Louvel, A. F. Forti, J. P. D. van Veldhoven, H. de Vries, T. Mulder-Krieger, F. M. McRobb, A. Negri, J. Goose, R. Abel, H. W. T. van Vlijmen, L. Wang, E. Harder, W. Sherman, A. P. IJzerman and T. Beuming, Predicting Binding Affinities for GPCR Ligands Using Free-Energy Perturbation, *ACS Omega*, 2016, **1**(2), 293–304.
36. D. A. Goldfeld, R. Murphy, B. Kim, L. Wang, T. Beuming, R. Abel and R. A. Friesner, Docking and Free Energy Perturbation Studies of Ligand Binding in the Kappa Opioid Receptor, *J. Phys. Chem. B*, 2015, **119**(3), 824–835.

37. L. Pérez-Benito, H. Keränen, H. van Vlijmen and G. Tresadern, Predicting Binding Free Energies of PDE2 Inhibitors. The Difficulties of Protein Conformation, *Sci. Rep.*, 2018, **8**(1), 4883.
38. M. Ciordia, L. Pérez-Benito, F. Delgado, A. A. Trabanco and G. Tresadern, Application of Free Energy Perturbation for the Design of BACE1 Inhibitors, *J. Chem. Inf. Model.*, 2016, **56**(9), 1856–1871.
39. C. Schindler, F. Rippmann and D. Kuhn, Relative Binding Affinity Prediction of Farnesoid X Receptor in the D3R Grand Challenge 2 Using FEP+, *J. Comput.-Aided Mol. Des.*, 2017, **32**(1), 1–8.
40. H. Keränen, L. Pérez-Benito, M. Ciordia, F. Delgado, T. B. Steinbrecher, D. Oehlrich, H. W. Van Vlijmen, A. A. Trabanco and G. Tresadern, Acylguanidine Beta Secretase 1 Inhibitors: A Combined Experimental and Free Energy Perturbation Study, *J. Chem. Theory Comput.*, 2017, **13**(3), 1439–1453.
41. W. Shunzhou, A. P. Bhati, S. Skerratt, K. Omoto, V. Shanmugasundaram, S. K. Bagal and P. V. Coveney, Evaluation and Characterization of Trk Kinase Inhibitors for the Treatment of Pain: Reliable Binding Affinity Predictions from Theory and Computation, *J. Chem. Inf. Model.*, 2017, **57**(4), 897–909.
42. A. de Ruiter, S. Boresch and C. Oostenbrink, Comparison of Thermodynamic Integration and Bennett Acceptance Ratio for Calculating Relative Protein-ligand Binding Free Energies, *J. Comput. Chem.*, 2013, **34**(12), 1024–1034.
43. Z. Cournia, B. Allen and W. Sherman, Relative Binding Free Energy Calculations in Drug Discovery: Recent Advances and Practical Considerations, *J. Chem. Inf. Model.*, 2017, **57**(12), 2911–2937.
44. Å. Johan and J. Marelius, The Linear Interaction Energy Method for Predicting Ligand Binding Free Energies, *Comb. Chem. High Throughput Screening*, 2001, **4**(8), 613–626.
45. T. Hansson, J. Marelius and J. Åqvist, Ligand Binding Affinity Prediction by Linear Interaction Energy Methods, *J. Comput.-Aided Mol. Des.*, 1998, **12**(1), 27–35.
46. S. Genheden and U. Ryde, The MM/PBSA and MM/GBSA Methods to Estimate Ligand-binding Affinities, *Expert Opin. Drug Discovery*, 2015, **10**(5), 449–461.
47. T. Hou, J. Wang, Y. Li and W. Wang, Assessing the Performance of the MM/PBSA and MM/GBSA Methods. 1. The Accuracy of Binding Free Energy Calculations Based on Molecular Dynamics Simulations, *J. Chem. Inf. Model.*, 2010, **51**(1), 69–82.
48. S. Wold, M. Sjöström and L. Eriksson, PLS-regression: A Basic Tool of Chemometrics, *Chemom. Intell. Lab. Syst.*, 2001, **58**(2), 109–130.
49. R. Wang, L. Lai and S. Wang, Further Development and Validation of Empirical Scoring Functions for Structure-based Binding Affinity Prediction, *J. Comput.-Aided Mol. Des.*, 2002, **16**(1), 11–26.
50. Y. Cao and L. Li, Improved Protein-ligand Binding Affinity Prediction by Using a Curvature-dependent Surface-area Model, *Bioinformatics*, 2014, **30**(12), 1674–1680.

51. M. L. Verdonk, J. C. Cole, M. J. Hartshorn, C. W. Murray and R. D. Taylor, Improved Protein-ligand Docking using GOLD, *Proteins: Struct., Funct., Bioinf.*, 2003, **52**(4), 609–623.
52. M. P. Repasky, M. Shelley and R. A. Friesner, Flexible Ligand Docking with Glide, *Curr. Protoc. Bioinf.*, 2007, **18**(1), 8–12.
53. B. Hans-Joachim, Prediction of Binding Constants of Protein Ligands: a Fast Method for the Prioritization of Hits Obtained from De Novo Design or 3D Database Search Programs, *J. Comput.-Aided Mol. Des.*, 1998, **12**(4), 309.
54. M. Rarey, B. Kramer, T. Lengauer and G. Klebe, A Fast Flexible Docking Method Using an Incremental Construction Algorithm, *J. Mol. Biol.*, 1996, **261**(3), 470–489.
55. R. Wang, L. Liu, L. Lai and Y. Tang, SCORE: A New Empirical Method for Estimating the Binding Affinity of a Protein-ligand Complex, *Molecular Modeling Annual*, 1998, **4**(12), 379–394.
56. W. Deng, C. Breneman and M. J. Embrechts, Predicting Protein-ligand Binding Affinities Using Novel Geometrical Descriptors and Machine-learning Methods, *J. Chem. Inf. Comput. Sci.*, 2004, **44**(2), 699–703.
57. N. Artemenko, Distance Dependent Scoring Function for Describing Protein-ligand Intermolecular Interactions, *J. Chem. Inf. Model.*, 2008, **48**(3), 569–574.
58. P. J. Ballester and J. B. Mitchell, A Machine Learning Approach to Predicting Protein-ligand Binding Affinity with Applications to Molecular Docking, *Bioinformatics*, 2010, **26**(9), 1169–1175.
59. K.-Y. Hsin, S. Ghosh and H. Kitano, Combining Machine Learning Systems and Multiple Docking Simulation Packages to Improve Docking Prediction Reliability for Network Pharmacology, *PLoS One*, 2013, **8**(12), e83922.
60. D. Zilian and C. A. Sotriffer, SFCscore RF: A Random Forest-based Scoring Function for Improved Affinity Prediction of Protein-Ligand Complexes, *J. Chem. Inf. Model.*, 2013, **53**(8), 1923–1933.
61. S. Zhang, A. Golbraikh and A. Tropsha, Development of Quantitative Structure- Binding Affinity Relationship Models Based on Novel Geometrical Chemical Descriptors of the Protein- Ligand Interfaces, *J. Med. Chem.*, 2006, **49**(9), 2713–2724.
62. G.-B. Li, L.-L. Yang, W.-J. Wang, L.-L. Li and S.-Y. Yang, ID-Score: A New Empirical Scoring Function Based on a Comprehensive Set of Descriptors Related to Protein-ligand Interactions, *J. Chem. Inf. Model.*, 2013, **53**(3), 592–600.
63. T. Sato, T. Honma and S. Yokoyama, Combining Machine Learning and Pharmacophore-based Interaction Fingerprint for In Silico Screening, *J. Chem. Inf. Model.*, 2009, **50**(1), 170–185.
64. S. Das, M. P. Krein and C. M. Breneman, Binding Affinity Prediction with Property-encoded Shape Distribution Signatures, *J. Chem. Inf. Model.*, 2010, **50**(2), 298–308.

65. V. Chupakhin, G. Marcou, I. Baskin, A. Varnek and D. Rognan, Predicting Ligand Binding Modes from Neural Networks Trained on Protein-ligand Interaction Fingerprints, *J. Chem. Inf. Model.*, 2013, **53**(4), 763–772.
66. Z. Deng, C. Chuaqui and J. Singh, Structural Interaction Fingerprint (SIFt): A Novel Method for Analyzing Three-dimensional Protein-ligand Binding Interactions, *J. Med. Chem.*, 2004, **47**(2), 337–344.
67. C. de Graaf, C. Rein, D. Piwnica, F. Giordanetto and D. Rognan, Structure-based Discovery of Allosteric Modulators of Two Related Class BG-protein-coupled Receptors, *ChemMedChem*, 2011, **6**(12), 2159–2169.
68. O. Trott and A. J. Olson, AutoDock Vina: Improving the Speed and Accuracy of Docking with a New Scoring Function, Efficient Optimization, and Multithreading, *J. Comput. Chem.*, 2010, **31**(2), 455–461.
69. J. D. Durrant and J. A. McCammon, BINANA: a Novel Algorithm for Ligand-binding Characterization, *J. Mol. Graphics Modell.*, 2011, **29**(6), 888–893.
70. J. D. Durrant and J. A. McCammon, NNScore: A Neural-network-based Scoring Function For the Characterization of Protein-ligand Complexes, *J. Chem. Inf. Model.*, 2010, **50**(10), 1865–1871.
71. X. Ouyang, S. D. Handoko and C. K. Kwoh, Cscore: A Simple Yet Effective Scoring Function for Protein-ligand Binding Affinity Prediction Using Modified Cmac Learning Architecture, *J. Bioinf. Comput. Biol.*, 2011, **9**, 1–14.
72. L. Li, B. Wang and S. O. Meroueh, Support Vector Regression Scoring of Receptor-ligand Complexes for Rank-ordering and Virtual Screening of Chemical Libraries, *J. Chem. Inf. Model.*, 2011, **51**(9), 2132–2138.
73. Q. Liu, C. K. Kwoh and J. Li, Binding Affinity Prediction for Protein-ligand Complexes Based on β Contacts and B Factor, *J. Chem. Inf. Model.*, 2013, **53**(11), 3076–3085.
74. D. D. Nguyen and G.-W. Wei, Algebraic Graph Learning of Protein-ligand Binding Affinity, *J. Chem. Inf. Model.*, 2019, **59**(7), 3291–3304.
75. P. J. Ballester, A. Schreyer and T. L. Blundell, Does a More Precise Chemical Description of Protein-ligand Complexes Lead to More Accurate Prediction of Binding Affinity?, *J. Chem. Inf. Model.*, 2014, **54**(3), 944–955.
76. I. Goodfellow, Y. Bengio and A. Courville, *Deep Learning*, MIT Press, 2016.
77. A. Krizhevsky, I. Sutskever and G. E. Hinton. Imagenet Classification with Deep Convolutional Neural Networks, in*Advances in Neural Information Processing Systems*, 2012, pp. 1097–1105.
78. K. Xu, J. Ba, R. Kiros, K. Cho, A. Courville, R. Salakhudinov, R. Zemel, and and Y. Bengio. Show, Attend and Tell: Neural Image Caption Generation with Visual Attention, in *International Conference on Machine Learning*, 2015, pp. 2048–2057.

79. J. Devlin, M.-W. Chang, K. Lee, K. Toutanova, Bert: Pre-training of Deep Bidirectional Transformers for Language Understanding, 2018, arXiv preprint arXiv:1810.04805.
80. J. C. Pereira, E. R. Caffarena and C. N. dos Santos, Boosting Docking-based Virtual Screening with Deep Learning, *J. Chem. Inf. Model.*, 2016, **56**(12), 2495–2506.
81. M. Ragoza, J. Hochuli and E. Idrobo, Jocelyn Sunseri, and David Ryan Koes. Protein-ligand Scoring with Convolutional Neural Networks, *J. Chem. Inf. Model.*, 2017, **57**(4), 942–957.
82. J. Jiménez, M. Skalic, G. Martnez-Rosell and G. D. Fabritiis, K DEEP: Protein-Ligand Absolute Binding Affinity Prediction via 3D-Convolutional Neural Networks, *J. Chem. Inf. Model.*, 2018, **58**(2), 287–296.
83. D. R. Koes, M. P. Baumgartner and C. J. Camacho, Lessons Learned in Empirical Scoring with smina from the CSAR 2011 Benchmarking Exercise, *J. Chem. Inf. Model.*, 2013, **53**(8), 1893–1904.
84. G. M. Morris, R. Huey, W. Lindstrom, M. F. Sanner, R. K. Belew, D. S. Goodsell and A. J. Olson, AutoDock4 and AutoDockTools4: Automated Docking with Selective Receptor Flexibility, *J. Comput. Chem.*, 2009, **30**(16), 2785–2791.
85. J. Jiménez, S. Doerr, G. Martnez-Rosell, A. S. Rose and G. D. Fabritiis, DeepSite: Protein-binding Site Predictor Using 3D-convolutional Neural Networks, *Bioinformatics*, 2017, **33**(19), 3036–3042.
86. M. Skalic, A. Varela-Rial, J. Jiménez, G. Martnez-Rosell and G. D. Fabritiis, LigVoxel: Inpainting Binding Pockets Using 3D-convolutional Neural Networks, *Bioinformatics*, 2018, **35**(2), 243–250.
87. M. Skalic, J. Jiménez, D. Sabbadin and G. D. Fabritiis, Shape-Based Generative Modeling for de-novo Drug Design, *J. Chem. Inf. Model.*, 2019, **59**(3), 1205–1214.
88. J. Gilmer, S. S. Schoenholz, P. F. Riley, O. Vinyals and G. E. Dahl. Neural Message Passing for Quantum Chemistry. in *Proceedings of the 34th International Conference on Machine Learning*, **vol. 70**, 2017, pp. 1263–1272, JMLR. org.
89. E. N. Feinberg, D. Sur, Z. Wu, B. E. Husic, H. Mai, Y. Li, S. Sun, J. Yang, B. Ramsundar and V. S. Pande, PotentialNet for Molecular Property Prediction, *ACS Cent. Sci.*, 2018, **4**(11), 1520–1530.
90. H. H. Aghdam and E. J. Heravi. *Guide to Convolutional Neural Networks*, Springer, New York, NY, **vol. 10**, 2017, p. 978.
91. D. E. Dudgeon, Multidimensional digital signal processing, *Engewood Cliffs*, 1983.
92. I. Sergey and C. Szegedy. Batch Normalization: Accelerating Deep Network Training by Reducing Internal Covariate Shift, 2015, arXiv preprint arXiv:1502.03167.
93. Z. Gaieb, C. D. Parks, M. Chiu, H. Yang, C. Shao, W. P. Walters, M. H. Lambert, N. Nevins, S. D. Bembenek, M. K. Ameriks and T. Mirzadegan, D3R Grand Challenge 3: Blind Prediction of

Protein-ligand Poses and Affinity Rankings, *J. Comput.-Aided Mol. Des.*, 2019, **33**(1), 1–18.

94. J. Jiménez-Luna, L. Pérez-Benito, G. Martínez-Rosell, S. Sciabola, R. Torella, G. Tresadern, G. D. Fabritiis, *DeltaDelta Neural Networks for Lead Optimization of Small Molecule Potency*, (submitted, under review), 2019.
95. D. K. Duvenaud, D. Maclaurin, J. Iparraguirre, R. Bombarell, T. Hirzel, A. Aspuru-Guzik and R. P. Adams, Convolutional Networks on Graphs for Learning Molecular Fingerprints, *Advances in Neural Information Processing Systems*, 2015, 2224–2232.
96. P. Battaglia, R. Pascanu, M. Lai and D. J. Rezende, *et al.*, Interaction Networks for Learning About Objects, Relations and Physics, *Advances in Neural Information Processing Systems*, 2016, 4502–4510.
97. S. Kearnes, K. McCloskey, M. Berndl, V. Pande and P. Riley, Molecular Graph Convolutions: Moving Beyond Fingerprints, *J. Comput.-Aided Mol. Des.*, 2016, **30**(8), 595–608.
98. K. T. Schütt, F. Arbabzadah, S. Chmiela, K. R. Müller and A. Tkatchenko, Quantum-chemical Insights from Deep Tensor Neural Networks, *Nat. Commun.*, 2017, **8**, 13890.
99. J. Bruna, W. Zaremba, A. Szlam, Y. LeCun. Spectral Networks and Locally Connected Networks on Graphs, 2013, arXiv preprint arXiv:1312.6203.
100. M. Defferrard, X. Bresson and P. Vandergheynst, Convolutional Neural Networks on Graphs with Fast Localized Spectral Filtering, *Advances in Neural Information Processing Systems*, 2016, 3844–3852.
101. T. N. Kipf and M. Welling, Semi-supervised Classification with Graph Convolutional Networks, 2016, arXiv preprint arXiv:1609.02907, ICLR 2017.
102. Y. Li, D. Tarlow, M. Brockschmidt, R. Zemel, Gated Graph Sequence Neural Networks, 2015, arXiv preprint arXiv:1511.05493.
103. K. Cho, B. V. Merriënboer, C. Gulcehre, D. Bahdanau, F. Bougares, H. Schwenk, Y. Bengio, Learning phrase representations using rnn encoder-decoder for statistical machine translation, 2014, arXiv preprint arXiv:1406.1078.
104. L. Breiman, Statistical Modeling: The Two Cultures, *Statist. Sci.*, 2001, **16**(3), 199–231.
105. F. Doshi-Velez and B. Kim. Towards a Rigorous Science of Interpretable Machine Learning, 2017, arXiv preprint arXiv:1702.08608.
106. A. Vellido, J. D. Martín-Guerrero and P. J. Lisboa, Making Machine Learning Models Interpretable, *ESANN*, 2012, **vol. 12**, 163–172. Citeseer.
107. S. Bach, A. Binder, G. Montavon, F. Klauschen, K.-R. Müller and W. Samek, On Pixel-wise Explanations for Non-linear Classifier Decisions by Layer-wise Relevance Propagation, *PLoS One*, 2015, **10**(7), e0130140.
108. J. Hochuli, A. Helbling, T. Skaist, M. Ragoza and D. R. Koes, Visualizing Convolutional Neural Network Protein-ligand Scoring, *J. Mol. Graphics Modell.*, 2018, **84**, 96–108.

109. B. Zhou, A. Khosla, A. Lapedriza, A. Oliva and A. Torralba, Learning Deep Features for Discriminative Localization, *Proceedings of the IEEE Conference on Computer Vision and Pattern Recognition*, 2016, 2921–2929.
110. M. Lin, Q. Chen, S. Yan. Network in Network, 2013, arXiv preprint arXiv:1312.4400.
111. M. Skalic, G. Martnez-Rosell, J. Jiménez and G. D. Fabritiis, PlayMolecule BindScope: Large-scale CNN-based Virtual Screening on the Web, *Bioinformatics*, 2018, **35**(7), 1237–1238.
112. R. R. Selvaraju, M. Cogswell, A. Das, R. Vedantam, D. Parikh and D. Batra. Grad-cam: Visual Explanations from Deep Networks via Gradient-based Localization. in *Proceedings of the IEEE International Conference on Computer Vision*, 2017, pp. 618–626.
113. J. T. Springenberg, A. Dosovitskiy, T. Brox, M. Riedmiller. Striving for Simplicity: The All Convolutional Net, 2014, arXiv preprint arXiv:1412.6806.
114. S. Sonnenburg, M. L. Braun, C. S. Ong, S. Bengio, L. Bottou, G. Holmes, Y. LeCun, K.-R. MÃžller, F. Pereira and C. E. Rasmussen, *et al.*, The Need for Open Source Software in Machine Learning, *J. Mach. Learn. Res.*, 2007, **8**(Oct), 2443–2466.
115. P. Henderson, R. Islam, P. Bachman, J. Pineau, D. Precup and D. Meger. Deep Reinforcement Learning That Matters. in *Thirty-Second AAAI Conference on Artificial Intelligence*, 2018.
116. M. Su, Q. Yang, Y. Du, G. Feng, Z. Liu, Y. Li and R. Wang, Comparative Assessment of Scoring Functions: The CASF-2016 Update, *J. Chem. Inf. Model.*, 2018, **59**(2), 895–913.
117. T. Cheng, X. Li, Y. Li, Z. Liu and R. Wang, Comparative Assessment of Scoring Functions on a Diverse Test Set, *J. Chem. Inf. Model.*, 2009, **49**(4), 1079–1093.
118. Y. Li, L. Han, Z. Liu and R. Wang, Comparative Assessment of Scoring Functions on an Updated Benchmark: 2. Evaluation Methods and General Results, *J. Chem. Inf. Model.*, 2014, **54**(6), 1717–1736.
119. M. L. Benson, R. D. Smith, N. A. Khazanov, B. Dimcheff, J. Beaver, P. Dresslar, J. Nerothin and H. A. Carlson, Binding MOAD, a High-Quality Protein-ligand Database, *Nucleic Acids Res.*, 2007, **36**, D674–D678.
120. A. Ahmed, R. D. Smith, J. J. Clark, J. B. Dunbar Jr. and H. A. Carlson, Recent Improvements to Binding MOAD: A Resource for Protein-ligand Binding Affinities and Structures, *Nucleic Acids Res.*, 2014, **43**(D1), D465–D469.
121. M. M. Mysinger, M. Carchia, J. J. Irwin and B. K. Shoichet, Directory of Useful Decoys, Enhanced (DUD-E): Better Ligands and Decoys for Better Benchmarking, *J. Med. Chem.*, 2012, **55**(14), 6582–6594.
122. J. J. Irwin and B. K. Shoichet, ZINC, a Free Database of Commercially Available Compounds for Virtual Screening, *J. Chem. Inf. Model.*, 2005, **45**(1), 177–182.

123. T. Liu, Y. Lin, X. Wen, R. N. Jorissen and M. K. Gilson, BindingDB: A Web-accessible Database of Experimentally Determined Protein-ligand Binding Affinities, *Nucleic Acids Res.*, 2006, **35**, D198–D201.
124. A. Pérez, G. Martnez-Rosell and G. D. Fabritiis, Simulations Meet Machine Learning in Structural Biology, *Curr. Opin. Struct. Biol.*, 2018, **49**, 139–144.
125. J.-F. Truchon and C. I. Bayly, Evaluating Virtual Screening Methods: Good and Bad Metrics for the "Early Recognition" Problem, *J. Chem. Inf. Model.*, 2007, **47**(2), 488–508.
126. K. Christian and P. Gedeck, Leave-cluster-out Cross-validation is Appropriate for Scoring Functions Derived from Diverse Protein Data Sets, *J. Chem. Inf. Model.*, 2010, **50**(11), 1961–1969.
127. R. P. Sheridan, Time-split Cross-validation as a Method for Estimating the Goodness of Prospective Prediction, *J. Chem. Inf. Model.*, 2013, **53**(4), 783–790.
128. J. Sieg, F. Flachsenberg and M. Rarey, In Need of Bias Control: Evaluating Chemical Data for Machine Learning in Structure-Based Virtual Screening, *J. Chem. Inf. Model.*, 2019, **59**(3), 947–961.
129. O. Russakovsky, J. Deng, H. Su, J. Krause, S. Satheesh, S. Ma, Z. Huang, A. Karpathy, A. Khosla and M. Bernstein, *et al.*, ImageNet Large Scale Visual Recognition Challenge, *Int. J. Comput. Vis.*, 2015, **115**(3), 211–252.

CHAPTER 8

Virtual Screening with Convolutional Neural Networks

FERGUS IMRIE,[a] ANTHONY R. BRADLEY[b] AND CHARLOTTE M. DEANE*[a]

[a] Oxford Protein Informatics Group, Department of Statistics, University of Oxford, Oxford OX1 3LB, UK; [b] Exscientia Ltd, 36 St. Giles', Oxford OX1 3LD, UK
*Email: deane@stats.ox.ac.uk

8.1 Introduction

In this chapter, we examine how convolutional neural networks (CNN), a type of deep learning methodology, can be applied to the evaluation of protein–ligand complexes *in silico*. CNNs have been the primary driver of the enormous progress in computational analysis of digital images and videos (computer vision) over the last few years. This success has sparked interest in, and the use of, CNNs across a diverse range of application areas. Recently there have been several applications of CNNs to the scoring of protein–ligand complexes, in particular to assess whether a ligand will bind (virtual screening), its binding mode (pose prediction), and the strength of binding (binding affinity prediction). The focus of this chapter is on the first of these, virtual screening, but the techniques and problem formulation are transferable to other tasks.

Drug Discovery Series No. 75
Artificial Intelligence in Drug Discovery
Edited by Nathan Brown

Published by the Royal Society of Chemistry, www.rsc.org

8.1.1 Virtual Screening

Experimental screening of compounds for binding is financially expensive and time consuming; in addition, hit rates are often extremely low.[1,2] Virtual screening can be used to improve the efficiency and effectiveness of experimental screens.[3] In particular, it can be used to refine the compound library for experimental screening, and inform optimisation of compounds.[4]

Virtual screening is very quick and cheap to perform, allowing orders of magnitudes more compounds to be assessed than is possible through experimental screens,[5] and has been shown to lead to enrichment of actives in follow-up experimental screens. As such, it has been heavily involved in many successful drug discovery projects.[6–10]

Virtual screening has several limitations over experimental approaches. Virtual screening requires prerequisite knowledge about the factors responsible for the desired biological response (*e.g.* binding to a protein target). In addition, the accuracy of virtual screening approaches is not sufficient to draw definitive conclusions about compounds: false negatives can lead to promising compounds being missed, while false positives still represent the majority of highly scored compounds in most cases, even when substantial enrichment has been achieved.[11] Given these limitations, it is still the case that experimental verification is needed to confirm a hit.

There are two fundamentally different methodologies for virtual screening: (i) ligand-based virtual screening and (ii) structure-based virtual screening.

8.1.1.1 Ligand-based Virtual Screening

Ligand-based virtual screening (LBVS) uses information about only the small molecule compounds, without any explicit information about the protein target. Using a set of ligands that have known binding properties (*i.e.* active or inactive), features of these ligands can be used to predict whether a new compound will bind. There are numerous methodologies that can be broadly grouped into pharmacophoric models,[12] chemical similarity search,[13] and machine learning models.[14,15] LBVS can be applied without a structure for the protein target, and without knowledge of the binding mode. A major limitation of LBVS is that models are built for a specific biological response, and are not generalisable to new targets. To build LBVS models, interaction data for the target in question is required, so LBVS cannot be used in advance of experimental work.

8.1.1.2 Structure-based Virtual Screening

In contrast, structure-based virtual screening (SBVS) uses information about the protein target in question, typically a 3D structure, to help determine if a compound will bind to the protein.[3] Unlike LBVS, SBVS can be used in advance of any experimental binding data for the target of interest.

SBVS is a multi-step process, with the following representing a typical pipeline. First a suitably prepared[16] 3D structure for the protein target is needed. Preferentially this is determined experimentally, for example using X-ray crystallography, or alternatively a model of the protein can be constructed from its amino acid sequence. Once a 3D structure is available, binding pockets, locations on the target where small molecules can bind, need to be established.[3] These can also either be determined experimentally, or by computational methods.[17] Prospective ligands are then docked into the binding pockets of the protein target. Docking consists of two components: first possible binding modes, or poses, of the ligand within the active site of the protein are proposed; next, poses are ranked in order to determine the most likely representation of a possible true binding mode (pose prediction). Ranking of poses has traditionally been achieved by using an estimate of the binding affinity (*e.g.* AutoDock Vina[18]); however this need not be the case and machine learning methods have also been used to train scoring functions specifically for this task.[19–21]

SBVS utilises one, or several, potential poses for a protein–ligand complex to assess the likelihood that the compound will bind to the protein based on the poses generated.[4] In this chapter, we will focus on the design and development of scoring functions for this final step, as opposed to the docking protocol more generally.

8.1.2 Traditional Approaches to Virtual Screening

Traditional approaches have typically used experimental data to parametrize a physically inspired function.[18,22–28] These functions are then used to guide the search for the binding mode and to assess the strength of interactions between the protein and ligand, often providing an estimate of the binding affinity.

Scoring functions used in docking protocols can generally be categorised as one of four types depending on the information and method used: force-field, knowledge-based, empirical, or hybrid.[29]

Force-field scoring functions estimate the change in free energy upon binding by summing the terms of a molecular mechanics force field, such as van der Waals and electrostatic interactions between the two molecules. Other contributions to the interaction energy are also often included, for example the solvent effects on exposed hydrophilic and hydrophobic atoms, along with the intramolecular energies, or strain energies, of the two binding partners. This approach has both the advantage and disadvantage of adopting physical models which do not depend on experimental data.[30]

Knowledge-based scoring functions, or statistical potentials, are usually derived from the distribution of pairwise distances between atoms (or another feature) observed in a corpus of protein structures, such as the Protein Data Bank.[31] These are then converted into an energy function that describes the preferred geometries of the pairs of atoms.[32,33]

Empirical scoring functions aim solely to reproduce experimental affinity data.[34] They consist of a weighted sum of largely uncorrelated physically-meaningful terms, for example the number of hydrogen bonds present, the molecular weight of the ligand, or entropic terms related to the size and flexibility of the ligand. The weighting of these features is either determined using experimental binding data for known protein–ligand complexes or assigned manually.

Scoring functions which combine elements of these approaches are known as hybrid scoring functions. Common to all of these approaches is the *a priori* assumption of the functional form of the relationship between the input features and the binding affinity (or other quantity being predicted). While this results in these techniques being interpretable, the use of rigid functional forms inherently limits their ability to capture complex relationships and is constrained to our current understanding of inter- and intra-molecular interactions.

8.1.3 Machine Learning Scoring Functions

Machine learning techniques have begun to be used over the last decade to develop more accurate scoring functions for pose prediction,[20] virtual screening,[19,35] and affinity prediction.[36,37] Machine learning approaches differ fundamentally in that they do not assume the functional form of the relationship between descriptive features and binding affinity, or other target quantity such as active or inactive. Instead, machine learning methods are able to infer relationships between features and the target quantity directly from the data.

Many models, particularly early attempts, reuse the features of traditional approaches,[38–40] but exploit the greater flexibility in model structure to produce better representations of the same input data.[41] For example, Li *et al.*[40] demonstrated the ability of a random forest model to continue to learn from new data and additional features long after the performance of a multiple linear regression model plateaus.

Other machine learning approaches have either designed specific features[36,38] or used more general features,[15] such as molecular fingerprints, as inputs. Two notable examples of such approaches are RF-Score[36] and NNScore.[38] We will briefly outline the method of both approaches to help highlight the difference in approach that CNNs offer.

RF-Score. In 2010, Ballester and Mitchell published RF-Score,[36] a structure-based method for predicting protein–ligand binding affinity using random forest regression. They represented protein–ligand complexes using the number of protein–ligand atom pairs with a separation of less than 12 Å. Four protein atom types (C, N, O, S) and nine ligand atom types (C, N, O, F, P, S, Cl, Br, I) combine to give 36 distinct atom pairs, whose counts for a given protein–ligand complex serve as the input features to the model.

The authors trained a random forest to predict binding affinity using a set of 1105 complexes from PDBbind.[42] In a benchmark of RF-Score against the state-of-the-art traditional scoring functions, including those found in GOLD,[43] and Glide,[24,44] RF-Score outperformed these functions on a test set of 195 complexes. In a later publication,[39] Wójcikowski *et al.* trained an updated version of RF-Score for virtual screening, performing roughly in line with AutoDock Vina[18] in a cross-target split of DUD-E.[45]

NNScore. Around the same time, Durrant and McCammon published NNScore,[38,46] a structure-based approach which used an ensemble of neural networks, each with a single hidden layer, to predict binding affinity. Similar to RF-Score, the representation of the protein–ligand complex is primarily based on atom pairs, in this case AutoDock atom types,[18] while the authors distinguished between atom pairs within 2 Å and pairs within 4 Å. In addition, they also included an estimate of the total electrostatic energy for each atom pair within 4 Å, basic counts of 13 different ligand atom types, and the number of rotatable bonds in the ligand as additional features, resulting in a total of 194 distinct input features. In virtual screening tests on influenza N1 neuraminidase, the authors reported slightly enriched performance over AutoDock Vina,[46] with similar levels of performance reported in a follow-up publication.[38]

8.1.4 Rationale for Deep Learning Approaches

For both traditional approaches and standard machine learning models, the particular features are chosen prior to fitting parameters. This biases the model to the choice of features and leads to an unnecessary loss of information through the elimination or approximation of the raw structural data. Considering the above examples, RF-score loses the counts of any atoms pairs not included as features, or counts further than 12 Å apart, while making the approximate assumption that all atom pairs within 12 Å are equivalent and the number of occurrences is all that matters. The features of NNScore are more descriptive, but make similar approximations.

Deep learning approaches are able to minimise the initial featurisation process by taking as input a format that more closely mimics the underlying protein–ligand complex. The model is then able to learn directly from the data the features that explain a given property, such as whether a compound binds, rather than the features being defined in advance.

There have been several recent deep learning approaches described that have learnt features for virtual screening or binding affinity prediction in an end-to-end manner and have been shown to outperform approaches that take as input precomputed features.[19,37,47,48] In particular, Ragoza *et al.*[19] developed a CNN-based approach for SBVS that minimised initial featurisation of input data, using only a 3D voxel grid of atomic coordinates with basic atom typing.

We will first describe the data sets and methods for training and testing SBVS techniques. Then we will give an overview of CNNs, their use and development in image recognition. Finally, we will describe their application into tools for SBVS, highlight their strengths and limitations, and discuss the outlook for such tools.

8.2 Virtual Screening

8.2.1 Data Sets for Structure-based Virtual Screening

There are many possible databases upon which to train and test virtual screening methods, and the most appropriate choice will depend on the goals of the specific project. A common goal, in particular within the academic research community, is to develop a universal scoring function. Universal in this sense means not directly optimised for a particular group or family of proteins, with the goal of being able to make accurate predictions for arbitrary proteins. The majority of data sets are designed for this purpose and contain a diverse range of proteins. Robust assessment of scoring functions is essential for the continued development of virtual screening, and thus suitable benchmarks are critically important. The following represent the most commonly used benchmarking sets. We discuss their construction, strengths, and limitations.

DUD-E. A widely used virtual screening benchmark is the Database of Useful Decoys: Enhanced (DUD-E)[45] data set. DUD-E consists of 102 targets across eight protein categories, more than 20 000 active molecules, and over one million decoy molecules. Active molecules were extracted from ChEMBL09[49] if their activity/affinity (Ki, Kd, IC50, EC50) was ≤1μM. Ideally, inactive molecules should also be selected based on experimental evidence.[50] However, this is typically not possible. As a result, compounds that are presumed to be inactives are used as decoys due to the very low chance for an arbitrary molecule to bind to a given target. Due to the scarcity of true inactive molecules, the majority of virtual screening benchmarks include putative inactives, or decoys, instead. Mysinger *et al.* took inactives from ChEMBL where available (with affinity ≥30 μM), but the majority of decoy compounds were selected from ZINC.[51]

Decoys were selected to have similar physicochemical properties to active compounds while being topological dissimilar. This is done to unbias properties that, while important, should not distinguish binders from non-binders (*e.g.* molecular weight), while the topological filter was applied to avoid including active compounds in the decoy sets. In total, 50 decoys were chosen per active compound. In an effort to reduce analogous bias, active ligands were clustered by their Bemis–Murcko[52] atomic frameworks, with only the most active ligand included, subject to a minimum number of actives per target. Despite these efforts, active compounds in some of the targets remain highly similar, and substantial physicochemical property biases remain,[53,54] although this is not a problem unique to DUD-E.

DEKOIS. In 2011, Vogel *et al.* proposed a new generator of decoy compounds sets called Demanding Evaluation Kits for Objective In Silico Screening (DEKOIS).[55] This workflow was further refined in 2013, giving rise to DEKOIS 2.0,[56] a virtual screening database with 81 targets across 11 target classes. Active molecules were taken from BindingDB,[57] with the activity cut-off selected dynamically on a per target basis. Bauer *et al.* removed compounds possessing a 1000-fold weaker target affinity than the most potent inhibitor for the respective target. Decoys were taken exclusively from ZINC,[51] with 30 decoys selected per active compound.

Similarly to DUD-E, the authors designed their tool to avoid the introduction of well-known and described biases into the decoy sets, *i.e.*, analog bias and artificial enrichment. Physicochemical properties of both active ligand and decoys are matched to limit the analog bias, while the authors proposed a quantitative measure to assess the risk of including false-negative compounds as decoys, known as the latent actives in the decoy set (LADS) score,[55,56] based functional class fingerprints. Property similarity and LADS score are then simultaneously optimised to produce the final decoy set.

MUV. Whilst the use of putative inactives is widespread and the methods adopted to ensure decoys are not binders are fairly effective, experimentally verified inactives remain the gold standard.[50] However, using such inactives only overcomes one of the problems for a benchmarking set. The Maximum Unbiased Validation Sets (MUV)[58] is designed with these two factors in mind. MUV is based on PubChem[59] bioactivity data from 17 targets, each with 30 actives and 15 000 decoys. This presents an order of magnitude increase in the number of decoys per active compound compared to the other data sets discussed. Actives were selected from confirmatory screens and were chosen to be maximally spread based on simple descriptors (such as basic atom counts) and embedded in decoys, which were selected from a primary screen for the same target.

The MUV data sets were designed to avoid analog bias and artificial enrichment, which produce overly optimistic predictions of virtual screening performance. The result is a very challenging benchmark that minimises the effects of artificial enrichment in evaluating virtual screening methods, in particular ligand-based methods. Some papers have questioned the appropriateness of using MUV as a structure-based virtual screening benchmark,[19,60] due to the use of cell-based assays for some of the targets and limited attention to structure-based considerations, among other factors. Indeed, the only approaches that have shown meaningful predictive power on MUV are ligand-based methods. However, these methods have included substantial amounts of target-specific data in the training set, and have often used large external data sets without any regard to the overlap between training and test sets.[14,15]

MUBD. The Maximal Unbiased Benchmarking Data Sets (MUBD) seeks to extend the principles of the MUV data set to new targets, in particular where putative, rather than true, inactives must be used. As such, the principles adopted can be used to create a data set suitable for both LBVS and SBVS, simultaneously addressing the relative lack of LBVS data sets and minimising the bias often found in SBVS benchmark sets. In particular, Xia *et al.*[61] built decoy data sets targeting the G Protein-Coupled Receptors (GPCRs). They have since expanded the concept to other families.[62,63] In the case of MUBD-GPCR,[61] active ligands were taken from GLL,[64] an existing virtual screening benchmark for GPCRs. Diversity between active ligands was ensured by enforcing a maximum similarity based on MACCS fingerprints.[65]

Decoys from ZINC were matched with six physicochemical properties such that no compound has physicochemical properties outside of the bounds of the active ligands, while a maximum fingerprint similarity to any active compounds was enforced to prevent the introduction of potential active structures into the decoy set. Compounds were further filtered to ensure the similarity of physicochemical properties was maximised while maintaining a random spatial distribution of the decoys around the active ligands, coupled with targeting an optimal embedding of actives in the decoys by fingerprints. For each active 39 decoys were chosen.

In contrast to the other data sets described above, each MUBD only includes targets for a specific family or collection of proteins, resulting in a highly targeted scoring function. Such a set can be very helpful when the goal is to learn a model for a novel target within that family or to assess selectivity of compounds within a protein family.

ChEMBL set. The final SBVS data set we will discuss is one generated from ChEMBL by Riniker and Landrum,[66] following Heikamp and Bajorath.[67] They selected a set of 50 human targets from ChEMBL version 14. They chose actives that had at least 10 μM potency, had a molecular weight under 700 $g\,mol^{-1}$, and did not have metal ions. The actives were down-sampled using the RDKit[68] diversity picker to select the 100 most diverse compounds for each target. For each active, two decoys were randomly selected from the ZINC database subject to a minimum similarity using a simple atom-count fingerprint (ECFC0). This yielded a total of 10 000 decoys that were shared across all targets. This data set has been analysed in less detail than DUD-E, but it is likely to suffer from artificial enrichment due to the lack of physicochemical property matching of decoys to actives, and the use of the same decoys across all targets.

8.2.2 Appropriate Train/Test Splits for SBVS

Once a suitable data set has been selected, there are numerous possible ways of splitting data to form training and validation sets. Ideally, in any machine learning experiment one should keep a held-out test set of data that most closely represents the real-world challenge for the data set. In the majority of machine learning applications, a random split of data within each class is appropriate. However, for structure-based virtual screening, more careful consideration needs to be given in light of both the specific problem (binding to one target is not the same as binding to another) and the use-cases (often a prospective screen for a novel target).

Intra-target splitting. Intra-target splitting in its basic form is comparable to a random split of data within each class. Both training and test sets contain data from all targets (or a single target), *i.e.* each target has its ligands both in the training and test sets. Such an approach mimics experiments on targets for which there are already known ligands. Assessing models in this way does not allow conclusions to be made about the ability for a model to generalise to novel targets, which is highly relevant in most use-cases, and is likely to lead to an overly optimistic assessment of scoring function due to the known-biases that exist within many benchmarks unless further precautions are taken, such as chemotype splitting or time splitting (described below).

Cross-target splitting. In contrast, in cross-target splitting there is no overlap of targets between training and test data. As a result, all testing is performed on unseen targets, more closely mimicking many real-world scenarios. This is more challenging, as different proteins typically have different binding modes. This normally results in substantial differences between the active compounds between the targets, while substantial commonalities will normally exist for a collection of actives for a single target that bind in the same pocket.

Several publications have demonstrated the additional challenge of cross-target splitting compared to intra-target splits. Wu *et al.* achieved almost perfect performance on DUD-E when performing a random training/test split across all targets (intra-target splitting),[15] allowing training and test sets to contain examples from the same targets. Wójcikowski *et al.* saw performance in cross-validation almost triple when splitting their data either randomly across all targets (intra-target splitting) or on a single target basis, as opposed to keeping all examples for a given target in the same fold (cross-target splitting).[39]

Many proteins are highly similar and, as a result, have substantial overlap of active ligands. Consequently, cross-target splitting alone is insufficient to ensure that one has a robust method to assess the predictive power of models. There are several different ways of splitting data to ensure dissimilarity between training and validation/test sets (Figure 8.1).

Protein sequence similarity. Proteins with high sequence similarity often have similar structure and function, and hence similar ligands will display activity. To remove this source of bias, targets can be clustered during cross validation, ensuring highly similar targets are in the same fold, or can be removed from a potential test set if a maximum similarity threshold is met with any target in the training set. In the experimental setup of Ragoza *et al.*,[19] a threshold of 80% sequence similarity from a global sequence alignment was set for both cross-validation and the construction of test sets.

Binding site similarity. Binding sites can be extremely similar, even for targets with relatively low sequence similarity. In most cases, this would result in substantial overlap in activity between two targets that would not be identified by protein sequence similarity. To avoid this, Ragoza *et al.*[19] assessed their CNN protocol by performing ProBiS[69] structural alignment on the binding sites of all pairs of targets from the training and proposed test sets, removing any targets for which a significant alignment was found using the default ProBiS parameters. To illustrate the level of redundancy between commonly used benchmarks, these two steps combined resulted in a

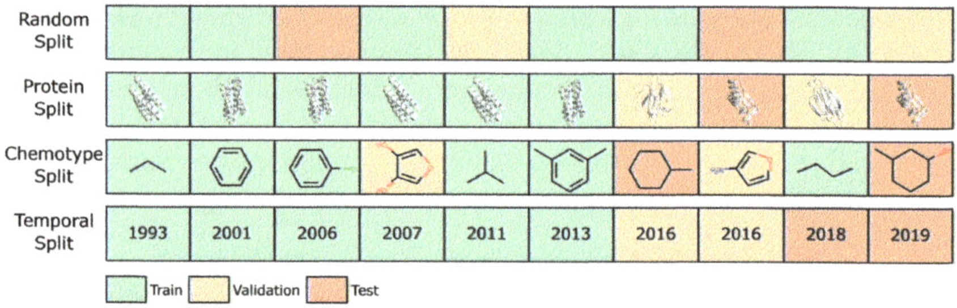

Figure 8.1 Representation of different methodologies for splitting a data set into train, validation, and test sets.

reduction in potential independent test sets from 50 initially, to a 13 target subset of the Riniker and Landrum ChEMBL set, and from 17 initially to a 9 target subset of the MUV set.[19]

Chemotype splitting. Scaffold, or chemotype, splitting separates the samples based on the 2D structural frameworks of the ligands, such as Bemis–Murko scaffolds.[52] Since scaffold splitting attempts to separate structurally different molecules into different subsets, it offers a greater challenge for learning algorithms than the random split.[15] Scaffold splits provide a stronger test of a given model's ability to generalise compared with random splitting,[48] and is most applicable for intra-target splitting.

Temporal splitting. Finally, if time information is known, training, validation, and test sets can be constructed on the basis of this information (temporal or time splitting).[15,70] This allows retrospective studies to more accurately reflect prospective ones, and is more challenging than a random split.[48] However, it is frequently not possible to easily incorporate, since time information is often not available. This idea is most readily applicable to intra-target splitting, but could also be adopted in the case of a cross-target split, where all data from training targets must have been published before any of the ligands from the test targets.

8.2.3 Evaluation Measures

Ensuring performance metrics match the goals of the screening process is crucial when building models, since model performance will be optimised to these metrics. In general, the goal of virtual screening is to prioritise binders above non-binders, with most emphasis being placed on the compounds with the highest scores, given that in the majority of scenarios it is only practical to experimentally test a very small subset of molecules screened *in silico*.

Receiver-operating characteristic (ROC) curves are a commonly used method of evaluation, with the area under such a curve a primary scoring measure. ROC curves plot True Positive Rate (TPR) on the y-axis, and False Positive Rate (FPR) on the x-axis. The area under the ROC curve is easily interpretable as the probability of a randomly chosen active being ranked above a randomly chosen inactive.

While informative, ROC curves place equal weight across all test cases; as a result, many studies report metrics designed to place more weight on the early part of the curve, such as BEDROC,[71] pROC,[72] or RIE.[73] However, these introduce additional parameters and are harder to interpret. Another common measure is ROC enrichment.[74,75] ROC enrichment measures the ratio of TPR to FPR at a given FPR. Common enrichment factor (EF) thresholds are 0.5%, 1%, 2%, and 5%. Note that the maximum possible ROC enrichment depends on the FPR threshold (*e.g.* at an FPR of 2%, the highest possible ROC enrichment is 50), while random performance has an expected enrichment factor of 1 at any FPR threshold. In contrast to other metrics discussed thus far that are global in nature, EFs are local metrics since they are assessed at a specific FPR, rather than across all FPRs.

Precision-Recall Curves (PRC) are an alternate global method of evaluating binary classifiers that can offer advantages in class imbalanced situations. PRCs plot precision on the y-axis against the recall, or TPR, on the x-axis. As noted in previous literature,[15,76] ROC curves and PRCs are highly correlated but perform significantly differently in the case of high-class imbalance. In particular, PRCs put more emphasis on the low recall side in the case of highly imbalanced data. When positive samples form only a small proportion of all samples, false-positive predictions exert a much greater influence on precision than FPR, amplifying the difference between PRC and ROC curves. The underlying data for virtual screening experiments have extremely low true positive rates, suggesting that the correct metric to analyse may depend on the experiment at hand. Several recent publications have recommended using precision-recall as the primary performance metric for virtual screening,[15,35] with Wu *et al.* proposing specific recommendations about when to use PRCs.[15] In particular they recommend that area under a PRC is used for data sets with positive rates less than 2%, otherwise ROC should be used instead. Random performance has an expected Area-Under-Curve (AUC) ROC of 0.5, whereas the expected AUC PRC of a random classifier is equal to the class imbalance (*e.g.* if there were 50 decoys for each active, the expected AUC PRC for a random classifier would be 0.02).

8.3 Convolutional Neural Networks

8.3.1 CNNs: A Primer

CNNs represent a class of deep learning methods that are designed to capture local and spatial connectivity. CNNs have been extensively applied in computer vision for analysing images; they have also been applied to other areas, such as neural machine translation[77] and latent semantic modelling.[78]

The precise architecture of CNNs can vary greatly, even for models designed for use on the same data set; however, CNNs share many commonalities and are constructed from smaller building blocks, known as layers. All CNNs contain an input and an output layer, as well as multiple hidden layers (Figure 8.2). The hidden layers of CNNs typically consist of convolutional

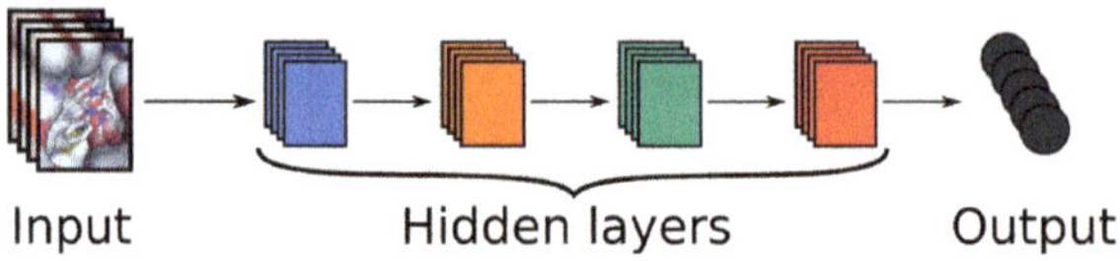

Figure 8.2 Overview of the basic format of a neural network architecture. The input is processed by a number of hidden layers to produce the output.

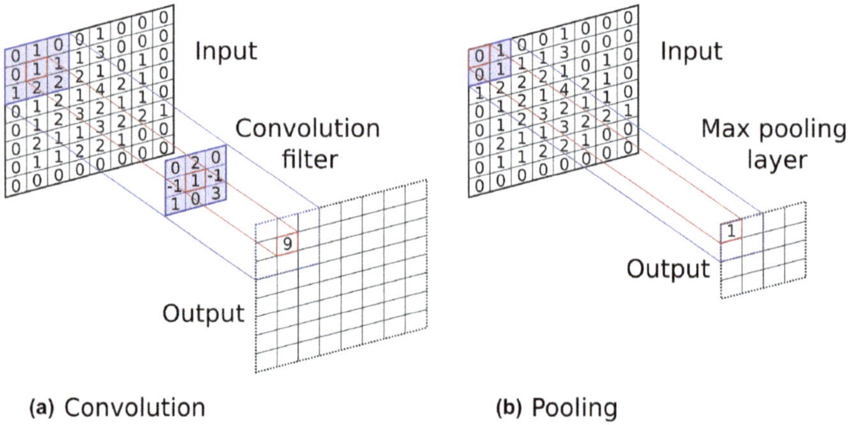

Figure 8.3 Illustration of the two most common layers used in CNNs. (a) A 3×3 convolution is passed across the input to the layer. The output pixel is a weighted sum of the source pixel and the elements around it. (b) A 2×2 max pooling operation with stride 2 results in a downsampling of the input from 8×8 to 4×4, taking the maximum element in each 2×2 grid.

layers, pooling layers, and fully connected layers, together with activation functions and normalisation layers. It is the different combinations of these layers that lead to the differences between networks.

Input layer. The input layer is where data is fed into the model. CNNs take input with fixed spatial dimensions (*e.g.* the height and width of an image) and a fixed depth dimension (*e.g.* 1 for a greyscale image or 3 for a standard RGB image).

Convolutional layers. A convolutional layer applies a series of convolution operations to its input to produce an output. Convolutional layers consist of convolutional filters, or kernels, which are convolved with the input to produce feature maps. Mathematically, a convolution is an operation on two functions returning a function that represents how the shape of one is modified by the other. In a CNN, the convolution operation is represented by matrix operations, specifically the dot products of portions of the input to a convolutional layer and each of the layer's filters (Figure 8.3(a)).

Convolutional filters are small, learnt, matrices, with dimensions equal to the size of the receptive field of the layer. The dimension of a convolutional filter determines how much of the input the filter acts upon. For example, a

3×3 filter acts on a square of pixels with side length 3. Each filter is passed over the input in an independent manner, producing an activation map or feature map. These activation maps are then concatenated to form the output of the convolutional layer. An example of the convolution operation can be seen in Figure 8.3(a).

Pooling layers. Pooling layers are used to reduce the spatial size of the representation passing through the network. This is done both to reduce the number of parameters and computation in the network, and also to control overfitting. In addition, this reduction of resolution helps to make the representation invariant to small translations of the input.[79] Pooling layers are commonly inserted between successive convolutional layers, or after a block of layers. The pooling layer operates independently on every depth slice of the input and resizes it spatially, typically using either the maximum or average operation of the pooling layer's receptive field. The most common form of pooling layer uses the maximum operation, and has filters of size 2×2 with a stride of 2 (the size of the step the filter moves each time); this results in a downsampling of every depth slice in the input by 2 along both width and height, discarding 75% of the activations by taking only the maximum in each 2×2 square of the input (Figure 8.3(b)). A standard pooling operation does not affect the depth dimension and acts on individual feature maps independently. It is possible to use any function to perform pooling, with common examples maximum, average and sometimes L2-norm pooling. Average pooling was often used historically but has recently fallen out of favour compared to the max pooling operation, which has been shown to work better in practice.[80] Several alternative pooling methods have been proposed that claim improvement over max and average pooling,[80,81] however they do not appear to have been generally adopted.

Other Layer Types. In addition to convolutional and pooling layers, there are several other types of commonly used layers. Here, we will briefly discuss activation and normalisation layers. More detail can be found in Goodfellow *et al.*[79] or Gu *et al.*[82] Activation layers are used to introduce non-linearities into the model. Without activation layers, neural networks would simply be linear functions of the input features. The most common form is the rectified linear unit, or ReLU,[83] which for every number in the input, takes the maximum of that number and zero as its output. Other commonly used activation functions are sigmoid, hyperbolic tangent, and leaky ReLU.[84] Choosing an activation function is normally determined by a hyperparameter search, or is selected to produce output within a specific range.

The initial input to machine learning models is often normalised in some way to control for the range of possible inputs. Normalisation layers can also be incorporated within a model to standarise the input to the subsequent layer in a particular way. Normalisation layers are not always included in CNNs; however, one commonly used example is batch normalisation.[85] Batch normalisation aims to stabilise the distribution (over a mini-batch) of inputs to a given network layer during training, in order to make the optimisation landscape smoother.[86] Batch normalisation layers are typically placed between convolutional layers, either before or after the non-linearity.

Training. CNNs are trained through backpropagation, which updates the parameters of the convolutional filters based on errors during the training process. The number of parameters in CNNs can be extremely large (>1 000 000) and parameters are not independent. Thus, finding optimal parameters is challenging. Stochastic gradient descent,[87] or other optimisers such as Adam,[88] are used to update the weights in the neural network based on small batches of data, known as minibatches. These optimisation techniques are stochastic in nature, thus the parameters of the trained model will depend on the random seed used during the training process. We refer the interested reader to a recent overview of optimisation techniques for training neural networks by Ruder.[89]

8.3.2 ImageNet

While CNNs were first introduced in the 1980s[90] and found some early use cases (for example LeNet for reading numbers on cheques[91,92]) broad adoption has only occurred much more recently. Before 2012, traditional image processing and analysis techniques were used, such as pyramid matching[93] and saliency maps.[94] In contrast to CNNs, these often required substantial pre-processing of images and did not learn the underlying features from scratch, but instead learnt how to combine and interpret pre-computed features. However, following their successful application as part of the annual ImageNet competition in 2012, CNNs have become the *de facto* method for almost all computer vision tasks.

The annual ImageNet Large Scale Visual Recognition Challenge (ILSVRC, or ImageNet)[95] was an annual competition in object category classification and detection of hundreds of object categories and millions of images that was introduced in 2010 and ran until 2017. Since its introduction, the competition has been the source of many of the substantial advances not just in computer vision, but in machine learning more broadly. Crucial to the development of more advanced algorithms was the release of the then largest annotated corpus of images that could be used for training, consisting of over 1.4 million images across 1000 different classes. This compares with around 20 000 images from 20 classes in the previous largest data set.

The competition has led to substantial improvement in several tasks including image classification. For example, all entries in the first two years of the challenge had top five error rates of over 25% at the image classification task, while the winner of the 2017 edition, SENets,[96] achieved a 2.3% error rate (Figure 8.4). A key shift occurred in 2012, with the introduction of convolutional neural networks. This caused a drop in classification error rate from 26% to 16%. Since then, all winning entries have been based on convolutional architectures, each with their own algorithmic advances, some of which will be described below.

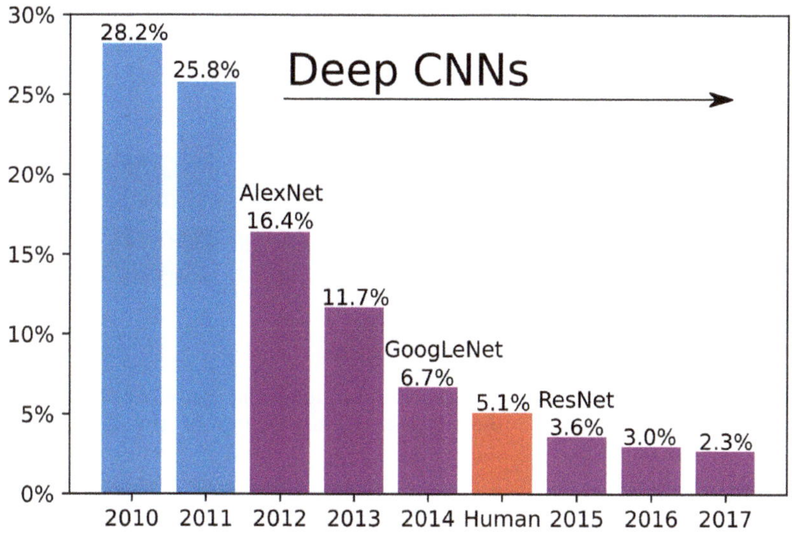

Figure 8.4 Top 5 error rate of the winning entries in the annual ImageNet competition since 2010. Deep CNNs have been responsible for the winning entry each year since 2012, with expert human performance surpassed in 2015.

8.3.3 Modern CNN Architectures

The majority of winning entries since 2012 have incorporated substantial algorithmic advances, as opposed to either hyperparameter or model architecture optimisation, or increases in the quantity or indeed quality of training data. We will briefly highlight several of the architectural advancements that have allowed more powerful networks to be trained.

AlexNet.[97] The first work that popularised CNNs in computer vision was AlexNet. AlexNet was submitted to ILSVRC in 2012 and significantly outperformed the second placed method, achieving a top 5 error of 16% compared to the runner-up which had 26% error. The network had a basic convolutional architecture, but was deeper (more layers) and wider (more filters per layer) than previously used networks. The model also stacked convolutional layers on top of each other whereas previously it was common to have a pooling layer immediately following each convolutional layer.

GoogLeNet.[98] The ILSVRC 2014 winning entry utilised a novel way of combining convolutional layers, developing the Inception Module (Figure 8.5) that dramatically reduced the number of parameters in the network (4 M, compared to AlexNet with 60 M). By combining convolutions with different kernel sizes, their model is able to extract features of different sizes at each point. Additionally, the authors used average pooling instead of fully connected layers at the end of their model, eliminating a large number of parameters that did not seem to impact performance.

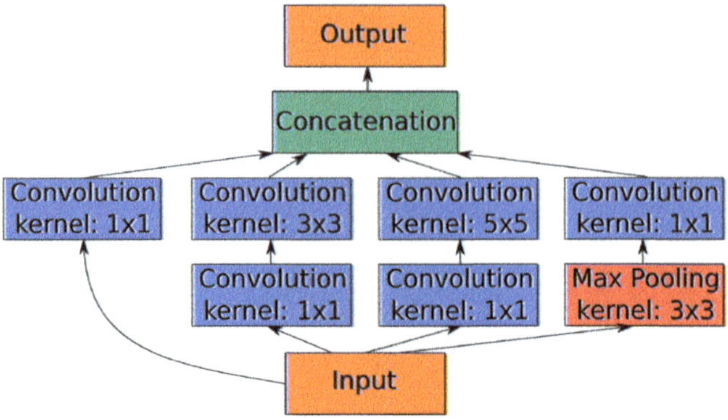

Figure 8.5 Architecture of the Inception Module, the main building block of the GoogLeNet[98] model, the winning entry of the 2014 ImageNet competition.

There have been several subsequent versions of GoogLeNet, most recently Inception-v4.[99]

ResNet[100] and DenseNet.[101] Residual networks (ResNets) developed by He *et al.* were the winner of ILSVRC 2015, while the DenseNet architecture was introduced by Huang *et al.* in 2016 and achieved state-of-the-art performance on several benchmarks at the time of release. Both featured novel skip connections between layers and made substantial use of batch normalisation. The key architectural difference in ResNet and DenseNet compared with other convolutional networks is the fashion in which layers are connected (Figure 8.6). Normally, each layer will receive as input solely the output of the previous layer (standard connectivity). In a ResNet, a layer receives as input the sum of the outputs from the previous two layers (residual connectivity), while in a DenseNet, within each dense block, a layer receives the output of all prior layers within that block (dense connectivity). The ability for the output of a layer to skip the next allows feature maps of different depths and hence complexities to form, and for new feature maps to be learnt from a combination of existing maps of differing complexities. Furthermore, the residual and dense connections improve gradient flow during backpropagation,[87] allowing deeper models to be trained effectively while exhibiting substantially improved parameter efficiency.[101]

8.4 CNN Applications for Virtual Screening

There have been two recent studies demonstrating the power and promise of formulating structure-based virtual screening as a computer vision problem and applying CNNs.[19,35] We first outline the input format utilised by both methods, which differs substantially from other machine-learning methods, before discussing the performance of these methods on several SBVS data sets.

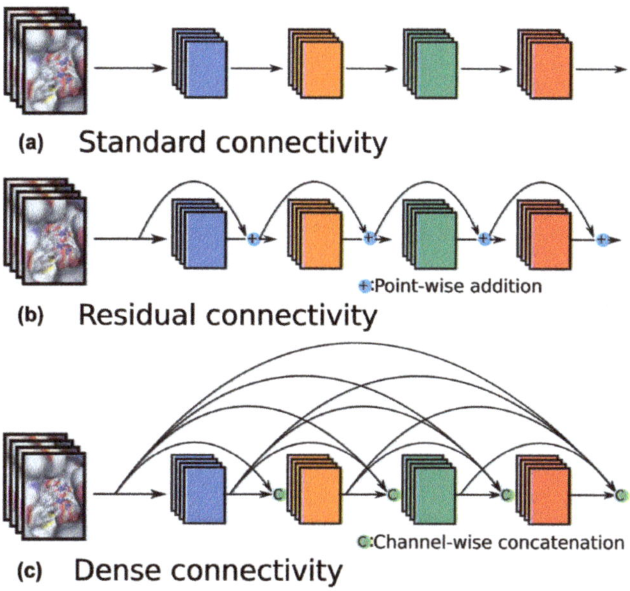

Figure 8.6 Illustration of connectivity in a standard convolutional neural network (a) compared to a ResNet (b) and DenseNet (c).
Adapted from ref. 35, https://doi.org/10.1021/acs.jcim.8b00350, under the terms of a CC BY 4.0 license, https://creativecommons.org/licenses/by/4.0/.

8.4.1 Input Format for CNN Structure-based Virtual Screening

Determination of binding can be reframed as a computer vision problem, due to the importance of spatial configurations for physical interactions. It is possible to visualise a protein–ligand complex as a 3D image using tools such as PyMOL.[102] The input format adopted by the CNN-based SBVS methods so far have attempted to mimic as closely as possible such a 3D image. The precise formulation varies slightly, and in some cases is readily customisable by the user, but all representations to date have been based on a grid discretisation of the binding site and an atom typing scheme. Here we will describe in detail the input format adopted by the gnina framework[19] as an example of how this is achieved.

The gnina featuriser overlays a fixed-size grid over the binding site, centred around the ligand. Instead of the RGB colour channels of a regular image, a set of common atom types are used, with separate channels for atoms of the protein and the ligand. A schematic of the input featurisation process is shown in Figure 8.7. In particular, Ragoza *et al.*[19] used a grid that was a 24 $Å^3$ cube, with a resolution of 0.5 Å. Each point of the 3D grid has 34 information channels, corresponding to distinct heavy atoms on either the protein (16 channels) or ligand (18). Permitted atom types are based on smina atom types (Table S1 in Ragoza *et al.*[19]), but are readily customisable. Atoms are converted from a point representation to a Gaussian representation within the van der Waals radius that is quadratically smoothed to zero at 1.5× the van der Waals radius from the input coordinates of a given atom. This input format provides no additional information beyond spatial coordinates and atom type, and does not explicitly include bonds, bond order, or hydrogens; the information provided in the input format is a

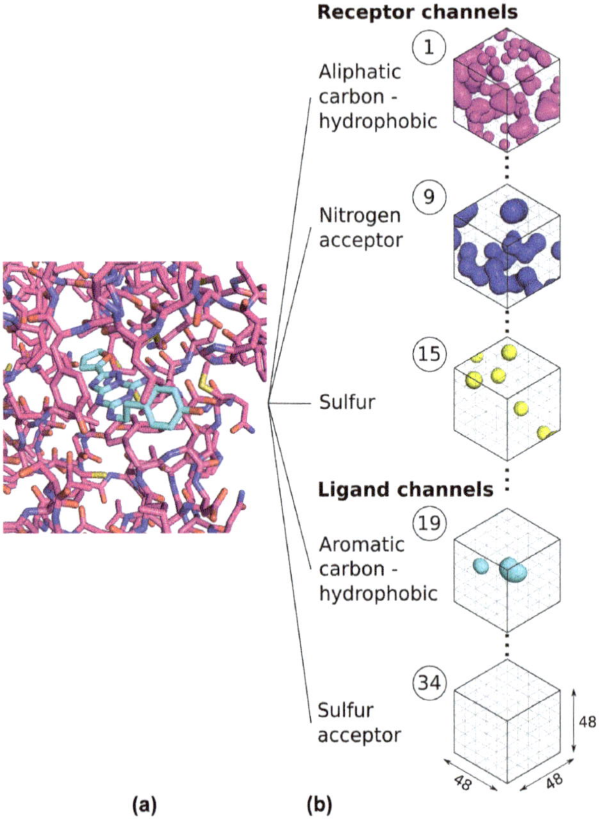

Figure 8.7 Input featurisation for PDB ID 3EML (ligand ZMA). (a) The protein–ligand complex is cropped to a 24 Å^3 box, centered on the ligand. The ligand is shown with carbons in cyan, the receptor with carbons in magenta, and the heteroatoms are coloured with standard colouring. (b) The complex is decomposed into information channels, one for each atom type, and divided into voxels with a resolution of 0.5 Å. Atoms within each channel have a Gaussian representation. This is shown for five of the 34 channels. The resulting atom type channels are concatenated to produce the (34, 48, 48, 48) input tensor for the CNNs.

comprehensive representation of the binding site of a single, static, docked protein–ligand complex, up to the chosen grid resolution and atom typing scheme.

8.4.2 Outline and Performance of CNN-based Methods

Ragoza *et al.*[19] proposed a neural network for protein–ligand scoring that took as input the format described in Section 1.4.1. The CNN model they employed consisted of three 3×3×3 convolutional layers (with 32, 64, and 128 filters respectively), each preceded by a 2×2×2 max pooling layer (Figure 8.8). They designed and trained their protocol using a 3-fold clustered cross-validation on the DUD-E.[45] Crystal structures do not exist for almost all of the ligands in DUD-E. As a result, ligand poses for all actives and decoys were generated using AutoDock Vina,[18] specifically the smina[25] implementation. Ligands were docked against a reference receptor within a box centered around a reference ligand with 8 Å of padding using smina's default arguments for exhaustiveness and sampling. The top-ranked pose for each protein–ligand pair was used for training.

Their final protocol was then evaluated on two independently constructed test sets, a subset of the ChEMBL database[103] curated by Riniker and Landrum,[66] following Heikamp and Bajorath,[67] and the maximum unbiased validation (MUV)[58] data set, which is based on PubChem bioactivity data. At test time, to evalutate a new ligand, they scored all docked poses using a single, universal model, and took the maximum as the final score. They demonstrated that this approach offered substantial improvement over both AutoDock Vina,[18] as well as two other machine learning scoring functions, RF-Score[36] and NNScore[38] (Table 8.1).

More recently, we extended this work in Imrie *et al.*[35] We adopted the same input format but proposed several improvements that each had an independent effect on performance. Firstly, a densely-connected neural network architecture was utilised. The model employed contained three dense blocks, with four convolutional blocks within each (Figure 8.8). The convolutional blocks consisted of a batch normalisation layer, a convolutional layer with a 3×3×3 kernel, followed by a ReLU activation. Between each dense block, a 1×1×1 convolutional layer was followed by a 2×2×2 max pooling layer. The first convolutional layer contained 32 filters, after which each 3×3×3 convolutional layer contained 16 filters, while the 1×1×1 convolutional layers matched the number of input features. A schematic of the network architecture can be seen in Figure 8.8. Adopting a DenseNet architecture instead of the more traditional CNN of Ragoza *et al.*[19] resulted in a 20.8–36.5% increase in AUC PRC.

Secondly, at test time the model scored each protein–ligand complex by averaging over an ensemble of docked poses. While training, the model only used the top-ranked Vina pose. Many of these poses will be incorrect and restricting evaluation to only the top-ranked pose at test time would result in

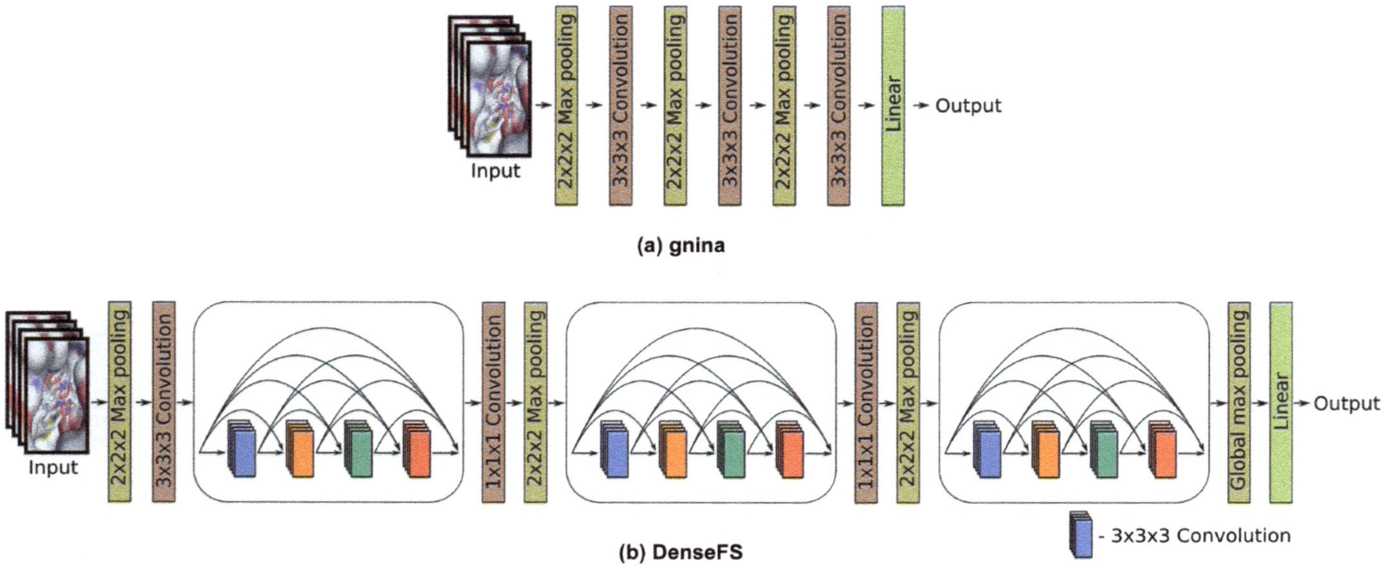

Figure 8.8 Illustration of model architectures of the models employed by (a) Ragoza *et al.*[19] and (b) Imrie *et al.*[35] Adapted from ref. 35, https://doi.org/10.1021/acs.jcim.8b00350, under the terms of a CC BY 4.0 license, https://creativecommons.org/licenses/by/4.0/.

Table 8.1 Mean AUC ROC, AUC PRC and ROC enrichment across targets in the DUD-E data set for gnina and DenseFS, compared to the AutoDock Vina scoring function and two machine learning scoring functions, RF-Score and NNScore. Results for Vina, gnina and DenseFS are taken from Imrie *et al.*,[35] results for RF-Score and NNScore from Ragoza *et al.*[19] Performance metrics are described in Section 1.2.3.

Metric	Vina	RF-Score	NNScore	gnina	DenseFS
AUC ROC	0.703	0.622	0.584	0.862	0.917
AUC PRC	0.093	—	—	0.263	0.443
0.5% EF	15.017	5.628	4.166	44.521	79.321
1% EF	10.383	4.274	2.980	30.652	47.986
2% EF	7.135	3.499	2.460	19.724	28.408
5% EF	4.726	2.678	1.891	10.595	13.744

scoring only inaccurate poses for many of the active molecules. Determining the final score for a compound by averaging the scores of the top n ranked poses, where n was determined through cross-validation, was compared to using the maximum. This meant that the score for a compound was averaged over a group of poses that on average should be more likely to be close to a native pose than a single pose. Overall, adopting average scoring instead of a max scoring policy led to improvements in AUC PRC of between 7.6% and 14.4%.

Thirdly, transfer learning was used to construct protein family-specific models for each of the four major protein classes in DUD-E. It has been shown that in most cases a scoring function constructed for a specific protein family will outperform a universal model.[104,105] As such, protein family-specific models were constructed using transfer learning by fine tuning a universal model on data from targets belonging to the target's family. Adopting protein family-specific models instead of a universal model led to average improvements in AUC PRC of 18.3% to 24.0%.

Finally, an ensemble of models was employed. Ensemble methods exploit the predictions of multiple models to improve performance. The predictions of the three replicas (each trained with a different random seed) were combined by averaging the scores produced by each of the models. This provided a small, but appreciable, improvement, with an increase in average AUC PRC of 3.4% to 11.0%.

The performance of CNN-based methods for virtual screening far exceeds conventional methods, such as docking, and other machine learning methods (Table 8.1). In retrospective testing, DenseFS identified more than 5× as many binders at a 0.5% false-positive threshold than Vina, both in cross validation on DUD-E and on an independent test set.[35] The performance of CNN-based methods has not yet been extensively verified in prospective studies, although some initial results are promising, with a CNN-based method providing best-in-class performance on several virtual screening tasks in the D3R 2017 community challenge.[21] Given the nascence of the field, we expect the relative performance of CNN-based methods to improve further.

8.5 Other Closely Related Tasks

8.5.1 Pose Prediction

Pose prediction is the task of selecting the correct binding mode from a set of (docked) structures of a ligand in complex with a protein structure. Identifying a pose that is close to native is a crucial step in SBDD in the absence of crystal data, and while docking software is able to sample the range of possible binding modes effectively in most cases, ranking the binding modes is more challenging.

Pose prediction performance can be assessed both in terms of single-complex ranking, and multi-complex ranking (sometimes referred to as inter-target ranking and intra-target ranking, respectively[19]). Multi-complex ranking scores all poses across a range of protein–ligand complexes to create a global ranking of all poses that can be assessed with a ROC curve. This form of evaluation is similar to the training protocol of most methods but does not reflect the typical docking scenario. In single-complex ranking, the goal is to select the lowest RMSD pose among poses generated for each individual protein–ligand pair. A scoring function can do well in single-complex ranking even if the low RMSD pose has a poor score as long as all other poses for that target have worse scores. This better represents the typical docking scenario where the goal is to identify a low RMSD pose among many poses for the same complex.

Ragoza *et al.*[19] developed CNN models that performed well in multi-complex pose prediction evaluation; however they performed worse than Vina at single-complex ranking. One possible explanation for this is the limited data set size used to train their CNN models, which was derived from 337 co-crystals from the CSAR-NRC HiQ data set, with the addition of the CSAR HiQ Update.[106] Hochuli *et al.*[20] trained a similar CNN model with a data set derived from the 2016 PDBbind refined set[107] an order of magnitude larger in size. This resulted in a substantial improvement in multi-complex pose prediction, although the authors did not report single target results. It is likely that the single-complex ranking performance of CNN methods now exceeds classical methods, such as Vina, although there is currently no systematic study demonstrating this.

8.5.2 Binding Affinity Prediction

Assessing the strength of binding is of crucial importance in the drug design process. Being able to predict accurately the binding affinity of a small molecule to a protein would enable optimisation of the specific small molecule for binding. While virtual screening allows for the identification of initial binders, these then need to be progressed into lead molecules by increasing both the affinity and specificity to a particular target.

There have been several recent studies using CNNs for binding affinity prediction.[20,21,37,48] Hochuli *et al.*[20] and Sunseri *et al.*[21] adopt the same input format as described in Section 1.4.1, while refining their model architecture and training objectives towards predicting binding affinity and pose. Jiménez *et al.*[37] utilise a similar input format but adopt a different typing scheme that captures property types (*e.g.* hydrophobic, hydrogen-bond donor or acceptor, aromatic, positive or negative ionisable, metallic) rather than atom types directly.

These methods have reported state-of-the-art performance when evaluated on the 2016 PDBbind core set,[37] as well as performing strongly in the D3R Grand Challenge 3[108] with one CNN-based entry placing highly for several targets.[21]

8.6 Visualisation

Deep learning methods have frequently been criticised for being uninterpretable.[109] In addition to prioritising compounds for experimental screening, an important potential use of scoring functions is for molecular optimisation. This requires knowledge about the parts of the molecule that are responsible for binding. As such, high predictive performance alone is not enough; it is crucial to understand why predictions are being made.

Visualisation techniques can help examine the drivers for the predictions of a CNN. Ragoza *et al.*[19] proposed a method based on masking atoms and residues in the protein and ligand, and using the change in score from their CNN as a proxy for the importance of a given atom or residue for binding. Hochuli *et al.*[20] examined both the masking method and a gradient-based method in further detail, while introducing a third method, conserved layer-wise relevance propagation (CLRP), a refinement of layer-wise relevence propagation.[110,111] The authors claim these three methods are somewhat orthogonal. They claim masking shows how the network scores changes to a molecule, gradients how the network wants to change the molecule, and CLRP how the network scores the unchanged molecule. Finally, we took a machine learning approach to visualisation and replaced the final layer of the model (the classifier) with a $1 \times 1 \times 1$ convolutional layer with the same parameters as the trained classifier.[35] This transforms the final layer from a global classifier into a regional classifier and allows specific regions of the image to be assessed. A score is produced for each sub-section, rather than a single score for the entire complex.

All methods are able to highlight important atoms of the ligand, which can then be targeted for modification by medicinal chemistry. However, the different visualisation techniques often do not identify the same atoms as important, despite using the same trained model. In addition, the visualisations only provide an indication that given atoms are important, rather than directly providing guidance as to why those atoms were important and what modifications should be made. As such, substantial work remains in order to understand the predictions of CNNs. This is an open problem for the broader machine learning community, and represents an active area of research.[112]

Finally, this is not a problem that is isolated to CNN-based methods. Indeed Sheridan discusses several issues that arise from using ligand-based QSAR models to produce atom-by-atom contributions,[113] as well as speculating that any atom-by-atom view of activity could be too granular.

8.7 Outlook

There have been several recent applications of CNNs to the scoring of protein–ligand complexes, following the progress in computer vision achieved through the development of CNNs. This has resulted in state-of-the-art performance on existing benchmarks in virtual screening,[19,35] pose prediction,[20] and binding affinity prediction.[37] This is very promising as the application of these techniques is nascent and substantial room for improvement exists.

In particular, we believe that the development of CNNs for protein–ligand scoring will rely on integrating domain-specific knowledge with the advances in computer vision. An early example of this was provided in Imrie *et al.*[35] where we utilised knowledge of the differences between protein families and an understanding of the current limitations of docking, together with improvements to the CNN architecture utilised.

The prominence of machine learning methods in cheminformatics will necessitate new and improved training and benchmarking sets in order to train more powerful models and more robustly assess methods. Two recent studies have suggested that inadequacies in benchmarks have allowed machine learning methods to perform strongly, despite not utilising features that are known to be important.[54,114] In particular, both studies show limited degradation in performance from removing all information about the protein receptor, both in training and at test time.

There are many potential challenges for deep learning virtual screening methods. Here we will highlight three factors that differentiate virtual screening from traditional computer vision tasks. Firstly, there are typically very limited amounts of data available for a single protein target. The methods described in this chapter have focused on making predictions for novel targets, utilising no data for the target of interest. This is highly challenging and can be the case at the start of a drug discovery project. However, frequently a small amount of data will be available. Utilising this requires methods that are able to exploit limited data. Some methods have been proposed to address this challenge,[115] but further developments are needed to more effectively harness both off- and on-target data. Secondly, learning to predict on a novel protein target is considered a tougher challenge that requires a greater amount of generalisation than learning to identify a novel image category.[115] Finally, the cost of labeling a new example is much higher than in typical computer vision tasks, such as image recognition. This reduces the ability to increase the number of training examples quickly or on a large scale and requires extensive experimental setup to label any new data point. This increases the importance of learning from limited amounts of data and selecting the most informative data to be labelled.

The true test of methods is in prospective trials rather than retrospective experiments. There has been some use of CNNs in prospective experiments, such as by Sunseri *et al.*[21] in the Drug Design Data Resource (D3R) Grand Challenge 3.[108] They found that their CNN-based scoring function outperformed AutoDock Vina on most tasks, but that the pose prediction target chosen for the challenge, Cathepsin S, was particularly challenging for *de novo* docking. However, the CNN provided best-in-class performance on several virtual screening tasks, underscoring the relevance of deep learning to the field of drug discovery. The true potential of CNNs will only be understood following real-world prospective screens in live drug discovery projects.

References

1. R. Lahana, How many leads from HTS?, *Drug Discovery Today*, 1999, **4**, 447–448.
2. C. S. Ramesha, How many leads from HTS? – Comment, *Drug Discovery Today*, 2000, **5**, 43–44.
3. S. P. Leelananda and S. Lindert, Computational methods in drug discovery, *Beilstein J. Org. Chem.*, 2016, **12**, 2694–2718.
4. G. Klebe, Virtual ligand screening: strategies, perspectives and limitations, *Drug Discovery Today*, 2006, **11**, 580–594.
5. J. Lyu, S. Wang, T. E. Balius, I. Singh, A. Levit, Y. S. Moroz, M. J. O'Meara, T. Che, E. Algaa, K. Tolmachova, A. A. Tolmachev, B. K. Shoichet, B. L. Roth and J. J. Irwin, Ultra-large library docking for discovering new chemotypes, *Nature*, 2019, **566**, 224–229.
6. P. Siedlecki, R. G. Boy, T. Musch, B. Brueckner, S. Suhai, F. Lyko and P. Zielenkiewicz, Discovery of two novel, small-molecule inhibitors of DNA methylation, *J. Med. Chem.*, 2006, **49**, 678–683.
7. R. Kiss, B. Kiss, A. Konczol, F. Szalai, I. Jelinek, V. Laszlo, B. Noszal, A. Falus and G. M. Keseru, Discovery of novel human histamine H4 receptor ligands by large-scale structure-based virtual screening, *J. Med. Chem.*, 2008, **51**, 3145–3153.
8. N. Odolczyk, J. Fritsch, C. Norez, N. Servel, M. F. Da Cunha, S. Bitam, A. Kupniewska, L. Wiszniewski, J. Colas, K. Tarnowski, D. Tondelier, A. Roldan, E. L. Saussereau, P. Melin-Heschel, G. Wieczorek, G. L. Lukacs, M. Dadlez, G. Faure, H. Herrmann, M. Ollero, F. Becq, P. Zielenkiewicz and A. Edelman, Discovery of novel potent ΔF508-CFTR correctors that target the nucleotide binding domain. *EMBO, Mol. Med.*, 2013, **5**, 1484–1501.
9. D. Gau, T. Lewis, L. Mcdermott, P. Wipf, D. Koes and P. Roy, Structure-based virtual screening identifies small molecule inhibitor of the profilin1-actin interaction, *J. Biol. Chem.*, 2017, **293**, 2606–2616.

10. M. Bollini, E. S. Leal, N. S. Adler, M. G. Aucar, G. A. Fernández, M. J. Pascual, F. Merwaiss, D. E. Alvarez and C. N. Cavasotto, Discovery of Novel Bovine Viral Diarrhea Inhibitors Using Structure-Based Virtual Screening on the Envelope Protein E2, *Front. Chem.*, 2018, **6**, 1–10.
11. G. Schneider and H.-J. Böhm, Virtual screening and fast automated docking methods, *Drug Discovery Today*, 2002, **7**, 64–70.
12. A. R. Leach, V. J. Gillet, R. A. Lewis and R. Taylor, Three-Dimensional Pharmacophore Methods in Drug Discovery, *J. Med. Chem.*, 2010, **53**, 539–558.
13. P. Willett, J. M. Barnard and G. M. Downs, Chemical Similarity Searching, *J. Chem. Inf. Model.*, 1998, **38**, 983–996.
14. B. Ramsundar, S. Kearnes, P. Riley, D. Webster, D. Konerding and V. Pande, Massively Multitask Networks for Drug Discovery, 2015, arXiv preprint:1502.02072.
15. Z. Wu, B. Ramsundar, E. N. Feinberg, J. Gomes, C. Geniesse, A. S. Pappu, K. Leswing and V. Pande, MoleculeNet: A Benchmark for Molecular Machine Learning, *Chem. Sci.*, 2018, **9**, 513–530.
16. G. Madhavi Sastry, M. Adzhigirey, T. Day, R. Annabhimoju and W. Sherman, Protein and ligand preparation: parameters, protocols, and influence on virtual screening enrichments, *J. Comput.-Aided Mol. Des.*, 2013, **27**, 221–234.
17. X. Zheng, L. Gan, E. Wang and J. Wang, Pocket-Based Drug Design: Exploring Pocket Space, *AAPS J.*, 2013, **15**, 228–241.
18. O. Trott and A. Olson, AutoDock Vina: improving the speed and accuracy of docking with a new scoring function, efficient optimization and multithreading, *J. Comput. Chem.*, 2010, **31**, 455–461.
19. M. Ragoza, J. Hochuli, E. Idrobo, J. Sunseri and D. R. Koes, Protein-Ligand Scoring with Convolutional Neural Networks, *J. Chem. Inf. Model.*, 2017, **57**, 942–957.
20. J. Hochuli, A. Helbling, T. Skaist, M. Ragoza and D. R. Koes, Visualizing convolutional neural network protein–ligand scoring, *J. Mol. Graphics Modell.*, 2018, **84**, 96–108.
21. J. Sunseri, J. E. King, P. G. Francoeur and D. R. Koes, Convolutional neural network scoring and minimization in the D3R 2017 community challenge, *J. Comput.-Aided Mol. Des.*, 2019, **33**, 19–34.
22. H. J. Böhm, The development of a simple empirical scoring function to estimate the binding constant for a protein–ligand complex of known three-dimensional structure, *J. Comput.-Aided Mol. Des.*, 1994, **8**, 243–256.
23. M. D. Eldridge, C. W. Murray, T. R. Auton, G. V. Paolini and R. P. Mee, Empirical scoring functions: I. The development of a fast empirical scoring function to estimate the binding affinity of ligands in receptor complexes, *J. Comput.-Aided Mol. Des.*, 1997, **11**, 425–445.
24. R. A. Friesner, J. L. Banks, R. B. Murphy, T. A. Halgren, J. J. Klicic, D. T. Mainz, M. P. Repasky, E. H. Knoll, M. Shelley, J. K. Perry, D. E. Shaw, P. Francis and P. S. Shenkin, Glide: A New Approach for

Rapid, Accurate Docking and Scoring. 1. Method and Assessment of Docking Accuracy, *J. Med. Chem.*, 2004, **47**, 1739–1749.
25. D. R. Koes, M. P. Baumgartner and C. J. Camacho, Lessons learned in empirical scoring with smina from the CSAR 2011 benchmarking exercise, *J. Chem. Inf. Model.*, 2013, **53**, 1893–1904.
26. H. Gohlke, M. Hendlich and G. Klebe, Knowledge-based scoring function to predict protein–ligand interactions, *J. Mol. Biol.*, 2000, **295**, 337–356.
27. Y. Huang, B. Niu, Y. Gao, L. Fu and W. Li, CD-HIT Suite: A web server for clustering and comparing biological sequences, *Bioinformatics*, 2010, **26**, 680–682.
28. H. Zhou and J. Skolnick, GOAP: A generalized orientation-dependent, all-atom statistical potential for protein structure prediction, *Biophys. J.*, 2011, **101**, 2043–2052.
29. R. Wang, Y. Lu and S. Wang, Comparative Evaluation of 11 Scoring Functions for Molecular Docking, *J. Med. Chem.*, 2003, **46**, 2287–2303.
30. N. Huang, C. Kalyanaraman, K. Bernacki and M. P. Jacobson, Molecular mechanics methods for predicting protein–ligand binding, *Phys. Chem. Chem. Phys.*, 2006, **8**, 5166–5177.
31. H. M. Berman, J. Westbrook, Z. Feng, G. Gilliland, T. N. Bhat, H. Weissig, I. N. Shindyalov and P. E. Bourne, The Protein Data Bank, *Nucleic Acids Res.*, 2000, **28**, 235–242.
32. S. Miyazawa and R. L. Jernigan, Estimation of effective interresidue contact energies from protein crystal structures: quasi-chemical approximation, *Macromolecules*, 1985, **18**, 534–552.
33. M. J. Sippl, Calculation of conformational ensembles from potentials of mena force: An approach to the knowledge-based prediction of local structures in globular proteins, *J. Mol. Biol.*, 1990, **213**, 859–883.
34. L. P. Pason and C. A. Sotriffer, Empirical Scoring Functions for Affinity Prediction of Protein-ligand Complexes, *Mol. Inf.*, 2016, **35**, 541–548.
35. F. Imrie, A. R. Bradley, M. van der Schaar and C. M. Deane, Protein Family-specific Models using Deep Neural Networks and Transfer Learning Improve Virtual Screening and Highlight the Need for More Data, *J. Chem. Inf. Model.*, 2018, **58**, 2319–2330.
36. P. J. Ballester and J. B. O. Mitchell, A machine learning approach to predicting protein–ligand binding affinity with applications to molecular docking, *Bioinformatics*, 2010, **26**, 1169–1175.
37. J. Jiménez, M. Škalič, G. Martínez-Rosell and G. De Fabritiis, KDEEP: Protein-Ligand Absolute Binding Affinity Prediction via 3D-Convolutional Neural Networks, *J. Chem. Inf. Model.*, 2018, **58**, 287–296.
38. J. D. Durrant and J. A. McCammon, NNScore 2.0: A neural-network receptor-ligand scoring function, *J. Chem. Inf. Model.*, 2011, **51**, 2897–2903.

39. M. Wójcikowski, P. J. Ballester and P. Siedlecki, Performance of machine-learning scoring functions in structure-based virtual screening, *Sci. Rep.*, 2017, 7, 46710.
40. H. Li, K.-S. Leung, M.-H. Wong and P. J. Ballester, Substituting random forest for multiple linear regression improves binding affinity prediction of scoring functions: Cyscore as a case study, *BMC Bioinf.*, 2014, **15**, 291.
41. H. Li, K.-S. Leung, M.-H. Wong and P. J. Ballester, The Importance of the Regression Model in the Structure-Based Prediction of Protein-Ligand Binding. *Computational Intelligence Methods for Bioinformatics and Biostatistics. CIBB 2014*, *Lecture Notes Comput. Sci.*, 2015, **8623**, 219–230.
42. T. Cheng, X. Li, Y. Li, Z. Liu and R. Wang, Comparative Assessment of Scoring Functions on a Diverse Test Set, *J. Chem. Inf. Model.*, 2009, **49**, 1079–1093.
43. G. Jones, P. Willett, R. C. Glen, A. R. Leach and R. Taylor, Development and validation of a genetic algorithm for flexible docking, *J. Mol. Biol.*, 1997, **267**, 727–748.
44. R. A. Friesner, R. B. Murphy, M. P. Repasky, L. L. Frye, J. R. Greenwood, T. A. Halgren, P. C. Sanschagrin and D. T. Mainz, Extra Precision Glide: Docking and Scoring Incorporating a Model of Hydrophobic Enclosure for Protein-Ligand Complexes, *J. Med. Chem.*, 2006, **49**, 6177–6196.
45. M. M. Mysinger, M. Carchia, J. J. Irwin and B. K. Shoichet, Directory of useful decoys, enhanced (DUD-E): Better ligands and decoys for better benchmarking, *J. Med. Chem.*, 2012, **55**, 6582–6594.
46. J. D. Durrant and J. A. McCammon, NNScore: A Neural-Network-Based Scoring Function for the Characterization of Protein-Ligand Complexes, *J. Chem. Inf. Model.*, 2010, **50**, 1865–1871.
47. I. Wallach, M. Dzamba and A. Heifets, AtomNet: A Deep Convolutional Neural Network for Bioactivity Prediction in Structure-based Drug Discovery, 2015, arXiv preprint:1510.02855.
48. J. Gomes, B. Ramsundar, E. N. Feinberg and V. S. Pande, Atomic Convolutional Networks for Predicting Protein-Ligand Binding Affinity, 2017, arXiv preprint:1703.10603.
49. A. P. Bento, A. Gaulton, A. Hersey, B. Al-Lazikani, D. Michalovich, J. Chambers, L. J. Bellis, M. Davies, S. McGlinchey, Y. Light and J. P. Overington, ChEMBL: a large-scale bioactivity database for drug discovery, *Nucleic Acids Res.*, 2011, **40**, D1100–D1107.
50. N. Lagarde, J.-F. Zagury and M. Montes, Benchmarking Data Sets for the Evaluation of Virtual Ligand Screening Methods: Review and Perspectives, *J. Chem. Inf. Model.*, 2015, **55**, 1297–1307.
51. J. J. Irwin and B. K. Shoichet, ZINC – A Free Database of Commercially Available Compounds for Virtual Screening, *J. Chem. Inf. Model.*, 2005, **45**, 177–182.
52. G. W. Bemis and M. A. Murcko, The Properties of Known Drugs. 1. Molecular Frameworks, *J. Med. Chem.*, 1996, **39**, 2887–2893.

53. L. Chaput, J. Martinez-Sanz, N. Saettel and L. Mouawad, Benchmark of four popular virtual screening programs: construction of the active/decoy dataset remains a major determinant of measured performance, *J. Cheminf.*, 2016, **8**, 56.
54. J. Sieg, F. Flachsenberg and M. Rarey, In Need of Bias Control: Evaluating Chemical Data for Machine Learning in Structure-Based Virtual Screening, *J. Chem. Inf. Model.*, 2019, **59**, 947–961.
55. S. M. Vogel, M. R. Bauer and F. M. Boeckler, DEKOIS: Demanding Evaluation Kits for Objective in Silico Screening – A Versatile Tool for Benchmarking Docking Programs and Scoring Functions, *J. Chem. Inf. Model.*, 2011, **51**, 2650–2665.
56. M. R. Bauer, T. M. Ibrahim, S. M. Vogel and F. M. Boeckler, Evaluation and Optimization of Virtual Screening Workflows with DEKOIS 2.0 – A Public Library of Challenging Docking Benchmark Sets, *J. Chem. Inf. Model.*, 2013, **53**, 1447–1462.
57. T. Liu, Y. Lin, X. Wen, R. N. Jorissen and M. K. Gilson, BindingDB: a web-accessible database of experimentally determined protein–ligand binding affinities, *Nucleic Acids Res.*, 2006, **35**, D198–D201.
58. S. G. Rohrer and K. Baumann, Maximum unbiased validation (MUV) data sets for virtual screening based on PubChem bioactivity data, *J. Chem. Inf. Model.*, 2009, **49**, 169–184.
59. A. Souvorov, D. Landsman, D. J. Lipman, D. M. Church, D. A. Benson, D. R. Maglott, E. Sequeira, E. Yaschenko, G. D. Schuler, G. Starchenko, J. Ostell, K. Sirotkin, K. Canese, K. D. Pruitt, L. Y. Geer, L. Wagner, M. Shumway, M. DiCuccio, M. Feolo, O. Khovayko, R. L. Tatusov, R. Edgar, S. Federhen, S. H. Bryant, S. T. Sherry, T. Barrett, T. A. Tatusova, T. L. Madden, V. Miller, V. Chetvernin, W. Helmberg, Y. Kapustin and D. L. Wheeler, Database resources of the National Center for Biotechnology Information, *Nucleic Acids Res.*, 2007, **36**, D13–D21.
60. P. Tiikkainen, P. Tiikkainen, P. Markt, P. Markt, G. Wolber, G. Wolber, J. Kirchmair, J. Kirchmair, S. Distinto, S. Distinto, A. Poso, A. Poso, O. Kallioniemi and O. Kallioniemi, Critical comparison of virtual screening methods against the MUV data set, *J. Chem. Inf. Model.*, 2009, **49**, 2168–2178.
61. J. Xia, H. Jin, Z. Liu, L. Zhang and X. S. Wang, An Unbiased Method To Build Benchmarking Sets for Ligand-Based Virtual Screening and its Application To GPCRs, *J. Chem. Inf. Model.*, 2014, **54**, 1433–1450.
62. J. Xia, E. L. Tilahun, E. H. Kebede, T.-E. Reid, L. Zhang and X. S. Wang, Comparative Modeling and Benchmarking Data Sets for Human Histone Deacetylases and Sirtuin Families, *J. Chem. Inf. Model.*, 2015, **55**, 374–388.
63. J. Xia, T.-E. Reid, S. Wu, L. Zhang and X. S. Wang, Maximal Unbiased Benchmarking Data Sets for Human Chemokine Receptors and Comparative Analysis, *J. Chem. Inf. Model.*, 2018, **58**, 1104–1120.
64. E. A. Gatica and C. N. Cavasotto, Ligand and Decoy Sets for Docking to G Protein-Coupled Receptors, *J. Chem. Inf. Model.*, 2012, **52**, 1–6.

65. M. J. McGregor and P. V. Pallai, Clustering of Large Databases of Compounds: Using the MDL "Keys" as Structural Descriptors, *J. Chem. Inf. Model.*, 1997, **37**, 443–448.
66. S. Riniker and G. A. Landrum, Open-source platform to benchmark fingerprints for ligand-based virtual screening, *J. Cheminf.*, 2013, **5**, 26.
67. K. Heikamp and J. Bajorath, Large-scale similarity search profiling of ChEMBL compound data sets, *J. Chem. Inf. Model.*, 2011, **51**, 1831–1839.
68. G. Landrum, *RDKit: Open-source cheminformatics*, http://www.rdkit.org/.
69. J. Konc and D. Janežič, ProBiS algorithm for detection of structurally similar protein binding sites by local structural alignment, *Bioinformatics*, 2010, **26**, 1160–1168.
70. R. P. Sheridan, Time-Split Cross-Validation as a Method for Estimating the Goodness of Prospective Prediction, *J. Chem. Inf. Model.*, 2013, **53**, 783–790.
71. J.-F. Truchon and C. I. Bayly, Evaluating Virtual Screening Methods: Good and Bad Metrics for the "Early Recognition" Problem, *J. Chem. Inf. Model.*, 2007, **47**, 488–508.
72. R. D. Clark and D. J. Webster-Clark, Managing bias in ROC curves, *J. Comput.-Aided Mol. Des.*, 2008, **22**, 141–146.
73. R. P. Sheridan, S. B. Singh, E. M. Fluder and S. K. Kearsley, Protocols for Bridging the Peptide to Nonpeptide Gap in Topological Similarity Searches, *J. Chem. Inf. Model.*, 2001, **41**, 1395–1406.
74. A. N. Jain and A. Nicholls, Recommendations for evaluation of computational methods, *J. Comput.-Aided Mol. Des.*, 2008, **22**, 133–139.
75. A. Nicholls, What do we know and when do we know it?, *J. Comput.-Aided Mol. Des.*, 2008, **22**, 239–255.
76. J. Davis and M. Goadrich, The relationship between Precision-Recall and ROC curves, *ICML*, 2006, 233–240.
77. J. Gehring, M. Auli, D. Grangier, D. Yarats and Y. N. Dauphin, Convolutional Sequence to Sequence Learning, *Proc. 34th Int. Conf. Machine Learning (ICML)*, 2017, **70**, 1243–1252.
78. Y. Shen, X. He, J. Gao, L. Deng and G. Mesnil, Learning Semantic Representations Using Convolutional Neural Networks for Web Search, *Proc. 23rd Int. Conf. World Wide Web*, 2014, 373–374.
79. I. Goodfellow, Y. Bengio and A. Courville, *Deep Learning*, MIT Press, http://www.deeplearningbook.org, 2016.
80. D. Yu, H. Wang, P. Chen and Z. Wei, *Mixed Pooling for Convolutional Neural Networks*, Rough Sets and Knowledge Technology, Cham, 2014, pp. 364–375.
81. C.-Y. Lee, P. W. Gallagher and Z. Tu, Generalizing Pooling Functions in Convolutional Neural Networks: Mixed, Gated, and Tree, *Proceedings of the 19th International Conference on Artificial Intelligence and Statistics*, Cadiz, Spain, 2016, pp. 464–472.

82. J. Gu, Z. Wang, J. Kuen, L. Ma, A. Shahroudy, B. Shuai, T. Liu, X. Wang, G. Wang, J. Cai and T. Chen, Recent advances in convolutional neural networks, *Pattern Recogn.*, 2018, **77**, 354–377.
83. V. Nair and G. E. Hinton, Rectified Linear Units Improve Restricted Boltzmann Machines, *Proceedings of the 27th International Conference on International Conference on Machine Learning*, USA, 2010, pp. 807–814.
84. A. L. Maas, A. Y. Hannun and A. Y. Ng, Rectifier nonlinearities improve neural network acoustic models, in *ICML Workshop on Deep Learning for Audio, Speech and Language Processing*, 2013.
85. S. Ioffe and C. Szegedy, Batch Normalization: Accelerating Deep Network Training by Reducing Internal Covariate Shift, *Proceedings of the 32nd International Conference on International Conference on Machine Learning – Volume 37*, 2015, pp. 448–456.
86. S. Santurkar, D. Tsipras, A. Ilyas and A. Madry, in Advances in Neural Information Processing Systems 31, ed. S. Bengio, H. Wallach, H. Larochelle, K. Grauman, N. Cesa-Bianchi and R. Garnett, Curran Associates, Inc., 2018, pp. 2483–2493.
87. D. E. Rumelhart, G. E. Hinton and R. J. Williams, Learning representations by back-propagating errors, *Nature*, 1986, **323**, 533–536.
88. D. P. Kingma and J. Ba Adam, A Method for Stochastic Optimization, *3rd International Conference on Learning Representations, ICLR 2015, May 7–9, 2015*, Conference Track Proceedings, San Diego, CA, USA, 2015.
89. S. Ruder, An overview of gradient descent optimization algorithms, *CoRR*, 2016, abs/1609.04747.
90. K. Fukushima, Neocognitron: A self-organizing neural network model for a mechanism of pattern recognition unaffected by shift in position, *Biol. Cybern.*, 1980, **36**, 193–202.
91. Y. Lecun, L. Bottou, Y. Bengio and P. Haffner, Gradient-based learning applied to document recognition, *Proc. IEEE*, 1998, **86**, 2278–2324.
92. Y. LeCun, P. Haffner, L. Bottou and Y. Bengio Object Recognition with Gradient-Based Learning. *Shape, Contour and Grouping in Computer Vision*, London, UK, 1999, p. 319.
93. S. Lazebnik, C. Schmid and J. Ponce, Beyond Bags of Features: Spatial Pyramid Matching for Recognizing Natural Scene Categories, *IEEE Conference on Computer Vision and Pattern Recognition (CVPR)*, 2006, pp. 2169–2178.
94. L. Itti, C. Koch and E. Niebur, A model of saliency-based visual attention for rapid scene analysis, *IEEE Trans. Pattern Anal. Mach. Intell.*, 1998, **20**, 1254–1259.
95. O. Russakovsky, J. Deng, H. Su, J. Krause, S. Satheesh, S. Ma, Z. Huang, A. Karpathy, A. Khosla, M. Bernstein, A. C. Berg and L. Fei-Fei, ImageNet Large Scale Visual Recognition Challenge, *Int. J. Comput. Vis.*, 2015, **115**, 211–252.
96. J. Hu, L. Shen and G. Sun Squeeze-and-Excitation Networks, *The IEEE Conference on Computer Vision and Pattern Recognition (CVPR)*, 2017.

97. A. Krizhevsky, I. Sutskever and G. E. Hinton, in Advances in Neural Information Processing Systems 25, ed. F. Pereira, C. J. C. Burges, L. Bottou and K. Q. Weinberger, *Curran Associates, Inc.*, 2012, pp. 1097–1105.
98. C. Szegedy, W. Liu, Y. Jia, P. Sermanet, S. E. Reed, D. Anguelov, D. Erhan, V. Vanhoucke and A. Rabinovich, Going Deeper with Convolutions. *IEEE Conference on Computer Vision and Pattern Recognition (CVPR)*, abs/1409.4842.
99. C. Szegedy, S. Ioffe, V. Vanhoucke and A. Alemi, *Inception-v4, Inception-ResNet and the Impact of Residual Connections on Learning*, 2017.
100. K. He, X. Zhang, S. Ren and J. Sun, Deep Residual Learning for Image Recognition, *The IEEE Conference on Computer Vision and Pattern Recognition (CVPR)*, 2016.
101. G. Huang, Z. Liu, K. Q. Weinberger and L. van der Maaten, Densely Connected Convolutional Networks, *The IEEE Conference on Computer Vision and Pattern Recognition (CVPR)*, 2017.
102. L. Schrödinger, *The PyMOL Molecular Graphics System*, Version 1.2r3pre.
103. A. P. Bento, A. Gaulton, A. Hersey, L. J. Bellis, J. Chambers, M. Davies, F. A. Krüger, Y. Light, L. Mak, S. McGlinchey, M. Nowotka, G. Papadatos, R. Santos and J. P. Overington, The ChEMBL bioactivity database: An update, *Nucleic Acids Res.*, 2014, **42**, 1083–1090.
104. G. A. Ross, G. M. Morris and P. C. Biggin, One size does not fit all: The limits of structure-based models in drug discovery, *J. Chem. Theory Comput.*, 2013, **9**, 4266–4274.
105. Y. Wang, Y. Guo, Q. Kuang, X. Pu, Y. Ji, Z. Zhang and M. Li, A comparative study of family-specific protein–ligand complex affinity prediction based on random forest approach, *J. Comput.-Aided Mol. Des.*, 2015, **29**, 349–360.
106. J. B. Dunbar, R. D. Smith, C.-Y. Yang, P. M.-U. Ung, K. W. Lexa, N. A. Khazanov, J. A. Stuckey, S. Wang and H. A. Carlson, CSAR Benchmark Exercise of 2010: Selection of the Protein–Ligand Complexes, *J. Chem. Inf. Model.*, 2011, **51**, 2036–2046.
107. Z. Liu, M. Su, L. Han, J. Liu, Q. Yang, Y. Li and R. Wang, Forging the Basis for Developing Protein–Ligand Interaction Scoring Functions, *Acc. Chem. Res.*, 2017, **50**, 302–309.
108. Z. Gaieb, C. D. Parks, M. Chiu, H. Yang, C. Shao, W. P. Walters, M. H. Lambert, N. Nevins, S. D. Bembenek, M. K. Ameriks, T. Mirzadegan, S. K. Burley, R. E. Amaro and M. K. Gilson, D3R Grand Challenge 3: blind prediction of protein–ligand poses and affinity rankings, *J. Comput.-Aided Mol. Des.*, 2019, **33**, 1–18.
109. Z. C. Lipton, The Mythos of Model Interpretability, *Queue*, 2018, **16**, 30:31–30:57.
110. S. Bach, A. Binder, G. Montavon, F. Klauschen, K.-R. Müller and W. Samek, On Pixel-Wise Explanations for Non-Linear Classifier Decisions by Layer-Wise Relevance Propagation, *PLoS One*, 2015, **10**, 1–46.

111. S. Lapuschkin, A. Binder, G. Montavon, K.-R. Müller and W. Samek, TheLRP Toolbox for Artificial Neural Networks, *J. Mach. Learn. Res.*, 2016, **17**, 1–5.
112. S. Carter, Z. Armstrong, L. Schubert, I. Johnson and C. Olah, Activation Atlas, *Distill 2019*, https://distill.pub/2019/activation-atlas.
113. R. P. Sheridan, Interpretation of QSAR Models by Coloring Atoms According to Changes in Predicted Activity: How Robust Is It?, *J. Chem. Inf. Model.*, 2019, **59**, 1324–1337.
114. L. Chen, A. Cruz, S. Ramsey, C. J. Dickson, J. S. Duca, V. Hornak, D. R. Koes and T. Kurtzman, Hidden bias in the DUD-E Dataset Leads to Misleading Performance of Deep Learning in Structure-Based Virtual Screening, 2019, https://chemrxiv.org/articles/Hidden_Bias_in_the_DUD-E_Dataset_Leads_to_Misleading_Performance_of_Deep_Learning_in_Structure-Based_Virtual_Screening/7886165/1.
115. H. Altae-Tran, B. Ramsundar, A. S. Pappu and V. Pande, Low Data Drug Discovery with One-Shot Learning, *ACS Cent. Sci.*, 2017, **3**, 283–293.

CHAPTER 9

Machine Learning in the Area of Molecular Dynamics Simulations

SHUZHE WANG AND SEREINA RINIKER*

Department of Chemistry and Applied Biosciences, ETH Zürich, Vladimir-Prelog-Weg 2, 8093 Zürich, Switzerland
*Email: sriniker@ethz.ch

9.1 Introduction

9.1.1 Basics of Molecular Dynamics

Molecular dynamics (MD) is a simulation technique used to study the time evolution of complex systems and obtain insights that are not accessible experimentally. With increasing computational power, MD has become a useful tool in drug-discovery pipelines to explore both thermodynamics (*e.g.* free energy of binding) and kinetics (*e.g.* on/off rates).[1–5]

Classical MD involves the numerical integration of Newton's equation of motion,

$$\vec{f}_i = m_i \vec{a}_i, \tag{9.1}$$

where $\vec{f}_i$ is the force acting on atom i, m_i is its mass and $\vec{a}_i$ its acceleration. Atoms are thereby treated as point charges in space with an associated mass and size. The forces acting on the atoms are recalculated in small time intervals (usually 1–2 femtoseconds) and are derived from a pre-defined

Drug Discovery Series No. 75
Artificial Intelligence in Drug Discovery
Edited by Nathan Brown

Published by the Royal Society of Chemistry, www.rsc.org

potential-energy function U, which is a function of the position of all particles $\vec{R}$,

$$\begin{aligned} U(\vec{R}) = & \underbrace{\sum_{\text{bonds}} k_i^{\text{bond}}(r_i - r_{i,0})^2}_{U_{\text{bonds}}} + \underbrace{\sum_{\text{angles}} k_i^{\text{angle}}(\theta_i - \theta_{i,0})^2}_{U_{\text{angles}}} \\ & + \underbrace{\sum_{\text{dihederal}} k_i^{\text{dihederal}}(1 + \cos(m_i\phi_i + \delta_i))}_{U_{\text{dihederals}}} \\ & + \underbrace{\sum_i \sum_{j \neq i} 4\varepsilon_{ij}^{LJ}\left(\left(\frac{\sigma_{ij}^{LJ}}{r_{ij}}\right)^{12} - \left(\frac{\sigma_{ij}^{LJ}}{r_{ij}}\right)^{6}\right) + \sum_i \sum_{j \neq i} \frac{q_i q_j}{4\pi\varepsilon_0 r_{ij}}}_{U_{\text{non-bonded}}} \end{aligned} \tag{9.2}$$

where k^{bond} and r_0 are the force constant and ideal value of the bond-stretching terms, k^{angle} and θ_0 the force constant and ideal value of the bond-angle bending terms, k^{dihedral}, m and Φ the parameters of the dihedral-angle torsion terms, ε^{LJ} and σ^{LJ} the Lennard–Jones (LJ) parameters to describe the van der Waals interactions, and q the partial charges. As can be seen from eqn (9.2), a fixed-charge force field with this form involves many parameters that need to be fitted empirically, either against quantum-mechanical (QM), experimental data, or a combination thereof. Different force-field families have been developed over the past decades for common biologically relevant molecules such as amino acids, nucleic acids, lipids, carbohydrates, water and organic solvents. Due to the vastness of chemical space, the parametrisation of small organic molecules is more difficult. AMBER,[6–8] CHARMM,[9–11] and OPLS[12–14] are all-atom force fields, where all hydrogens are modelled explicitly. The GROMOS[15–18] force fields on the other hand still use a united-atom approach for aliphatic hydrogens, *i.e.* these are combined with their carbon atom into a larger atom. An overview of the four major force-field families, their history and parametrisation strategies are given in ref. 19, further reviews on force fields are *e.g.* ref. 20,21.

Using a classical fixed-charge force field in an MD simulation is a trade-off between speed and accuracy. Running microseconds of simulations of relatively complex biomolecular systems with *ab initio* QM methods is still not feasible in the near future. Nevertheless, researchers are actively exploring various avenues for a more accurate description of the physical interactions. This includes, for example, the improvement of the non-bonded interactions with a better dispersion description,[22] the introduction of off-site charges to account for anisotropic charge distributions[23] or the replacement of the point charges all together with polarizable multipoles,[23,24] as well as accounting for chemical reactions with reactive force fields[25–27] or with QM/MM simulations.[28,29]

9.1.2 Machine-learning Applications

The current wave of machine-learning (ML) applications started with AlexNet in 2012,[30] which significantly outperformed all other algorithms in the ImageNet Large Scale Visual Recognition Challenge.[31] It revitalised the connectism approach and all variants of artificial neural-network (ANNs) architectures, collectively known as deep-learning (DL) approaches. To this day, image processing is still the most highly studied DL application, followed by natural language processing. In both instances, state-of-the-art methods have been successfully adapted to biomedical research, such as disease diagnostics with imaging data[32] or automated representation learning from medical records.[33,34]

Long before DL, ML methods such as Bayesian inference, tree-based models and genetic algorithms have been applied in different areas of cheminformatics and computer-aided drug discovery, and are still widely used today, *e.g.* for quantitative structure–activity relationship (QSAR) and quantitative structure–property relationship (QSPR) models,[35,36] the design of scoring functions for docking,[37,38] or protein folding.[39,40] DL approaches can in some cases improve the performance of ML models in these tasks.[41–44] In addition, DL led to innovations in chemical-reaction planning with reinforcement learning,[45] or focused compound library design with generative models.[46,47] Some of these topics are discussed in other chapters of this book.

The use of ML in computational chemistry in general is an interdisciplinary endeavour, where appropriate domain-specific knowledge needs to be fused with an understanding of the statistical-learning models.

9.1.3 MD and ML

Both ML and MD simulations benefit from the continuing increase in computer power. With the growing size of high-performance computing (HPC) clusters, hundreds of parallel simulations have become possible, and especially the advances in GPU computing permit MD simulations to reach longer time scales and larger systems. GPU computing has also enabled the age of DL since efficient forward/backward propagation on massive computational graphs has become feasible. The availability of ample computer resources allows both the application of more sophisticated ML methods and the performance of large numbers of QM calculations and MD simulations of practically relevant systems. Thus, it has become possible to produce rapidly large datasets for learning from QM or MD data.

In this chapter, we aim to give an overview of recent research exploring the synergies between ML and MD. The primary focus is thereby on approaches involving supervised ML methods. First, we will look at applications of ML in the area of force-field development (without neural network potentials,[48–51] which would require a chapter by themselves). Second, we discuss how ML can be exploited to improve enhanced sampling in MD simulations. Third, approaches are described that take MD data as input to learn thermodynamic and kinetic properties of molecular systems. Finally, we give an outlook on various aspects that make the combination of MD and ML a powerful tool in computational chemistry.

9.2 Using Machine Learning to Improve Force Fields

Parametrising a classical force field is a multi-objective optimisation as the different potential-energy terms contribute simultaneously to the thermodynamic (and kinetic) properties of a molecule. Different sources of reference data (*e.g.* QM calculations, experimental data of pure liquids and/or crystals) and parametrisation strategies have been employed in the past to develop force fields. Ref. 19 gives an overview of the history and parametrisation of the four main force-field families. All of them make use of the concept of "atom types", either only for the LJ parameters or additionally to encode bonded parameters. This means that a chemical element (carbon, oxygen, nitrogen, *etc.*) can have different atom types depending on its chemical environment. In the past, the development cycle of force fields was often multiple years,[19] involving varying degrees of manual work,[52] leading to potential sub-optimal human choices and errors.[53] There is thus an interest by the community to employ ML approaches to improve both the speed of force-field parametrisation as well as to remove as much manual input as possible. With the help of ML, force-field development is aimed at becoming more automatic, systematic and data-driven. In the following, recent work on the use of ML for force-field parameterisation is discussed, grouped by the ML approach that is employed: multi-variate linear regression, Bayesian inference, genetic algorithm, random forest regression, and artificial neural networks.

9.2.1 Multi-variate Linear Regression

Linear models are the simplest ML approaches but are typically easy to interpret and involve relatively few parameters. An example for the use of multi-variate linear regression in force-field parametrisation is ForceBalance,[54] where a regularised least-square regression fit is used to parametrise any of the force-field terms using multiple reference input data (*i.e.* experimental thermodynamic properties and QM calculations) by iteratively reducing the error between the calculated values and the reference data. An advantage of ForceBalance is its modularised design that provides easy-to-use setups for performing an optimisation routine and extensions to different MD and QM packages for carrying out simulations. It has been utilised for the refitting of the AMBER force field,[54] as well as parameter optimisations with novel functional forms.[55–58] ForceBalance is also used in the Open Force Field initiative,[59] which is a collaborative effort of academic and industry partners with the aim to produce more accurate force fields while standardising the workflows to do so along the way, including modules such as property mining,[60] property calculation,[61] and parameter assignment.[62] A GAFF-like force field based on direct chemical environment perception instead of atom types has been reported.[62] It will be further improved with ForceBalance and automated chemical perception exploration in a fully data-driven manner.[63]

9.2.2 Bayesian Inference

In Bayesian inference,[64] the ability of different parameter sets to accurately describe the reference data is estimated using,

$$p(\theta|D) = \frac{p(D|\theta)p(\theta)}{p(D)} \tag{9.3}$$

where θ is the parameter set, and D the available data to fit against. The posterior distribution $p(\theta|D)$ is proportional to the likelihood $p(D|\theta)$ and a prior distribution of the parameter set. Techniques from statistical mechanics such as Markov-chain Monte Carlo or Gibbs sampler can be used to sample the posterior distribution. Based on the maximum likelihood, a parameter set that fits the reference data well can then be obtained. It is also possible for the posterior to indicate more than one region in the parameter space with high probability density, suggesting different parameter combinations are equally performant given the input data. Quantification of uncertainty or confidence level for parameter sets is therefore also possible.[65,66] Such information is absent in other optimisation approaches such as linear regression or genetic algorithms. To efficiently sample the posterior distribution without performing costly simulations for every parameter set, reweighting of a reference simulation can be used to infer the outcome of other parameter sets. In configuration sampling-based reweighting, adequate phase-space overlap with the reference simulation is needed for accurate prediction.[67]

Two examples of using Bayesian inference for parametrising water models have been reported in the past few years.[68,69] Rizzi *et al.*[68] reported TIP4P parameters fitted to density, self-diffusion, and enthalpy, whereas Wu *et al.*[69] applied a hierarchical Bayesian model to fit the TIP5P parameters using experimental diffusivity, density, and radial distribution functions. They obtained a high model uncertainty from the posterior distribution, indicating that more than one parameter set is optimal in describing all three properties.

9.2.3 Genetic Algorithm

Genetic algorithms (GA) are widely used for optimising parameters.[70–75] A basic GA workflow starts with a pool of "reasonable" parameters as the first generation. The set of parameters is ranked by a chosen fitness function and only top-ranked sets survive, *i.e.* are taken to the next step. These sets are subjected to genetic operations, typically recombination and mutation, where the former combines parent sets to obtain different parameters and the latter can add elements of randomness into the child parameters. The ranking and selection of sets followed by performing genetic operations to yield a new generation of parameters are done in an iterative fashion until a chosen convergence criterion is fulfilled.

An application of GA in force-field parametrisation is the tool Paramfit,[76] which can be used to obtain bonded parameters compatible with the AMBER force-field family by optimising against QM reference data. As input for a given molecule, Paramfit requires a set of conformers and the set of bonded parameters to be optimised. First, physically reasonable parameter sets are randomly selected and ranked using a fitness function. In Paramfit, the fitness function is based on the difference between the force field and the QM single-point energy or forces. In the recombination operation, pairs of parameter sets are randomly picked between lower and upper bounds defined by the parent parameter values. In the mutation operation, a proportion of the parameters are altered randomly. Convergence of the GA is reached when the parameter values ranked best remain stable for several iterations. In addition to the GA formalism, Paramfit makes use of interpolation by a simplex algorithm, which was observed to speed up convergence. Paramfit is designed to fit multiple bonded parameters simultaneously instead of sequentially. The authors recovered almost identical torsion parameters for the alanine tetrapeptide system as the original AMBER ff99SB force field.[76] They also provided optimal values for the hyperparameters of the GA such as the recombination and mutation rates, claiming that these allow the global minimum to be found independent of the underlying functional form used. Paramfit is now a part of AmberTools[6] and was employed in the parameterisation of the AMBER lipid14 force field.[7]

In another application of GA, Li *et al.*[77] reported the optimisation of a polarisable force field for methanol. For this, they fitted the electrostatic parameters based on the reproduction of the QM-derived electrostatic potential (ESP) of methanol dimers, and subsequently trained the LJ parameters using the interaction energy between a central methanol molecule with its nearest neighbours. The converged parameters were used to simulate liquid methanol, yielding density and heat of vaporisation values that agreed better with experimental data than using the methanol model in the polarisable force field AMOEBA.[78] Recently, Fracchia *et al.*[79] employed the GA approach to improve the LJ parameters of five monovalent and bivalent ions by exploring the general functional form,

$$U_{\mathrm{LJ}} = \frac{C_1}{(r_{ij}+\theta)^a} + \frac{C_2}{(r_{ij})^b}, \tag{9.4}$$

where a and b are model choices, C_1 and C_2 are linear parameters to be fitted, and θ is the non-linear biasing term. An ensemble of configurations of a single ion in a water cluster was generated. Based on this ensemble, the parameters a, b, C_1 and C_2 were optimised using linear ridge regression, while a GA variant termed differential evolution[80] was employed to fit θ.

In coarse-grained (CG) force fields, multiple atoms (or even molecules) are grouped into spherical beads to reduce the number of particles in the system, and thus to enable more exhaustive sampling. Parametrisation is typically based on atomistic reference simulations and/or fitting to experimental data. As for atomistic force fields, this has involved a substantial amount of manual input in the past. Recently, a GA approach was described to optimise a CG water model with one-to-one mapping (*i.e.* one water molecule is represented by one CG bead) using experimental data.[81]

9.2.4 Random Forest Regression

Random forest (RF) classification and regression models[82] are often competitive in terms of performance to artificial neural networks, while being faster to train and easier to interpret. They have therefore been popular techniques in computer-aided drug discovery for the past decades (see *e.g.* ref. 35, 36, 83–86). A recent application of RF regression in force-field parametrisation is the prediction of partial charges.[87] In most general force fields, partial charges of small organic molecules are extracted from individual QM calculations, which is computationally relatively demanding. In addition, the resulting partial charges are conformation-dependent and with low transferability between similar substructures in different molecules. Bleiziffer *et al.*[87] performed high-level QM calculations for more than 130 000 lead-like molecules selected such that they represent the diverse chemical environments observed in lead-like molecules in CHEMBL[88] and ZINC.[89] The molecules contained the organic elements C, H, N, O, S, P and the four halogens. Two sets of calculations were performed for each molecule, one in a low dielectric environment ($\varepsilon = 4$) and one in a high dielectric environment ($\varepsilon = 80$). Reference partial charges were extracted from the electron density using the density-derived electrostatic and chemical (DDEC) method[90,91] to partition the electron density into point charges. It was found that the partial charges obtained with DDEC were least dependent on the conformation, which is highly desirable for non-polarisable force fields. The environment around an atom was encoded by an atom-centered atom-pairs fingerprint, which was used as input feature vector in the training of the RF models (one for each element and dielectric constant). The partial charges of a new molecule can then be predicted with high speed and transferability. The method scales favourably with molecular size compared to *ab initio* or semi-empirical charging methods.

9.2.5 Artificial Neural Network

Artificial neural networks (ANNs)[92] take inspiration from the connectism approach in animal brains. In ANNs, perceptron units form multiple layers, whereby the input layer (with input features) is connected to various types of hidden layers (some can be non-linear activation layers), which in turn are linked to the output layer. The output is compared to a reference value and the error is back propagated to update all the weights in each layer. By the universal approximation theorem,[93] a finite ANN is capable of approximating continuous functions of arbitrary form. The resurgence of ANNs brought with it ideas to reduce the number of parameters involved, such as the convolutional layer in image processing and recurrent layers for time-series data. ANN models are suitable for representing non-linear, many-body interactions.

Continuing the topic of partial-charge assignment, Nebgen *et al.*[94] trained ANNs to fit partial charge, albeit for a different purpose. Their hierarchical interacting particle neural network (HIP-NN) uses only the atomic number of each atom and the pairwise distances between atoms as input features. The network architecture is based on an interspersed atom-specific layer and inter-atomic interaction layers. The fitting is done against DFT calculations post-processed with a number of charge partitioning schemes (DDEC not included). HIP-NN can be used to predict atomic charges for neutral organic molecules made of carbon, hydrogen, nitrogen, and oxygen. In contrast to the RF-based charge predictor in ref. 87, the charges assigned are dependent on the molecular conformation and thus, they would need to be dynamically updated if used in MD simulations (which has not been tried so far). Overall, the best HIP-NN model is able to predict partial charges with a mean absolute error not exceeding 0.004 e.

The work of Friederich *et al.*[95] illustrates the potential utility of energies predicted by ANN models. They investigated a series of tertiary amines with large substituents by generating conformations in a stochastic manner using random sets of dihedral angles (and removing conformers with clashes) and calculating semi-empirical energies of each conformer. They observed a strong non-linear and non-local dependence of the dihedral angles, which affects the accuracy of the torsional-angle terms if they are assumed to be simply additive in a force field. Classical force fields typically account for some of the correlations by (scaled) 1,4-non-bonded interactions. In ref. 95, the correlated dihedral potential-energy function was approximated using a shallow ANN with 2–3 hidden layers that were trained on QM derived dihedral energies and torsional angles as input. The fully connected nature of the network infers correlations between torsions. It was shown that the ANN-predicted dihedral energies in Monte Carlo simulations were more accurate than when the same molecule was simulated with MD and the GROMOS 54A7 force field[16] (parametrised with ATB[96]).

ANN models have also been used to obtain a more accurate CG representation from atomistic reference simulations. Wang *et al.*[97] developed the CGnet, which is a fully connected ANN trained to reproduce the free-energy surface of the atomistic system. As the free-energy surface is not known *a priori*, the loss function in the ANN is based on the forces, obtained from the ANN-approximated free energy by auto-differentiating with respect to the input coordinates. This ensures that the ANN represents a conservative force field. The authors also discussed the importance of regularising the ANN model by adding high-energy base potentials at unphysical states. This is needed because high-energy configurations are absent from the training dataset and thus, if such configurations are encountered in the simulation, the extrapolation by the ANN will give unreasonable results. With a trained ANN model at hand, the system propagation in the CG phase space is performed with,

$$\vec{x}_{t+\tau}^{CG} = \vec{x}_t^{CG} - \tau \frac{D}{k_B T} \nabla U(\vec{x}_t^{CG}) + \sqrt{2\tau D}\varepsilon, \tag{9.5}$$

where $\vec{x}$ are the coordinates, τ is the timestep, D the diffusion constant of the system, and ε the Gaussian random noise (non-overdamped Langevin dynamics). The authors used the alanine dipeptide in implicit water as test system, whereby the dipeptide was modelled with five backbone atoms needed for the φ/ψ torsional angles. The regularised CGnet successfully reproduced the equilibrium distribution of the backbone torsional angles. Nevertheless, training such a network requires a sufficient amount of training data to sample the free-energy surface (FES) for a particular system. Moreover, a new network needs to be trained for every new system. The ability to transfer the learned knowledge to similar systems would thus be highly beneficial.

Another application of ANN in the area of CG representation is the work of Lemke *et al.*,[98] where the authors attempted to solve the transferability problem in CG models by employing a convolutional network architecture. The test system in this case consisted of a series of glutamic/aspartic acid oligomers, whereby only the carbon atoms of the carboxy group were retained in the CG representation of a monomer. The convolutional layer takes in tetrameric blocks starting from one end of the oligomer, and sliding one monomer at a time to the other end of the oligomer. The aim of the ANN approach was to distinguish reference conformations generated by MD, which obey the Boltzmann distribution (labelled 1), from conformations obtained from a uniform distribution (labelled 0). The ANN outputs a probabilistic value between zero and one, which can be interpreted as the free-energy difference ΔG between the same conformer coming from the Boltzmann distribution and the unphysical uniform distribution. As the unphysical distribution is uniform, the absolute free energy only differs from ΔG by a constant. Thus, the trained ANN can be used in a Monte Carlo scheme, where a new move of a bead will be accepted or rejected based on the difference in absolute free energy between the old and new configuration *via* the Metropolis criterion. It was found that the models trained using shorter oligomers were able to predict the distribution of the end-to-end distance of longer oligomers.

9.2.6 Remarks

The different applications of ML in force-field development discussed above indicate that there is a high potential for such approaches in this area, but they also highlight challenges. One such challenge is the identification of a suitable descriptor for training, which needs to be rotation and translation invariant, and also to be constructed in an efficient manner. A similar challenge arises from the generation of a suitably diverse and large training dataset. These aspects increase in difficulty when extending the approaches beyond simple test systems.

9.3 Improving Sampling in MD Simulations

Next to force-field accuracy, the second major challenge in MD simulations is sampling. The time resolution in classical MD simulations is 1–2 femtoseconds, resulting in millions of time steps to be calculated in order to reach time scales of nanoseconds or microseconds. Many important biological processes occur, however, on even longer time scales of milliseconds to seconds. Therefore, numerous methods have been developed in the past to enhance sampling in MD simulations (see *e.g.* ref. 99–101). Many of these involve system-dependent parameters, calling for approaches to efficiently determine them. In recent years, ML techniques have been explored as a potential solution to address this need.

Apart from methods such as temperature replica exchange[102,103] or accelerated MD,[104] which enhance sampling globally, most techniques focus the sampling in a subspace of the entire phase space of the system. This is typically done along a "reaction coordinate" (also called order parameter or coupling parameter), which can consist of multiple input features or degrees of freedom. The choice of the reaction coordinate is system (and problem) dependent. For example, it could be a series of torsional angles in a molecule or distances of atoms between two molecules. In essence, enhanced sampling aims to explore the free-energy landscape as a function of the reaction coordinate with the help of a biasing force, such that transitions between free-energy minima occur within reasonable simulation time. The true free-energy landscape is then recovered by reweighting. If many degrees of freedom contribute to a reaction coordinate, the resulting reduced phase space may still be too vast to be effectively explored. This can be addressed by the use of a collective variable that maps the contributing degrees of freedom to a lower manifold.

A detailed overview of all existing enhanced-sampling approaches and their variants is beyond the scope of this section. In the following, three of the fundamental approaches are summarised. In replica-exchange (RE) methods, parallel simulations (replicas) of the system are carried out, with attempts to swap the configurations of neighbouring replicas (Monte Carlo moves) at fixed time intervals. The replicas differ from another in a single parameter. This can, for example, be the temperature,[105] which leads to a global sampling enhancement as mentioned above. The larger thermal fluctuations at the higher temperatures allow the system to cross barriers and visit other free-energy minima, which is not possible at lower replicas in reasonable simulation time. If the replicas differ in a parameter of the

Hamiltonian, the approach is called Hamiltonian RE.[106–111] The sampling is thereby focused on the chosen degree(s) of freedom. Examples here are the force constant of a force-field term,[108,110] or the λ-coupling parameter as used in the free-energy methods thermodynamic integration[112,113] and free-energy perturbation.[114,115]

In umbrella sampling (US),[116] a continuous reaction coordinate is split into discrete points, and separate simulations are performed at each point. To ensure that the simulation samples the chosen value of the reaction coordinate well, a restraining potential is applied. Subsequently, the free-energy profile is obtained by reweighting in order to withdraw the effect of the restraining force (WHAM analysis[117]). Examples are the permeation of a small molecule through a lipid membrane,[118] where simulations are performed at discrete points along the bilayer normal, or host–guest binding,[14] where the guest molecule is simulated at different distances from the host.

Another widely used approach to enhance sampling along a reaction coordinate is local elevation[119] or metadynamics,[120] where a biasing force along the chosen reaction coordinate is dynamically built up by adding Gaussian potentials. The resulting biasing potential is approximating the free-energy profile. The sampling performance can be further improved by a subsequent US simulation with the obtained biasing potential (*i.e.* LEUS[121]).

In the following, approaches are discussed that use ML techniques to either enhance sampling globally or along a given reaction coordinate, or to estimate an optimal collective variable of a system/problem.

9.3.1 General Sampling Enhancement

Sakae *et al.*[122] applied the recombination idea in GA to enhance phase-space exploration of amino-acid oligomers. Parallel simulations were started from a set of seed conformations, and after a predefined number of time steps recombination was performed on pairs of parent conformations by randomly swapping consecutive sets of backbone dihedrals to obtain new child structures. A "propagation stage" first relaxed the dihedral restraints in the new structures before they were being accepted or rejected based on the comparison to the potential energy of the parent structures. As in most GA approaches, the design of the recombination step is highly system-dependent and not straightforward to transfer to other systems.

Another approach proposed in the literature is called COCO-MD,[123] where parallel simulations are performed followed by principal component analysis (PCA), a dimension-reduction method, to identify poorly sampled regions after projecting all the frames to a lower-dimensional manifold. Subsequently, simulations are initiated in these regions and a new iteration round is started.

The use of Bayesian inference was explored for the inclusion of "weak information" to steer the sampling in protein-folding simulations.[124] As the timescale for folding processes is typically beyond the reach of MD, most successful approaches are currently information-centric, *i.e.* based on statistical learning from existing crystal structures in databases. Perez *et al.*[124] took a physics-centric approach instead and tested it on globular proteins. The statistical information from the PDB[125] was thereby included as "weak information" in MD simulations, *i.e.* as long-range (*e.g.* the proportion of hydrophobic contacts) and short-range (*e.g.* from secondary-structure motifs) harmonic restraints. The decision to apply the restraints for a given configuration was made through Bayesian inference: the probability of a protein fold given a set of weak information is proportional to the joint probability of the fold with the set of weak information. The procedure was set up as an iterative process. First, diverse sampling was obtained by using Hamiltonian-RE (with biasing force from "weak information") and temperature-RE simulations. After a chosen number of time steps, the set of weak information that would add the least amount of strain to the system was activated and a new set of RE simulations was carried out. The authors tested the approach on 20 small globular proteins containing up to 96 residues. In all cases, they recovered the native fold within 500 ns of simulation, and for 15 of the 20 proteins the native fold was within the top five most stable clusters. Force-field defects were found to be a major source of error, as some native structures were deemed to be unstable by the force field. In addition, it is not possible to observe the formation of disulfide bridges with classical force fields.

9.3.2 Estimating the Biasing Potential for a Given Reaction Coordinate

Galvelis and Sugita[126] developed a sophisticated workflow to learn the optimal biasing potential for a given set of input features. First, a subset of frames from an initial simulation were extracted at regular intervals and nearest-neighbour density clustering of the frames was performed to estimate the free energy as a function of the input features. Next, an ANN was trained to approximate the free-energy surface. Based on this, a new biasing potential was derived and another ANN was trained to approximate the biasing potential. The latter step was needed as a way to regularise erroneous extrapolation of the free-energy surface by the first ANN. Finally, the output of the second ANN (*i.e.* biasing forces given the input features) was employed in a new simulation. The whole procedure has been wrapped in an iterative scheme where the ANNs are retrained after a chosen number of time steps. The workflow was tested on six small polypeptides with selected torsions as the reaction coordinate. Accurate free energies were obtained up to the alanine pentapeptide, for which an eight-dimensional biasing potential was used (*i.e.* all backbone torsions). On the other hand, the free-energy profile for the tryptophan tripeptide was poorly converged, even though the same amount of torsions (eight) was selected for biasing. The authors hypothesised that this was due to the system having more minima separated by higher energy barriers. In general, the construction of the biasing potential becomes less efficient with increasing number of dimensions. Moreover, the scheme requires the optimisation of additional hyperparameters such as number of nearest neighbours, a maximum value of the biasing potential, and a minimum value of the free energy. The alternative methods of Sidky *et al.*[127] and Guo *et al.*[128] learn the adaptive biasing forces in a similar way. In contrast to the method of Galvelis *et al.*,[126] both alternative techniques utilise a network architecture that includes Bayesian regularisation to prevent overfitting. While the former relies on gradual updated estimate of free energy to bias the simulation, the latter instead relies on on-the-fly estimates of the free-energy derivative.

The reinforced dynamics scheme of Zhang *et al.*[129] takes inspiration from reinforcement learning, which aims to find the optimal policy function: the best action at a given state. For sampling in MD simulations, the optimal policy function is the inverted FES. Applying this as the biasing force allows the exploration of all states evenly. The scheme works in an iterative fashion. First, multiple ANNs are independently trained (with different initial parameters) on the potential of mean force (PMF) at state space points obtained from biased simulations. The predictions of all ANNs are treated as a form of statistical confidence (similar to ideas in NN potential works[51]): state space points where the ANNs do not give similar predictions are deemed to be undersampled. Better sampling of these points is then achieved by adding more bias to regions where all ANN give similar prediction. Reinforced dynamics scheme was showcased on the polyalanine-10 system, where the final FES learnt by the ANN ensemble agreed well with values obtained by thermodynamic integration at metastable states of the peptide.

9.3.3 Estimating Optimal Collective Variables

As the performance of an enhanced-sampling approach depends heavily on the chosen reaction coordinate, the use of ML techniques is being explored to estimate the optimal collective variable of a system/problem to be used as reaction coordinate.

Sultan *et al.*[130] proposed a supervised ML method to accelerate sampling between known end states (*e.g.* two protein conformations). A support vector machine (SVM) classifier was trained using short MD simulations of the end-states, resulting in an optimal plane separating the end-states. The signed distance of a given conformation from the separation plane can then be used as an effective collective variable. Applying a biasing potential along this collective variable enhances transitions between the end-states. When more than two end-states need to be probed, the distance of a conformation to every separation plane can be used as the collective variable. The authors tested the approach on the (un)folded states of the decamer chignolin and found that 100 ns of biased simulations with the determined collective variable were sufficient to induce transitions. They also observed that the collective variables obtained with logistic regression (*i.e.* a linear model) or ANN models did not outperform SVM.

The variational dynamics encoder (VDE),[131] which is based on the autoencoder idea,[132] can be employed to learn the optimal collective variable.[133] An autoencoder consists of two connected networks: encoder and decoder.

The encoder takes the vector of input features and performs non-linear transformations in the hidden layers before outputting an encoded vector of lower dimension. The decoder then takes the encoded vector as input and aims to reconstruct the input-feature vector with minimum error. Thus, the aim of an autoencoder is to learn a lower dimensional representation (referred to as the latent representation) that best encapsulates the input features. The optimised network can generate encoded vectors of the input features with maximum information content. The VDE approach is an autoencoder variant, which incorporates the time information in the MD trajectories. First, the input features at time t are transformed into an encoded vector. But instead of having the decoder reconstruct the same snapshot, the decoder is trained to reproduce the input-feature vector at time $t+\tau$. This way, a low dimensional latent representation that describes the dynamics of the system can be obtained. The trained VDE was applied to seed conformations, from which biasing forces were obtained along the direction indicated by the low dimensional vector. The authors tested the VDE technique on a relatively large system, the FIP protein WW domains, for which long simulation trajectories were available from other studies.[134] They observed the unfolded, misfolded and folded states with faster sampling using only a 1D latent variable, which they obtained from TICA featurisation[135] of a set of 184 input features composed of backbone contact distances. In addition, they showed how the same VDE model can be used to enhance the sampling of a mutant analogue GTT[136] without the need for retraining.

Tiwary and co-workers[137] developed an approach using autoencoder termed reweighted autoencoded variational Bayes for enhanced sampling (RAVE). RAVE is an iterative scheme, which uses the autoencoder technique to learn the parameters of a latent Gaussian distribution $P(z;\mu,\sigma)$ that best reconstructs the Boltzmann probability distribution as a function of the input features: $P(x_1,x_2,\ldots)$. Instead of directly using the latent distribution as a 1D collective variable, it is transcribed into a more interpretable form. In order to do this, trial probability distributions from linear combinations of the input features $P(\chi=c_1x_1+c_2x_2+\ \ldots)$ are constructed with different weights $(c_1,c_2\ldots)$. The distribution with the optimal weights is the one that minimises the Kullback–Leibler divergence between $P(z)$ and $P(\chi)$. These weights can then be used to construct the collective variable, along which the biasing potential is applied for a chosen number of time steps. The procedure is iterated until the optimal weights are converged. RAVE was employed to estimate the binding free-energy profile of a hydrophobic ligand–substrate system in explicit water and showed at least twenty-fold speed-up compared to simulations with US or metadynamics.

Ribeiro *et al.*[138] extended the RAVE approach to multi-RAVE by using the factorisation of the conditional probability based on the previous collective variable. The Boltzmann probability distribution as a function of the input features $P(x_1,x_2,\ \ldots)$ is progressively better approximated by including more multiplicative terms $P_1(\chi_1)P_2(\chi_2)\ldots$, where χ_i indicates the ith collective variable and $P_i(\chi_i)$ is the ith probability distribution whose target phase space is not captured by $P_{i-1}(\chi_i - 1)$. The additional collective variables are obtained using alternative linear combinations of the input features (with different sets of weights) or using new input features. Multi-RAVE allows the incorporation of additional collective variables when the first collective variable fails to sample the desired rare event, even though sampling appears to be converged (*i.e.* the system is trapped at intermediate metastable state). The authors demonstrated the potential of the approach by inducing the unbinding of benzene from lysozyme within tens of nanoseconds of simulation instead of the actual expected timescale of hundreds of microseconds.

A different method by Tiwary and co-workers[139,140] was termed spectral gap optimization of order parameters (SGOOP), where the weights of the probability distribution (see discussion of RAVE above) are optimised using the spectral gap from Markov state models.[141] Employing the same conditional probability idea as in multi-RAVE, multi-SGOOP was proposed in 2018.[140] Another approach based on the autoencoder technique was proposed by Ferguson and co-workers.[142,143] The test system consisted of the alanine dipeptide and the Trp-cage miniprotein. An improved autoencoder architecture was used that involves three distinct aspects: (1) Use of circular bottleneck layers in the autoencoder network. This yielded significant computational savings when learning periodic collective variables. (2) Inclusion of system-specific knowledge into the design of the error function in the decoder reconstruction step. The authors observed a broadening of the configurational spaced that was explored. (3) Adaption of a hierarchical autoencoder, *i.e.* replacing the one-encoder-one-decoder with a one-encoder-multiple-decoder architecture, where each decoder sees different subsets of the input features. This allows the ranking of collective variables based on their physical importance.

9.4 Learning from MD Trajectories

When an MD simulation is performed, the output is a trajectory with the 3D coordinates and the energetics of the molecular system, with neighbouring frames being time-correlated. There is thus a vast amount of information about a system contained in such trajectories, which can be exploited by supervised ML methods. For this, the data needs to be extracted in such a manner that it can be used as input for an ML model. Note that this is typically done as a post-processing step, thus the MD engine and the ML software can be largely decoupled. This is in contrast to the previous sections, where often an efficient interface between the MD engine and the ML software is needed. In the following, we discuss different approaches proposed in the literature to extract the information from MD trajectories for ML, ordered by the amount and type of information used.

9.4.1 Application to Clustering

Clustering of frames in an MD trajectory is a widely used analysis technique that belongs to the unsupervised ML methods. A recent summary of unsupervised ML methods in MD simulations can be found in ref. 144. Numerous clustering algorithms have been proposed in the literature.[145–153] They can be broadly grouped into partitioning algorithms such as *K*-means[148] and *K*-centers,[149] hierarchical algorithms such as Ward or landmark agglomerative clustering,[150] and density-based algorithms such as JP (Jarvis–Patrick)[151] clustering, DBSCAN (density-based spatial clustering of applications with noise),[152] and CNN (common nearest neighbour)[153] clustering. The most common distance metric used for clustering in the context of MD simulations is the atom-positional root-mean-square deviation (RMSD). A discussion of all clustering algorithms is beyond the scope of this section, the focus is instead on approaches where supervised ML techniques are incorporated to improve clustering performance.

Brandt *et al.*[154] used a supervised XGBoost tree classifier to cluster conformations for two systems: villin headpiece (HP-35) with dihedral angles as input features, and T4 lysozyme with inter-residue distances as input features. The classifier was trained on structures from a set of predefined metastable states from density-based clustering. New MD frames were assigned to one of these metastable states using the classifier. In addition to the metastable state labels, the tree classifier indicated which features in the input vector were most important for the assignment to a specific metastable state.

Botlani *et al.*[155] used ML techniques to study allosteric modulations by quantifying the inter-residue correlations in 250 ns MD simulations of the apo form and the regulator-bound form of the PDZ2 domain. Using affinity propagation,[156] residues were first clustered based on their levels of correlation. The clusters obtained from both forms were found to be similar, indicating that the spatial arrangement of the strongly interacting residue clusters is not affected by binding. Next, the authors connected spatially close residues by weighted edges in a graph, where the weights were the inverse of motion correlation. By calculating the shortest weighted path between all pairs of residues in both forms, the authors could show that the global signaling paths changed significantly upon binding. Additionally, to quantify how shifts in the conformation space upon binding are correlated for pairs of residues, conformations common to both apo and regulator-bound forms have to be removed. This was achieved by training of an SVM classifier using all available data to distinguish apo and bound conformations, and subsequently removing all conformers that were predicted to be common to both states by the classifier. The results suggested that the regulatory signal weakens with distance from the regulatory site.

9.4.2 Application to Property Prediction

The aim of property prediction is to learn a generalisable model from a set of known compounds, which can then be applied to new molecules. This usually involves the design or selection of descriptors of the system to use as input for the training of a supervised ML model. Alternatively, it is also possible to use ANNs that take directly 3D structural data as input and determine hidden mappings to a given property.

9.4.2.1 Prediction of Binding Affinity

Ash and Fourches[157] developed an MD-based fingerprint to train a QSAR model to predict the activity of small-molecule inhibitors to kinase ERK2. The inhibitors were individually docked to ERK2 and each protein-inhibitor complex was simulated for 20 ns. For a given ligand pose obtained from the simulation, two 3D-shape descriptors were calculated: (i) 3D-D moments, which purely encode molecular geometrics *via* inter-atomic distances (ultrafast shape recognition[158]), and (ii) 3D-WHIM,[159] which is a weighted scheme based on different chemical information such as atomic mass, van der Waals volumes and electronegativities. These descriptors were calculated for 400 frames obtained at regular intervals from the simulation, and subsequently the average and standard deviation were calculated for each component to give the final fingerprints. From the set of 85 ERK2 inhibitors, the authors found that the first two principal components of the feature vector provided a good distinction of active and inactive molecules. This approach outperformed 2D descriptors or 3D descriptors calculated from a single ligand pose.

Most of the ANN approaches taking directly 3D structural data as input consider only a single static configuration (*e.g.* KDEEP,[44] multiscale weighted colored graph (MWCG),[160] or PotentialNet[43]). The use of trajectory data is still rare. One such example is reported by Yakovenko and co-workers.[161] The authors applied their approach to predict binding poses as part of the second D3R grand challenge.[162–164] The ANN model used was composed of a convolutional encoder unit linked to both a convolutional decoder unit and a fully connected network. The convolutional network takes preprocessed pixelated 3D coordinates as input. Given a snapshot of the protein–ligand complex (either after docking or from an MD simulation), 3D mesh grids are generated where each cell in the mesh grid records the density of a certain atomic property, *e.g.* the number of nitrogen atoms, hydrophobic motifs, or polar hydrogen density. The docked poses of 12 known ligands (to the same target protein) were used in the first round of training of the encoder. The aim of the encoder–decoder construct was to learn a condensed representation from the 3D coordinates that can best reconstruct the mesh grids of the protein–ligand complexes obtained from short equilibrium MD simulation. Multiple simulations were run for the same docked pose and were supplied to be reconstructed, in order to better account for the distribution of low-energy conformations of the complexes. The construct of encoder–fully connected network was used to predict the linear interaction energy (LIE)[165] of a docked pose. LIE was used in this work as the reference. The trained ANN model was then applied to all 102 ligands in the competition, where diverse poses were generated from ensemble docking. For each ligand, the pose with the lowest reconstruction error and best predicted LIE was chosen to have the reference LIE calculated and added to the training set. The process iterated for two rounds until no new optimal binding pose was observed for the 102 ligands. The submission achieved the second-best score in the pose prediction challenge of D3R.[161]

9.4.2.2 Prediction of Physicochemical Properties

An early simple approach for using information in MD data was described by Duffy and Jorgensen in 2000.[166] Solvation free energies in water, octanol, and hexadecane as well as octanol/water partition coefficients were estimated using a multi-variate linear regression model based on average properties obtained from Monte Carlo simulations.

A more sophisticated method to encode the information from MD in a fingerprint usable with ML techniques is called MDFPs.[167] The basic idea is to extract not only the average value of a feature from the MD trajectories but include higher statistical moments (*i.e.* variance, skew) in the form of standard deviation and median. The approach was first applied to predict the solvation free energy in different solvents from water to hexadecane as well as partition coefficients in different pairs of solvents. The MDFPs were constructed for a dataset of 643 small molecules, for which short simulations (5 ns) were performed in water. The resulting fingerprints were taken as input-feature vectors for an ensemble composed of a LASSO (*i.e.* linear model) and a gradient tree-boosting model, which were trained against experimental data. The idea of ensemble modelling exploits the fact that different ML methods tend to have different sources of error. Thus, combining different ML methods in heterogeneous fusion models[168] potentially improves the overall prediction accuracy (see *e.g.* ref. 169). The MDFP-ML approach outperformed ML models trained with topological fingerprints and gave a similar performance as rigorous MD-based methods. The underlying idea of MDFPs is highly versatile and can be explored for the prediction of other physicochemical properties as well as binding affinity.

For the prediction of liquid properties, Gong *et al.*[170] proposed a fully automated high-throughput simulation protocol to generate large amounts of liquid and vapour–liquid equilibrium data (*e.g.* isobaric heat capacities, liquid densities). The authors simulated a series of alkanes at multiple pressures and temperatures (using their temperature-tuned force field[171]) to obtain almost 50 000 data points. The subset of compounds, for which experimental data was available, was used to validate the accuracy of the simulated values. Subsequently, an ANN model was trained with all computed data points, which was then able to predict the properties of new alkanes at different temperatures and pressures. In this case, the input-feature vector for a molecule was based only on atom-type information and topological properties.

9.4.3 Application to Kinetic Models

A widely used approach to extract dynamical information from MD trajectories is to construct Markov state models (MSM).[141,172–177] It involves the Markovian assumption that the future does not depend on the past but only on the present. Thus, the transition to other metastable conformational states depends only on the current conformation. An advantage of this approach is that multiple parallel trajectories can be combined to give the final MSM, thus obtaining information about processes that are slower than those directly observed in one of the input trajectories. The construction of an MSM involves two clustering steps: first structural clustering into so-called microstates followed by kinetic clustering into the metastable sets (or macrostates). Both steps can be improved by the use of ML techniques. For the structural clustering, examples include the VDE[131] approach discussed above as well as the similar time-lagged autoencoder[178] method. The two methods differ in the design and dimension of the NN architecture and the latent variables. VDE utilises furthermore backpropagation of the network for saliency mapping, adding more interpretability to the model.

VAMPnets[179] offer a more streamlined and automated protocol to construct MSMs by encapsulating all stages of the process into one NN. VAMPnet aims to learn the optimal operator χ, which maps n-dimensional feature vectors to (usually smaller) m-dimensional probability vectors, where m is the desired number of metastable states. VAMPnet applies the variational approach for Markov processes (hence the acronym) by maximising the variational score based on the Koopman operator,[180] which is a function of χ. The network takes as input pairs of feature vectors x_t and $x_{t+\tau}$ (where τ is a user-defined parameter) and progressively optimises the variational score. Note that VAMPnet still requires tuning of the network architecture to the problem of interest as well as chemical intuition for selecting the number of metastable states. In addition, no statistical error estimation has yet been incorporated into the framework, and the black-box nature of ANNs makes the debugging of a VAMPnet challenging.

Since the first publication of VAMPnet in 2018, multiple efforts have been made to enhance the approach in various ways: Husic and Noé[181] have introduced the idea of deflation as a systematic way of removing uninteresting yet slow processes during the variational optimisation, allowing the subsequent MSM to focus on important modes of a system. Xie *et al.*[182] have incorporated graph convolutional architectures into VAMPnets, which improves the invariant encoding of local chemical environments, and they have applied it to material design. The work on state-free reversible VAMPnets by Chen *et al.*[183] led to networks with higher success rate in identifying eigenfunctions of the transfer operator. Wu *et al.*[184] extended the VAMPnet approach with a generator to form deep generative MSMs, with which it is possible to predict time-series of physically meaningful configurations unseen in the dataset. As the amount of data measured/simulated is continuing to grow with increasing speed, the data-driven approaches coupled with NNs have become popular to study complex dynamic systems.[185]

9.5 Perspectives and Challenges

Many crossovers between methodologies and techniques described in the different sections are possible. An example is the work of Brandt *et al.*,[154] where clustering of conformers with a RF model also determines key input features, which can be employed as reaction coordinates for enhanced sampling. A second example is the autoencoder approach, which is a popular tool for enhanced sampling as well as post-processing of MD trajectories. Furthermore, Schneider *et al.*[186] illustrated how ANN models can extract a high-dimensional free-energy surface from MD trajectories. Their approach shows similarity with the development of the CGnet.[97] While the focus of the former is to use the ANNs to further sample the systems, the emphasis of the latter is to derive a compact representation of the high-dimensional FES to obtain ensemble averages of different properties. In general, many of the ideas presented in this chapter aim to handle the curse of dimensionality by learning more compact representations of the input data. In the following, we discuss some of the future directions, opportunities, needs and challenges that we see after surveying the literature.

9.5.1 Datasets on Dynamics Information

A relatively large amount of "static" data is available in the literature, which is used *e.g.* for force-field development and enhanced sampling. This includes structures and energies from high-level QM calculations (*e.g.* Alexdria library,[187] Quantum-Machine,[188] or ANI-1[50]), as well as experimental structures and activity data (*e.g.* PDB,[125] CSD,[189] PDBBind,[190] BindingDB,[191] or ChEMBL[88]). In contrast, little computational or experimental data is available on dynamics. One exception is the microseconds MD trajectories of protein folding from DE Shaw research,[134,136,192] which have been used by other research groups for learning latent representation or testing MSM methods.[131,133,178,179] Such efforts to provide publicly available MD datasets are highly desirable for the community, as the time and resources used to generate trajectories can be better invested elsewhere. However, one difficulty with this idea is certainly the large storage space needed for a large number of long MD trajectories (*e.g.* many of the trajectories from DE Shaw Research are without the solvent coordinates). Additionally, it is crucial for utility that metadata accompanies every stored trajectory, *i.e.* information about the topology, starting coordinates, simulation protocol, input parameters and conditions. For this, standards for recording and parsing metadata fields will be needed.

9.5.2 Benchmarking

Different research groups often validate their new methods on different test systems, making a direct comparison between techniques difficult. One approach to directly compare methods is to organise a blind prediction challenge. Examples for such community-wide blind challenges are CASP[193] for crystal structure prediction, CAPRI[194] for protein–protein docking, the D3R challenges[164] for the prediction of binding poses, binding affinity for new protein targets as well as host–guest systems, and the SAMPL challenges[195–198] for the prediction of physicochemical properties of small molecules. While MD-based methods were rather weakly represented in such blind challenges in the past, we anticipate a wider participation in the future with the emergence of combined MD and ML approaches.

Another possibility for method comparison is the creation of standard benchmarking sets, as commonly done in other fields such as image and language deep learning.[31,199] Some efforts for benchmarking sets have been made in the last year (GuacaMol[200] and MoleculeNet[201]) for drug discovery related tasks. Time will tell if these are taken up by the community.

9.5.3 Open-source Implementation

Although the number of publications in computational chemistry that include source code is growing, it is not yet standardly done. If the original source code is not available, methods have to be re-implemented in order to reproduce the results and to apply the method on a new data set, which is time consuming and can be frustrating if important details were accidentally not reported. However, in many cases it is not sufficient to simply release source code, instead what is needed is re-usable code, which includes thorough documentation. In the area of applying deep learning to molecular systems, packages such as SchNetPack[202] and deeptime[179] are examples of good practices.

For the combination of MD and ML, a special challenge can be to create an efficient interface between the methods. Python is a suitable programming language for such a task due to the availability of a wide range of utility libraries. However, while many popular ML libraries have comprehensive Python APIs,[203–206] most MD software packages use their own file formats and work as command-line tools,[207–209] which complicates the intercommunication between ML tasks and MD tasks in a combined approach. One exception is the OpenMM MD engine,[210] which has extensive Python APIs for parsing various input formats, including analytical custom force expressions to define the biasing potential in simulations, and running the simulations. An additional plugin also enables the use of NNs *via* custom force expressions.[211] Efforts are currently being made to create Python handles for GROMACS.[212] Furthermore, developments of end-to-end differentiable MD engines are underway.[213,214]

The Molecular Sciences Software Institute (MOLSSI)[215] is actively improving the usability of computational chemistry programs. For selected software packages, MOLSSI designates experts to work alongside scientists to improve their code quality. We anticipate that this will positively impact the scientific software landscape in computational chemistry in the future.

9.5.4 Concluding Remarks

The advances in the field of ML will continue to impact computational chemistry and trigger approaches to learn from and for MD. ANN models may help to improve the accuracy of the description of the physical interactions in simulations, without adding unfeasible computational costs. Reinforcement learning ideas[216] can lend a hand to pathway exploration in enhanced-sampling simulations. Graph convolutional NN models[217,218] may offer a natural representation of MD trajectories for learning, while transfer learning[219] and multi-task learning[220] techniques could be employed when only small datasets are available for training. Facilitated by the easy implementation of new inductive biases,[221] alternative clever ideas may emerge that allow us to harness the information contained in MD simulations further and learn from it.

References

1. J. D. Durrant and J. A. McCammon, *BMC Biol.*, 2011, **9**, 71.
2. D. W. Borhani and D. E. Shaw, *J. Comput.-Aided Mol. Des.*, 2011, **26**, 15.
3. M. J. Harvey and G. D. Fabritiis, *Drug Discovery Today*, 2012, **17**, 1059.
4. M. D. Vivo, M. Masetti, G. Bottegoni and A. Cavalli, *J. Med. Chem.*, 2016, **59**, 4035.
5. A. Ganesan, M. L. Coote and K. Barakat, *Drug Discovery Today*, 2017, **22**, 249.
6. J. P. M. Jämbeck and A. P. Lyubartsev, *J. Phys. Chem. B*, 2014, **118**, 3793.
7. C. J. Dickson, B. D. Madej, Å. A. Skjevik, R. M. Betz, K. Teigen, I. R. Gould and R. C. Walker, *J. Chem. Theory Comput.*, 2014, **10**, 865.
8. V. Hornak, R. Abel, A. Okur, B. Strockbine, A. Roitberg and C. Simmerling, *Proteins: Struct., Funct., Bioinf.*, 2006, **65**, 712.
9. A. D. MacKerell, D. Bashford, M. Bellott, R. L. Dunbrack, J. D. Evanseck, M. J. Field, S. Fischer, J. Gao, H. Guo, S. Ha, D. Joseph-McCarthy, L. Kuchnir, K. Kuczera, F. T. K. Lau, C. Mattos, S. Michnick, T. Ngo, D. T. Nguyen, B. Prodhom, W. E. Reiher, B. Roux, M. Schlenkrich, J. C. Smith, R. Stote, J. Straub, M. Watanabe, J. Wiórkiewicz-Kuczera, D. Yin and M. Karplus, *J. Phys. Chem. B*, 1998, **102**, 3586.
10. K. Vanommeslaeghe, E. Hatcher, C. Acharya, S. Kundu, S. Zhong, J. Shim, E. Darian, O. Guvench, P. Lopes, I. Vorobyov and A. D. Mackerell, *J. Comput. Chem.*, 2009, **31**, 671.
11. K. Vanommeslaeghe and A. MacKerell, *Biochim. Biophys. Acta, Gen. Subj.*, 2015, **1850**, 861.
12. D. Kony, W. Damm, S. Stoll and W. F. van Gunsteren, *J. Comput. Chem.*, 2002, **23**, 1416.
13. E. Harder, W. Damm, J. Maple, C. Wu, M. Reboul, J. Y. Xiang, L. Wang, D. Lupyan, M. K. Dahlgren, J. L. Knight, J. W. Kaus, D. S. Cerutti, G. Krilov, W. L. Jorgensen, R. Abel and R. A. Friesner, *J. Chem. Theory Comput.*, 2015, **12**, 281.

14. K. Roos, C. Wu, W. Damm, M. Reboul, J. M. Stevenson, C. Lu, M. K. Dahlgren, S. Mondal, W. Chen, L. Wang, R. Abel, R. A. Friesner and E. D. Harder, *J. Chem. Theory Comput.*, 2019, **15**, 1863.
15. J. Hermans, H. J. C. Berendsen, W. F. van Gunsteren and J. P. M. Postma, *Biopolymers*, 1984, **23**, 1513.
16. N. Schmid, A. P. Eichenberger, A. Choutko, S. Riniker, M. Winger, A. E. Mark and W. F. van Gunsteren, *Eur. Biophys. J.*, 2011, **40**, 843.
17. M. M. Reif, P. H. Hünenberger and C. Oostenbrink, *J. Chem. Theory Comput.*, 2012, **8**, 3705.
18. B. A. C. Horta, P. T. Merz, P. F. J. Fuchs, J. Dolenc, S. Riniker and P. H. Hünenberger, *J. Chem. Theory Comput.*, 2016, **12**, 3825.
19. S. Riniker, *J. Chem. Inf. Model.*, 2018, **58**, 565.
20. P. Dauber-Osguthorpe and A. T. Hagler, *J. Comput.-Aided Mol. Des.*, 2018, **33**, 133.
21. A. T. Hagler, *J. Comput.-Aided Mol. Des.*, 2018, **33**, 205.
22. J. Hermann, R. A. DiStasio and A. Tkatchenko, *Chem. Rev.*, 2017, **117**, 4714.
23. G. A. Cisneros, M. Karttunen, P. Ren and C. Sagui, *Chem. Rev.*, 2013, **114**, 779.
24. H. Yu and W. F. van Gunsteren, *J. Chem. Phys.*, 2004, **121**, 9549.
25. A. C. T. van Duin, S. Dasgupta, F. Lorant and W. A. Goddard, *J. Phys. Chem. A*, 2001, **105**, 9396.
26. P. Pahari and S. Chaturvedi, *J. Mol. Model.*, 2011, **18**, 1049.
27. T. P. Senftle, S. Hong, M. M. Islam, S. B. Kylasa, Y. Zheng, Y. K. Shin, C. Junkermeier, R. Engel-Herbert, M. J. Janik, H. M. Aktulga, T. Verstraelen, A. Grama and A. C. T. van Duin, *npj Comput. Mater.*, 2016, **2**, 15011.
28. M. M. Islam, G. Kolesov, T. Verstraelen, E. Kaxiras and A. C. T. van Duin, *J. Chem. Theory Comput.*, 2016, **12**, 3463.
29. H. M. Senn and W. Thiel, *Angew. Chem., Int. Ed.*, 2009, **48**, 1198.
30. A. Krizhevsky, I. Sutskever and G. E. Hinton, Imagenet classification with deep convolutional neural networks, in *Advances in Neural Information Processing Systems 25*, ed. F. Pereira, C. J. C. Burges, L. Bottou and K. Q. Weinberger, Curran Associates, Inc., 2012, p. 1097.
31. J. Deng, W. Dong, R. Socher, L.-J. Li, K. Li and L. Fei-Fei, "ImageNet: A large-scale hierarchical image database", in *2009 IEEE Conference on Computer Vision and Pattern Recognition*. IEEE, 2009.
32. G. Litjens, T. Kooi, B. E. Bejnordi, A. A. A. Setio, F. Ciompi, M. Ghafoorian, J. A. van der Laak, B. van Ginneken and C. I. Sánchez, *Med. Image Anal.*, 2017, **42**, 60.
33. E. Choi, M. T. Bahadori, L. Song, W. F. Stewart and J. Sun, GRAM: Graph-based attention model for healthcare representation learning, *Proceedings of the 23rd ACM SIGKDD International Conference on Knowledge Discovery and Data Mining KDD '17*. ACM Press, 2017.
34. M. Rotmensch, Y. Halpern, A. Tlimat, S. Horng and D. Sontag, *Sci. Rep.*, 2017, **7**, 5994.

35. R. P. Sheridan, *J. Chem. Inf. Model.*, 2012, **52**, 814.
36. N. Basant, S. Gupta and K. P. Singh, *J. Chem. Inf. Model.*, 2015, **55**, 1337.
37. P. J. Ballester and J. B. O. Mitchell, *Bioinformatics*, 2010, **26**, 1169.
38. S. L. Kinnings, N. Liu, P. J. Tonge, R. M. Jackson, L. Xie and P. E. Bourne, *J. Chem. Inf. Model.*, 2011, **51**, 408.
39. I.-F. Chung, C.-D. Huang, Y.-H. Shen and C.-T. Lin, Recognition of structure classification of protein folding by NN and SVM hierarchical learning architecture, *Artificial Neural Networks and Neural Information Processing — ICANN/ICONIP 2003*, Springer, Berlin Heidelberg, 2003, p. 1159.
40. J. Cheng, A. Tegge and P. Baldi, *IEEE Rev. Biomed. Eng.*, 2008, **1**, 41.
41. T. Jo, J. Hou, J. Eickholt and J. Cheng, *Sci. Rep.*, 2015, **5**, 17573.
42. S. Wang, S. Sun, Z. Li, R. Zhang and J. Xu, *PLoS Comput. Biol.*, 2017, **13**, e1005324.
43. E. N. Feinberg, D. Sur, Z. Wu, B. E. Husic, H. Mai, Y. Li, S. Sun, J. Yang, B. Ramsundar and V. S. Pande, *ACS Cent. Sci.*, 2018, **4**, 1520.
44. J. Jiménez, M. Škalić, G. Martínez-Rosell and G. D. Fabritiis, *J. Chem. Inf. Model.*, 2018, **58**, 287.
45. M. H. S. Segler, M. Preuss and M. P. Waller, *Nature*, 2018, **555**, 604.
46. M. Olivecrona, T. Blaschke, O. Engkvist and H. Chen, *J. Cheminf.*, 2017, **9**, 48.
47. A. Gupta, A. T. Müller, B. J. H. Huisman, J. A. Fuchs, P. Schneider and G. Schneider, *Mol. Inf.*, 2017, **37**, 1700111.
48. J. Behler and M. Parrinello, *Phys. Rev. Lett.*, 2007, **98**, 146401.
49. J. Behler, *Int. J. Quantum Chem.*, 2015, **115**, 1032.
50. J. S. Smith, O. Isayev and A. E. Roitberg, *Chem. Sci.*, 2017, **8**, 3192.
51. K. Yao, J. E. Herr, D. W. Toth, R. Mckintyre and J. Parkhill, *Chem. Sci.*, 2018, **9**, 2261.
52. O. Krämer-Fuhrmann, J. Neisius, N. Gehlen, D. Reith and K. N. Kirschner, *J. Chem. Inf. Model.*, 2013, **53**, 802.
53. D. Mobley, C. C. Bannan, A. Rizzi, C. I. Bayly, J. D. Chodera, V. T. Lim, N. M. Lim, K. A. Beauchamp, D. R. Slochower, M. R. Shirts, M. K. Gilson and P. K. Eastman, *J. Chem. Theory Comput.*, 2018, **14**, 6076.
54. L.-P. Wang, T. J. Martinez and V. S. Pande, *J. Phys. Chem. Lett.*, 2014, **5**, 1885.
55. X. Mu, Q. Wang, L.-P. Wang, S. D. Fried, J.-P. Piquemal, K. N. Dalby and P. Ren, *J. Phys. Chem. B*, 2014, **118**, 6456.
56. S. I. L. K. Schumacher, E. G. Hohenstein, R. M. Parrish, L.-P. Wang and T. J. Martínez, *J. Chem. Theory Comput.*, 2015, **11**, 3042.
57. A. D. Wade, L.-P. Wang and D. J. Huggins, *J. Chem. Inf. Model.*, 2018, **58**, 1766.
58. K. A. McKiernan, L.-P. Wang and V. S. Pande, *J. Chem. Theory Comput.*, 2016, **12**, 5960.
59. "Open force field initiative," https://openforcefield.org/ (accessed March 2019).

60. K. A. Beauchamp, J. M. Behr, A. S. Rustenburg, C. I. Bayly, K. Kroenlein and J. D. Chodera, *J. Phys. Chem. B*, 2015, **119**, 12912.
61. G. D. R. Matos, D. Y. Kyu, H. H. Loeffler, J. D. Chodera, M. R. Shirts and D. L. Mobley, *J. Chem. Eng. Data*, 2017, **62**, 1559.
62. D. L. Mobley, C. C. Bannan, A. Rizzi, C. I. Bayly, J. D. Chodera, V. T. Lim, N. M. Lim, K. A. Beauchamp, D. R. Slochower, M. R. Shirts, M. K. Gilson and P. K. Eastman, *J. Chem. Theory Comput.*, 2018, **14**, 6076.
63. C. Zanette, C. C. Bannan, C. I. Bayly, J. Fass, M. K. Gilson, M. R. Shirts, J. D. Chodera and D. L. Mobley, *J. Chem. Theory Comput.*, 2018, **15**, 402.
64. J. Vallverdú, *Bayesians Versus Frequentists: A Philosophical Debate on Statistical Reasoning (SpringerBriefs in Statistics Book 0)*, Springer, 2015.
65. F. Cailliez and P. Pernot, *J. Chem. Phys.*, 2011, **134**, 054124.
66. R. A. Messerly, T. A. Knotts and W. V. Wilding, *J. Chem. Phys.*, 2017, **146**, 194110.
67. R. A. Messerly, S. M. Razavi and M. R. Shirts, *J. Chem. Theory Comput.*, 2018, **14**, 3144.
68. F. Rizzi, H. N. Najm, B. J. Debusschere, K. Sargsyan, M. Salloum, H. Adalsteinsson and O. M. Knio, *Multiscale Model. Simul.*, 2012, **10**, 1460.
69. S. Wu, P. Angelikopoulos, G. Tauriello, C. Papadimitriou and P. Koumoutsakos, *J. Chem. Phys.*, 2016, **145**, 244112.
70. J. Grefenstette, *IEEE Trans. Syst. Man Cybern.*, 1986, **16**, 122.
71. G. E. Liepins and M. R. Hilliard, *Ann. Oper. Res.*, 1989, **21**, 31.
72. H. Lam, S. Ling, F. Leung and P. Tam, Tuning of the structure and parameters of neural network using an improved genetic algorithm, in *IECON'01. 27th Annual Conference of the IEEE Industrial Electronics Society (Cat. No.37243)*, IEEE, Denver, CO, USA, 2002.
73. M. Gen and M. Yoo, Real time tasks scheduling using hybrid genetic algorithm, in *Studies in Computational Intelligence*, Springer, Berlin Heidelberg, 2008, p. 319.
74. Z.-Y. Yin, Y.-F. Jin, S.-L. Shen and H.-W. Huang, *Acta Geotech.*, 2016, **12**, 849.
75. M. Mohammadi, M. Lakestani and M. Mohamed, *Energy*, 2018, **143**, 56.
76. R. M. Betz and R. C. Walker, *J. Comput. Chem.*, 2015, **36**, 79.
77. Y. Li, H. Li, F. C. Pickard, B. Narayanan, F. G. Sen, M. K. Y. Chan, S. K. R. S. Sankaranarayanan, B. R. Brooks and B. Roux, *J. Chem. Theory Comput.*, 2017, **13**, 4492.
78. J. W. Ponder, C. Wu, P. Ren, V. S. Pande, J. D. Chodera, M. J. Schnieders, I. Haque, D. L. Mobley, D. S. Lambrecht, R. A. DiStasio, M. Head-Gordon, G. N. I. Clark, M. E. Johnson and T. Head-Gordon, *J. Phys. Chem. B*, 2010, **114**, 2549.
79. F. Fracchia, G. Del Frate, G. Mancini, W. Rocchia and V. Barone, *J. Chem. Theory Comput.*, 2018, **14**, 255.
80. R. Storn and K. Price, *J. Global Optim.*, 1997, **11**, 341.
81. H. Chan, M. J. Cherukara, B. Narayanan, T. D. Loeffler, C. Benmore, S. K. Gray and S. K. R. S. Sankaranarayanan, *Nat. Commun.*, 2019, **10**, 379.

82. C. Sheppard, *Tree-based Machine Learning Algorithms: Decision Trees, Random Forests, and Boosting*, CreateSpace Independent Publishing Platform, 2017.
83. A. Rusinko, M. W. Farmen, C. G. Lambert, P. L. Brown and S. S. Young, *J. Chem. Inf. Comput. Sci.*, 1999, **39**, 1017.
84. V. Svetnik, A. Liaw, C. Tong, J. C. Culberson, R. P. Sheridan and B. P. Feuston, *J. Chem. Inf. Comput. Sci.*, 2003, **43**, 1947.
85. W. Tong, H. Hong, H. Fang, Q. Xie and R. Perkins, *J. Chem. Inf. Comput. Sci.*, 2003, **43**, 525.
86. C. L. Bruce, J. L. Melville, S. D. Pickett and J. D. Hirst, *J. Chem. Inf. Model.*, 2007, **47**, 219.
87. P. Bleiziffer, K. Schaller and S. Riniker, *J. Chem. Inf. Model.*, 2018, **58**, 579.
88. A. Gaulton, L. J. Bellis, A. P. Bento, J. Chambers, M. Davies, A. Hersey, Y. Light, S. McGlinchey, D. Michalovich, B. Al-Lazikani and J. P. Overington, *Nucleic Acids Res.*, 2011, **40**, D1100.
89. J. J. Irwin and B. K. Shoichet, *J. Chem. Inf. Model.*, 2005, **45**, 177.
90. T. A. Manz and D. S. Sholl, *J. Chem. Theory Comput.*, 2012, **8**, 2844.
91. T. A. Manz and N. G. Limas, *RSC Adv.*, 2016, **6**, 47771.
92. W. S. McCulloch and W. Pitts, *Bull. Math. Biophys.*, 1943, **5**, 115.
93. B. C. Csáji, *Approximation with Artificial Neural Networks*, Faculty of Sciences, Etvs Lornd University, Hungary, **vol. 24**, 2001, p. 48.
94. B. Nebgen, N. Lubbers, J. S. Smith, A. E. Sifain, A. Lokhov, O. Isayev, A. E. Roitberg, K. Barros and S. Tretiak, *J. Chem. Theory Comput.*, 2018, **14**, 4687.
95. P. Friederich, M. Konrad, T. Strunk and W. Wenzel, *Sci. Rep.*, 2018, **8**, 2559.
96. J. Wang, W. Wang, P. A. Kollman and D. A. Case, *J. Mol. Graphics Modell.*, 2006, **25**, 247.
97. J. Wang, S. Olsson, C. Wehmeyer, A. Pérez, N. E. Charron, G. de Fabritiis, F. Noé and C. Clementi, *ACS Cent. Sci.*, 2019, **5**, 755.
98. T. Lemke and C. Peter, *J. Chem. Theory Comput.*, 2017, **13**, 6213.
99. R. C. Bernardi, M. C. Melo and K. Schulten, *Biochim. Biophys. Acta, Gen. Subj.*, 2015, **1850**, 872.
100. Y. Miao and J. A. McCammon, *Mol. Simul.*, 2016, **42**, 1046.
101. P. Tiwary and A. van de Walle, A review of enhanced sampling approaches for accelerated molecular dynamics, in *Multiscale Materials Modeling for Nanomechanics*, Springer International Publishing, 2016, p. 195.
102. R. H. Swendsen and J.-S. Wang, *Phys. Rev. Lett.*, 1986, **57**, 2607.
103. D. J. Earl and M. W. Deem, *Phys. Chem. Chem. Phys.*, 2005, **7**, 3910.
104. D. Hamelberg, J. Mongan and J. A. McCammon, *J. Chem. Phys.*, 2004, **120**, 11919.
105. Y. Sugita and Y. Okamoto, *Chem. Phys. Lett.*, 1999, **314**, 141.
106. W. L. Jorgensen, *Acc. Chem. Res.*, 1989, **22**, 184.
107. P. Kollman, *Chem. Rev.*, 1993, **93**, 2395.

108. Y. Sugita, A. Kitao and Y. Okamoto, *J. Chem. Phys.*, 2000, **113**, 6042.
109. H. Fukunishi, O. Watanabe and S. Takada, *J. Chem. Phys.*, 2002, **116**, 9058.
110. J. Hritz and C. Oostenbrink, *J. Chem. Phys.*, 2008, **128**, 144121.
111. C. D. Christ, A. E. Mark and W. F. van Gunsteren, *J. Comput. Chem.*, 2009, **31**, 1569.
112. J. G. Kirkwood, *J. Chem. Phys.*, 1935, **3**, 300.
113. J. Kästner, H. M. Senn, S. Thiel, N. Otte and W. Thiel, *J. Chem. Theory Comput.*, 2006, **2**, 452.
114. R. W. Zwanzig, *J. Chem. Phys.*, 1954, **22**, 1420.
115. D. Shivakumar, J. Williams, Y. Wu, W. Damm, J. Shelley and W. Sherman, *J. Chem. Theory Comput.*, 2010, **6**, 1509.
116. J. Kästner, *Wiley Interdiscip. Rev.: Comput. Mol. Sci.*, 2011, **1**, 932.
117. S. Kumar, J. M. Rosenberg, D. Bouzida, R. H. Swendsen and P. A. Kollman, *J. Comput. Chem.*, 1992, **13**, 1011.
118. D. Bochicchio, E. Panizon, R. Ferrando, L. Monticelli and G. Rossi, *J. Chem. Phys.*, 2015, **143**, 144108.
119. T. Huber, A. E. Torda and W. F. van Gunsteren, *J. Comput.-Aided Mol. Des.*, 1994, **8**, 695.
120. A. Laio and M. Parrinello, *Proc. Natl. Acad. Sci. U. S. A.*, 2002, **99**, 12562.
121. H. S. Hansen and P. H. Hünenberger, *J. Comput. Chem.*, 2010, **31**, 1.
122. Y. Sakae, J. E. Straub and Y. Okamoto, *J. Comput. Chem.*, 2018, **40**, 475.
123. A. Shkurti, I. Styliari, V. Balasubramanian, I. Bethune, C. Pedebos, S. Jha and C. Laughton, *J. Chem. Theory Comput.*, 2019, **15**, 2587.
124. A. Perez, J. L. MacCallum and K. A. Dill, *Proc. Natl. Acad. Sci. U. S. A.*, 2015, **112**, 11846.
125. H. M. Berman, *Nucleic Acids Res.*, 2000, **28**, 235.
126. R. Galvelis and Y. Sugita, *J. Chem. Theory Comput.*, 2017, **13**, 2489.
127. H. Sidky and J. K. Whitmer, *J. Chem. Phys.*, 2018, **148**, 104111.
128. A. Z. Guo, E. Sevgen, H. Sidky, J. K. Whitmer, J. A. Hubbell and J. J. de Pablo, *J. Chem. Phys.*, 2018, **148**, 134108.
129. L. Zhang, H. Wang and E. Weinan, *J. Chem. Phys.*, 2018, **148**, 124113.
130. M. M. Sultan and V. S. Pande, *J. Chem. Phys.*, 2018, **149**, 094106.
131. C. X. Hernández, H. K. Wayment-Steele, M. M. Sultan, B. E. Husic and V. S. Pande, *Phys. Rev. E*, 2018, **97**, 062412.
132. D. P. Kingma and M. Welling, 2013, arXiv:1312, 6114.
133. M. M. Sultan, H. K. Wayment-Steele and V. S. Pande, *J. Chem. Theory Comput.*, 2018, **14**, 1887.
134. D. E. Shaw, P. Maragakis, K. Lindorff-Larsen, S. Piana, R. O. Dror, M. P. Eastwood, J. A. Bank, J. M. Jumper, J. K. Salmon, Y. Shan and W. Wriggers, *Science*, 2010, **330**, 341.
135. Y. Naritomi and S. Fuchigami, *J. Chem. Phys.*, 2011, **134**, 065101.
136. S. Piana, K. Sarkar, K. Lindorff-Larsen, M. Guo, M. Gruebele and D. E. Shaw, *J. Mol. Biol.*, 2011, **405**, 43.
137. J. M. L. Ribeiro, P. Bravo, Y. Wang and P. Tiwary, *J. Chem. Phys.*, 2018, **149**, 072301.

138. J. M. L. Ribeiro and P. Tiwary, *J. Chem. Theory Comput.*, 2018, **15**, 708.
139. P. Tiwary and B. J. Berne, *Proc. Natl. Acad. Sci. U. S. A.*, 2016, **113**, 2839.
140. Z. Smith, D. Pramanik, S.-T. Tsai and P. Tiwary, *J. Chem. Phys.*, 2018, **149**, 234105.
141. *An Introduction to Markov State Models and Their Application to Long Timescale Molecular Simulations*, ed. G. R. Bowman, V. S. Pande and F. Noé, Springer, Netherlands, 2014.
142. W. Chen, A. R. Tan and A. L. Ferguson, *J. Chem. Phys.*, 2018, **149**, 072312.
143. W. Chen and A. L. Ferguson, *J. Comput. Chem.*, 2018, **39**, 2079.
144. M. Ceriotti, *J. Chem. Phys.*, 2019, **150**, 150901.
145. A. K. Jain, M. N. Murty and P. J. Flynn, *ACM Comput. Surv.*, 1999, **31**, 264.
146. R. Xu and D. C. Wunsch, *IEEE Rev. Biomed. Eng.*, 2010, **3**, 120.
147. M. R. Anderberg, *Cluster Analysis for Applications: Probability and Mathematical Statistics: A Series of Monographs and Textbooks (Probability & Mathematical Statistics Monograph)*, Academic Press, 2014.
148. S. Lloyd, *IEEE Trans. Inf. Theory*, 1982, **28**, 129.
149. T. F. Gonzalez, *Theor. Comput. Sci.*, 1985, **38**, 293.
150. D. Müllner, 2011, arXiv:1109, 2378.
151. R. Jarvis and E. Patrick, *IEEE Trans. Comput.*, 1973, **C-22**, 1025.
152. M. Ester, H.-P. Kriegel, J. Sander and X. Xu, A density-based algorithm for discovering clusters a density-based algorithm for discovering clusters in large spatial databases with noise, *Proceedings of the Second International Conference on Knowledge Discovery and Data Mining, ser. KDD'96*, AAAI Press, 1996, p. 226.
153. B. Keller, X. Daura and W. F. van Gunsteren, *J. Chem. Phys.*, 2010, **132**, 074110.
154. S. Brandt, F. Sittel, M. Ernst and G. Stock, *J. Phys. Chem. Lett.*, 2018, **9**, 2144.
155. M. Botlani, A. Siddiqui and S. Varma, *J. Chem. Phys.*, 2018, **148**, 241726.
156. B. J. Frey and D. Dueck, *Science*, 2007, **315**, 972.
157. J. Ash and D. Fourches, *J. Chem. Inf. Model.*, 2017, **57**, 1286.
158. P. J. Ballester and W. G. Richards, *J. Comput. Chem.*, 2007, **28**, 1711.
159. R. Todeschini and P. Gramatica, *SAR QSAR Environ. Res.*, 1997, **7**, 89.
160. D. D. Nguyen, Z. Cang, K. Wu, M. Wang, Y. Cao and G.-W. Wei, *J. Comput.-Aided Mol. Des.*, 2018, **33**, 71.
161. O. Yakovenko and S. J. M. Jones, *J. Comput.-Aided Mol. Des.*, 2017, **32**, 299.
162. S. Gathiaka, S. Liu, M. Chiu, H. Yang, J. A. Stuckey, Y. N. Kang, J. Delproposto, G. Kubish, J. B. Dunbar, H. A. Carlson, S. K. Burley, W. P. Walters, R. E. Amaro, V. A. Feher and M. K. Gilson, *J. Comput.-Aided Mol. Des.*, 2016, **30**, 651.
163. Z. Gaieb, S. Liu, S. Gathiaka, M. Chiu, H. Yang, C. Shao, V. A. Feher, W. P. Walters, B. Kuhn, M. G. Rudolph, S. K. Burley, M. K. Gilson and R. E. Amaro, *J. Comput.-Aided Mol. Des.*, 2017, **32**, 1.

164. Z. Gaieb, C. D. Parks, M. Chiu, H. Yang, C. Shao, W. P. Walters, M. H. Lambert, N. Nevins, S. D. Bembenek, M. K. Ameriks, T. Mirzadegan, S. K. Burley, R. E. Amaro and M. K. Gilson, *J. Comput.-Aided Mol. Des.*, 2019, **33**, 1.
165. J. Aqvist and J. Marelius, *Comb. Chem. High Throughput Screening*, 2001, **4**, 613.
166. E. M. Duffy and W. L. Jorgensen, *J. Am. Chem. Soc.*, 2000, **122**, 2878.
167. S. Riniker, *J. Chem. Inf. Model.*, 2017, **57**, 726.
168. S. Riniker, N. Fechner and G. A. Landrum, *J. Chem. Inf. Model.*, 2013, **53**, 2829.
169. S. Wang and S. Riniker, *J. Comput.-Aided Mol. Des.*, 2019, **34**, 393.
170. Z. Gong, Y. Wu, L. Wu and H. Sun, *J. Chem. Inf. Model.*, 2018, **58**, 2502.
171. Z. Gong, H. Sun and B. E. Eichinger, *J. Chem. Theory Comput.*, 2018, **14**, 3595.
172. C. Schütte, A. Fischer, W. Huisinga and P. Deuflhard, *J. Comput. Phys.*, 1999, **151**, 146.
173. W. C. Swope, J. W. Pitera and F. Suits, *J. Phys. Chem. B*, 2004, **108**, 6571.
174. J.-H. Prinz, H. Wu, M. Sarich, B. Keller, M. Senne, M. Held, J. D. Chodera, C. Schütte and F. Noé, *J. Chem. Phys.*, 2011, **134**, 174105.
175. B. Keller, P. Hünenberger and W. F. van Gunsteren, *J. Chem. Theory Comput.*, 2011, **7**, 1032.
176. J. D. Chodera and F. Noé, *Curr. Opin. Struct. Biol.*, 2014, **25**, 135.
177. B. E. Husic and V. S. Pande, *J. Am. Chem. Soc.*, 2018, **140**, 2386.
178. C. Wehmeyer and F. Noé, *J. Chem. Phys.*, 2018, **148**, 241703.
179. A. Mardt, L. Pasquali, H. Wu and F. Noé, *Nat. Commun.*, 2018, **9**, 5.
180. H. Wu and F. Noé, *J. Nonlinear Sci.*, 2010, **30**, 23.
181. B. E. Husic and F. Noé, *J. Chem. Phys.*, 2019, **151**, 054103.
182. T. Xie, A. France-Lanord, Y. Wang, Y. Shao-Horn and J. C. Grossman, *Nat. Commun.*, 2019, **10**, 1.
183. W. Chen, H. Sidky and A. L. Ferguson, *J. Chem. Phys.*, 2019, **150**, 214114.
184. H. Wu, A. Mardt, L. Pasquali and F. Noé, 2018, arXiv:1805, 07601.
185. B. Lusch, J. N. Kutz and S. L. Brunton, *Nat. Commun.*, 2018, 9.
186. E. Schneider, L. Dai, R. Q. Topper, C. Drechsel-Grau and M. E. Tuckerman, *Phys. Rev. Lett.*, 2017, **119**, 150601.
187. M. M. Ghahremanpour, P. J. van Maaren and D. van der Spoel, *Sci. Data*, 2018, **5**, 180062.
188. R. Ramakrishnan, P. O. Dral, M. Rupp and O. A. von Lilienfeld, *Sci. Data*, 2014, **1**, 140022.
189. C. R. Groom, I. J. Bruno, M. P. Lightfoot and S. C. Ward, *Acta Crystallogr., Sect. B: Struct. Sci., Cryst. Eng. Mater.*, 2016, **72**, 171.
190. R. Wang, X. Fang, Y. Lu, C.-Y. Yang and S. Wang, *J. Med. Chem.*, 2005, **48**, 4111.
191. T. Liu, Y. Lin, X. Wen, R. N. Jorissen and M. K. Gilson, *Nucleic Acids Res.*, 2007, **35**, D198.
192. K. Lindorff-Larsen, S. Piana, R. O. Dror and D. E. Shaw, *Science*, 2011, **334**, 517.

193. J. Moult, K. Fidelis, A. Kryshtafovych, T. Schwede and A. Tramontano, *Proteins: Struct., Funct., Bioinf.*, 2013, **82**, 1.
194. J. Janin, K. Henrick, J. Moult, L. T. Eyck, M. J. E. Sternberg, S. Vajda, I. Vakser and S. J. Wodak, *Proteins: Struct., Funct., Genet.*, 2003, **52**, 2.
195. M. T. Geballe, A. G. Skillman, A. Nicholls, J. P. Guthrie and P. J. Taylor, *J. Comput.-Aided Mol. Des.*, 2010, **24**, 259.
196. A. G. Skillman, *J. Comput.-Aided Mol. Des.*, 2012, **26**, 473.
197. D. L. Mobley, K. L. Wymer, N. M. Lim and J. P. Guthrie, *J. Comput.-Aided Mol. Des.*, 2014, **28**, 135.
198. C. C. Bannan, K. H. Burley, M. Chiu, M. R. Shirts, M. K. Gilson and D. L. Mobley, *J. Comput.-Aided Mol. Des.*, 2016, **30**, 927.
199. P. Rajpurkar, J. Zhang, K. Lopyrev and P. Liang, SQuAD: 100,000 questions for machine comprehension of text, in *Proceedings of the 2016 Conference on Empirical Methods in Natural Language Processing*, Association for Computational Linguistics, 2016.
200. N. Brown, M. Fiscato, M. H. S. Segler and A. C. Vaucher, *J. Chem. Inf. Model.*, 2019, **59**, 1096.
201. Z. Wu, B. Ramsundar, E. N. Feinberg, J. Gomes, C. Geniesse, A. S. Pappu, K. Leswing and V. Pande, *Chem. Sci.*, 2018, **9**, 513.
202. K. T. Schütt, P. Kessel, M. Gastegger, K. A. Nicoli, A. Tkatchenko and K.-R. Müller, *J. Chem. Theory Comput.*, 2018, **15**, 448.
203. M. Abadi, A. Agarwal, P. Barham, E. Brevdo, Z. Chen, C. Citro, G. S. Corrado, A. Davis, J. Dean, M. Devin, S. Ghemawat, I. Goodfellow, A. Harp, G. Irving, M. Isard, Y. Jia, R. Jozefowicz, L. Kaiser, M. Kudlur, J. Levenberg, D. Mané, R. Monga, S. Moore, D. Murray, C. Olah, M. Schuster, J. Shlens, B. Steiner, I. Sutskever, K. Talwar, P. Tucker, V. Vanhoucke, V. Vasudevan, F. Viégas, O. Vinyals, P. Warden, M. Wattenberg, M. Wicke, Y. Yu and X. Zheng, 2015, arXiv:1603, 04467. Software available at tensorflow.org.
204. A. Paszke, S. Gross, S. Chintala, G. Chanan, E. Yang, Z. DeVito, Z. Lin, A. Desmaison, L. Antiga and A. Lerer, Automatic differentiation in pytorch, *Proceedings of Neural Information Processing Systems*, 2017.
205. T. Chen, M. Li, Y. Li, M. Lin, N. Wang, M. Wang, T. Xiao, B. Xu, C. Zhang and Z. Zhang, 2015, arXiv:1512.01274.
206. F. Pedregosa, G. Varoquaux, A. Gramfort, V. Michel, B. Thirion, O. Grisel, M. Blondel, P. Prettenhofer, R. Weiss, V. Dubourg, J. Vanderplas, A. Passos, D. Cournapeau, M. Brucher, M. Perrot and E. Duchesnay, *J. Mach. Learn. Res.*, 2011, **12**, 2825.
207. H. J. C. Berendsen, D. van der Spoel and R. van Drunen, *Comput. Phys. Commun.*, 1995, **91**, 43.
208. D. A. Case, T. E. Cheatham, T. Darden, H. Gohlke, R. Luo, K. M. Merz, A. Onufriev, C. Simmerling, B. Wang and R. J. Woods, *J. Comput. Chem.*, 2005, **26**, 1668.
209. K. J. Bowers, D. E. Chow, H. Xu, R. O. Dror, M. P. Eastwood, B. A. Gregersen, J. L. Klepeis, I. Kolossvary, M. A. Moraes, F. D. Sacerdoti, J. K. Salmon, Y. Shan and D. E. Shaw, Scalable algorithms for molecular

dynamics simulations on commodity clusters, in *ACM/IEEE SC 2006 Conference (SC'06)*, IEEE, 2006.
210. P. Eastman, J. Swails, J. D. Chodera, R. T. McGibbon, Y. Zhao, K. A. Beauchamp, L.-P. Wang, A. C. Simmonett, M. P. Harrigan, C. D. Stern, R. P. Wiewiora, B. R. Brooks and V. S. Pande, *PLoS Comput. Biol.*, 2017, **13**, 1.
211. "OpenMM-MM", https://github.com/pandegroup/openmm-nn (accessed October 2018).
212. M. E. Irrgang, J. M. Hays and P. M. Kasson, *Bioinformatics*, 2018, **34**, 3945.
213. S. Schoenholz and E. Cubuk, 2019, arXiv:1912.04232.
214. "Time Machine" https://github.com/proteneer/timemachine (accessed May 2020).
215. "The Molecular Sciences Software Institute", https://molssi.org (accessed March 2019).
216. H. Van Hasselt, A. Guez and D. Silver, Deep reinforcement learning with double q-learning, *Thirtieth AAAI Conference on Artificial Intelligence*, 2016.
217. S. Kearnes, K. McCloskey, M. Berndl, V. Pande and P. Riley, *J. Comput.-Aided Mol. Des.*, 2016, **30**, 595.
218. M. Wang, L. Yu, D. Zheng, Q. Gan, Y. Gai, Z. Ye, M. Li, J. Zhou, Q. Huang, C. Ma, Z. Huang, Q. Guo, H. Zhang, H. Lin, J. Zhao, J. Li, A. Smola and Z. Zhang, *arXiv*, 2019, **1909**, 01315.
219. S. J. Pan and Q. Yang, *IEEE Trans. Knowl. Data Eng.*, 2010, **22**, 1345.
220. A. de la Vega de León, B. Chen and V. J. Gillet, *J. Cheminf.*, 2018, **10**, 26.
221. "Swift for Tensorflow," https://www.tensorflow.org/swift (accessed May 2020).

Section 5: Molecular Design

CHAPTER 10

Compound Design Using Generative Neural Networks

T. BLASCHKE AND J. BAJORATH*

Life Science Informatics, B-IT, University of Bonn, Endenicher Allee 19c, D-53115 Bonn, Germany
*Email: bajorath@bit.uni-bonn.de

10.1 Introduction

Drug discovery is challenging and often compared to finding needles in haystacks. A potential drug needs to satisfy multiple criteria including, among others, desired biological activity, metabolic stability, and efficacy. However, navigating chemical space for drug discovery is a non-trivial exercise, given its principal vastness with on the order of 10^{60} to 10^{100} possible small organic compounds.[1] Hence, only minute fractions of theoretically possible chemical space can be explored in practice, which largely depends on computational approaches. For example, biological screening of large compound collections is further expanded through *in silico* (virtual) screening techniques. Moreover, compound synthesis is complemented by *in silico* design.

Regardless of computational details, the interplay between experimental and *in silico* approaches is an integral part of early-phase drug discovery. Virtual compound libraries further extend accessible chemical space. There are different compound design strategies that can be applied. For example, reaction-based design produces large numbers of virtual candidates that can be further explored.[2] Another design strategy applies expert rules to generate structural analogs of template compounds. Alternative design strategies

Drug Discovery Series No. 75
Artificial Intelligence in Drug Discovery
Edited by Nathan Brown

Published by the Royal Society of Chemistry, www.rsc.org

have their pros and cons and intrinsic limitations, but they are essential for expanding existing compound collections *in silico* and further extending chemical space coverage.

Going beyond structure generation, *in silico* methods for the identification of new chemical entities with desired properties rely on sufficiently accurate prediction of qualitative structure-property/activity relationships (SP/ARs) or quantitative SP/ARs (QSP/ARs). In general, such approaches are statistical in nature and attempt property or activity prediction on the basis of chemical structure. Despite methodological differences, QSPR and QSAR approaches typically have in common that they use calculated molecular similarity as an indicator of property or activity similarity. Hence, activity is accounted for as a structure-dependent variable. Activity prediction plays a central role in computational medicinal chemistry and "*de novo* design" can also be guided by QSAR-type models.[2] By combining structure generation and activity prediction, *de novo* design offers opportunities for charting biologically relevant chemical space and provides pools of candidate compounds with desirable properties for further chemical exploration.[3–5] For property prediction, machine learning (ML) methods including deep learning (DL) are often applied.

10.2 Principles of Deep Learning

DL uses artificial neural networks (ANNs) for ML. The ANN concept has been known since 1943 when it was called threshold logic.[6] ANNs consist of multiple nodes that are called neurons in analogy to the central nervous system. Each node receives input data and transforms the data into a formatted output. A group of nodes receiving the same input data is called a layer. The output of a layer consists of the output of all individual neurons. Contemporary ANNs consist of multiple layers that are connected with each other such that the output of one layer becomes the input of the next one.

A basic ANN architecture is shown in Figure 10.1. It consists of three layers including the input layer, a hidden layer, and an output layer. Depending on the type of ANN, neurons in adjacent layers are either fully or partially connected. Input neurons accept variables that are transformed by neurons comprising the hidden layer and the resulting values are calculated in the output layer. The output of each node is calculated as follows:

$$Y_i = H\left(\sum_j W_{ij} * x_{ij} + b_i\right) \tag{10.1}$$

Y_i refers to the output value of the node i and x_j refers to the input variables. W_{ij} is the weight of input j of node i, b_i is the bias of the node and H is the activation function, which usually is a nonlinear function such as a sigmoid or hyperbolic tangent function.

Training of an ANN involves iterative modification of weight values in the network to minimize prediction errors through back-propagation.[7] ANNs

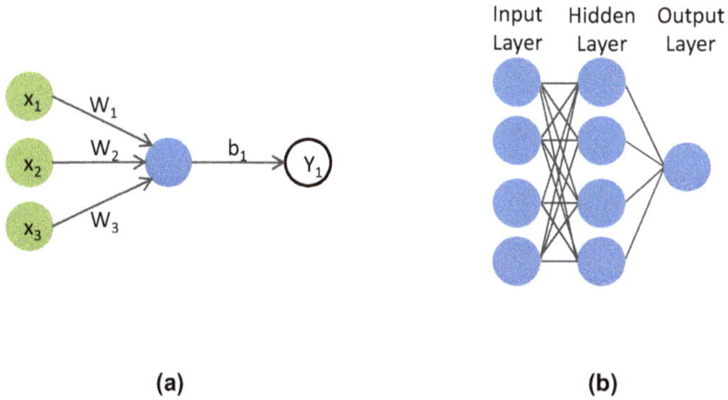

Figure 10.1 Simple illustration of ANNs. An ANN consists of multiple nodes. (a) The output value of a single node is calculated from input values *via* an activation function. (b) Multiple nodes are grouped into layers. A modern ANN is composed of input, hidden and output layers.

were found to be particularly prone to overfitting. Therefore, they were increasingly replaced by other ML approaches such as support vector machine (SVM)[8] or random forest (RF)[9] algorithms. However, more recently ANNs have experienced a renaissance as multi-layered architectures for DL, termed deep neural networks (DNNs). In the context of DL, the problem of overfitting is addressed, for example, by applying the dropout[6] and dropconnect[10] approaches that convert a subset of output values in each layer to zero (dropout) or set a random subset of weights throughout the network to zero (dropconnect). The resulting information loss across connected layers balances overfitting by DNNs. Another algorithmic improvement was the introduction of the rectified linear unit (ReLU)[11] to avoid gradient deterioration in the network. Furthermore, the introduction of convolutional neuron layers enabled the use of larger numbers of input variables for specific DNN architectures.

The first widely used ANN architecture for DL was a fully connected feedforward DNN, which represented an extension of original ANNs. DNNs contain multiple hidden layers with hundreds of nonlinear process units. The large number of hidden layers distinguishes DNNs from conventional ANNs.

Another widely used architecture for DL is the recurrent neural network (RNN), which is characterized by the presence of connections between neurons in a hidden layer. These connections result in the formation of directed circles as shown in Figure 10.2. Therefore, an RNN can accept sequential data as input and process these data sequentially, which makes this architecture very suitable for iterative or time-dependent modeling such as language processing.[12] The type of neuron typically used in RNNs is the so-called "long short term memory unit" (LSTM).[13] Alternatively, a more recently introduced, computationally more efficient variant termed "gated recurrent unit" (GRU) is used.[14] Both LSTMs and GRUs contain a memory cell that stabilizes value gradients when training RNNs for predicting sequential data dependencies.

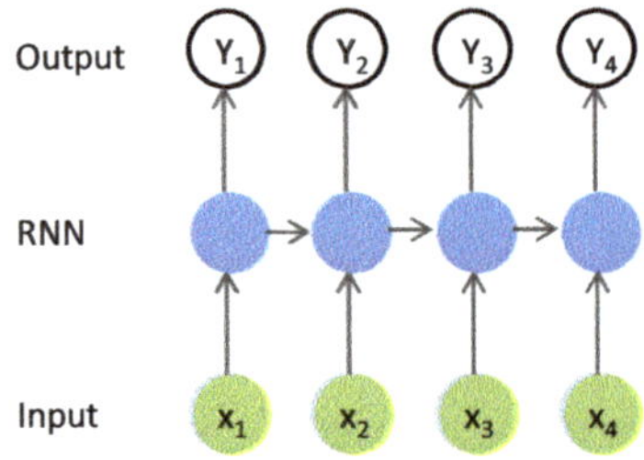

Figure 10.2 Sequential processing by RNN. The RNN receives a sequential input and generates an output value at each step.

Autoencoder (AE) is another increasingly popular architecture for unsupervized learning.[15] AE consists of two connected ANNs including an "encoder" and "decoder". The encoder distributes information received from the input layer over a limited number of hidden nodes. The decoder contains an input and output layer having the same number of nodes. Instead of predicting class labels of test instances, the decoder reconstructs input data from a limited number of hidden nodes. Thereby, the AE learns to convert input data to a lower dimensional representation. Thus, an AE is used for nonlinear dimensionality reduction of data encodings.

10.3 *De Novo* Design *via* Deep Learning

10.3.1 Molecular Representation

A critical parameter for *de novo* compound design and predictive modeling is the selection of molecular representations. Most generative modeling has so far been carried out on the basis of molecular graphs. Here, a molecule is represented as an undirected graph of vertices (atoms) and edges (bonds) connecting them. Graphs can be used for ML in different ways, for example, by storing atom types, bond types, and connectivity information in multidimensional arrays. However, this data structure makes searching for molecules in large databases computationally expensive. Textual or string encoding of molecules is more efficient. The most popular string encoding of molecular structure is the so-called simplified molecular-input line-entry system (SMILES) representation.[16] SMILES strings encode a molecular graph as a sequence of characters using depth-first graph traversal. There are different ways to represent a molecule using SMILES. Unique SMILES representations are produced by algorithmic canonicalization.[17]

For generative modeling, SMILES strings are usually first tokenized based on single characters (except for atom types comprising two characters) and then converted into a "one-hot" string representation, *i.e.*, a unique *N*-dimensional vector where only one element is set to 1 and the others to 0 (with *N* being the number of possible tokens). Generative models using one-hot string representations produce a categorical distribution of each token that is sampled to generate SMILES encoding new structures.

10.3.2 Recurrent Neural Networks

RNNs currently are the most popular models for processing and generating sequential data. Such models are often trained to predict only a single missing token in a sequence. Simultaneous prediction of multiple tokens is generally limited by combinatorial explosion. ML models trained to predict the next character for an input sequence are applied in a generative mode by concatenating the predicted token to the sequence and returning the extended sequence as input for the next iteration, as illustrated in Figure 10.3.

However, this autoregressive generation scheme has a high probability of failure because the model is trained on a given data distribution and not the generative distribution. Further improved generative models are derived to predict sequences by maximum likelihood estimation of a new token on the basis of tokens added in previous iterations. During each step, the model generates a probability distribution for maximizing the likelihood of the next token. This approach is also referred to as "teacher forcing".[18] Formally, a SMILES string is expressed as a sequence of tokens (x) of length T. During training, the model generates a probability distribution P for estimating the likelihood of next token x^t minimizing the following loss function:

$$J(\theta) = -\sum_{t=1}^{T} \log P(x^t \mid x^{t-1}, \ldots, x^1) \tag{10.2}$$

θ denotes the parameter settings of the network.

This approach was applied by Segler *et al.*[19] and Yuan *et al.*[20] to generate new compound structures. RNNs were trained on a large number of SMILES strings and successfully generated new SMILES representations not contained in training sets. RNNs created valid SMILES by learning the underlying probability distribution of characters in strings. Initial training of RNN models on a large collection of strings aims at learning the SMILES syntax.

Once this is accomplished, the resulting models are fine-tuned by subsequent training on smaller data sets to focus on sequence generation. This approach is termed "transfer learning". Segler *et al.*[19] have shown that transfer learning on the basis of SMILES strings makes it possible to

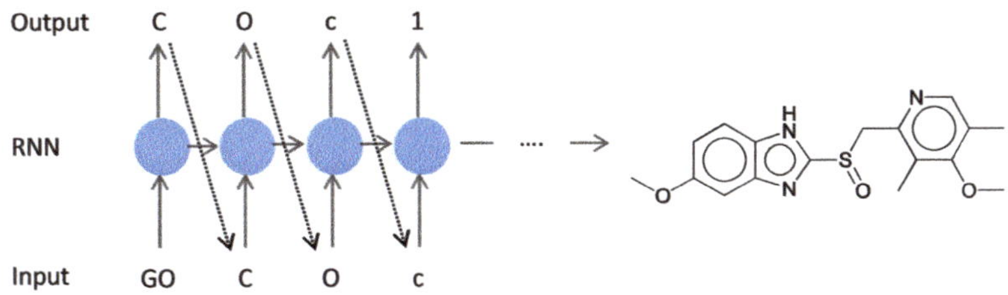

Figure 10.3 Structure generation using an RNN. At each step, a token is sampled based on the probability distribution of the model. The generated token is then used as the input for the next calculation step.

generate focused libraries containing novel compounds that are similar to training instances.

An alternative approach for generating a focused compound library is so-called "reinforcement learning" (RL) where a model is trained to generate molecules with desired properties. For example, Jaques *et al.*[21] trained a DNN following the so-called "Deep Q-learning" RL strategy to generate SMILES with desirable cLogP and drug-likeness parameter values. Here, structure generation required the inclusion of expert rules for penalizing chemically overly simplistic and other undesirable structures. Furthermore, Olivecrona *et al.*[22] developed a policy-based RL approach for fine-tuning pre-trained RNNs and generating compounds with user-defined properties. This approach relies on augmenting a reward R for the generation a molecule m with the likelihood of generating m *a priori*:

$$R'(m) = [R(m) + \log P_{\text{base}}(m) - \log P_{\text{fine-tune}}(m)]^2 \tag{10.3}$$

$R(m)$ denotes the reward for generating the molecule, $P_{\text{base}}(m)$ is the likelihood of generating the molecule with the original model prior to fine-tuning, and $P_{\text{fine-tune}}(m)$ the likelihood of generating m with the fine-tuned model. This methodology makes it possible to generate a focused compound library without the need for pre-defined expert rules. When attempting to predict dopamine receptor type 2 ligands, Olivecrona *et al.* detected known active compounds in the output library that were not used for modeling.[22]

10.3.3 Autoencoder Variants

As stated above, AE essentially is a dimensionality reduction approach. The encoder maps high-dimensional input data to a lower dimensional representation (latent space), while the decoder reconstructs the original input in lower dimensional space. AE training aims at minimizing the information loss associated with data reconstruction by extracting informative features from the high-dimensional input data. Dimensionality reduction performance is strongly influenced by the architecture of the encoder and decoder ANN. Generally, the AE maps a molecule m into a continuous data space z and the decoder reconstructs m from its continuous representation. During this process, the AE model is not required to learn a generalized numerical molecular representation, which has potential drawbacks. Importantly, the model likely learns an explicit mapping of the training set and thus the decoder might not be able to distinguish between different instances in continuous data space. To avoid training of an explicit mapping, modeling might be restricted by deriving a latent variable from the input data. To these ends, the "variational AE" (VAE), introduced by Kingma *et al.*[23] in 2013, formally interprets a continuous data representation as a latent variable for a probabilistic generative model. In 2016, VAE was first used for structure generation by Gómez-Bombarelli *et al.*[24]

Following the VAE methodology, the latent variable z is drawn from a given prior distribution $p(z)$ and passed on to the decoder $p_\theta(m/z)$:

$$p(m,z) = p_\theta\,(m|z)p(z) \tag{10.4}$$

VAE modeling maximizes the lower bound of the log-likelihood $p_\theta(m|z)$ to generate a sequence, while simultaneously minimizing the Kullback–Leibler divergence (D_{KL}) between the output of the encoder $q_\phi(z|m)$ and a set of independent unit normal distributions $N(0,1)$. A loss function is formulated:

$$L = -D_{KL}\,(q_\phi\,(z|m)||N(0,1)) + E[\log p_\theta\,(m|z)] \tag{10.5}$$

VAE training then reversibly transforms SMILES strings into a continuous numerical representation, as shown in Figure 10.4.

The generation of new structures with desirable properties *via* VAE is attempted by searching for latent solutions in the continuous latent space using an optimization method (such as Bayesian optimization) and decoding latent solutions into SMILES strings. The use of unit normal distributions leads to an uneven distribution of SMILES in latent space. Hence, the majority of SMILES strings are mapped close to 0 in the latent space while others are sparsely mapped into regions with values larger than 0. The uneven distribution of SMILES representations hinders the application of optimization methods, which can be addressed by selecting an alternative prior distribution. However, computing D_{KL} is computationally infeasible for many distributions.

To address this issue, Blaschke *et al.*[25] introduced the "adversarial AE" (AAE) that imports selected properties into latent space such that searching for optimal solutions in latent space becomes more feasible. The main difference between the AAE and VAE is that an additional discriminator D is added to the architecture. Instead of minimizing the D_{KL} metric on the basis of a prior distribution, D directs optimization to follow a specific target distribution while minimizing the reconstruction error of the decoder.[25]

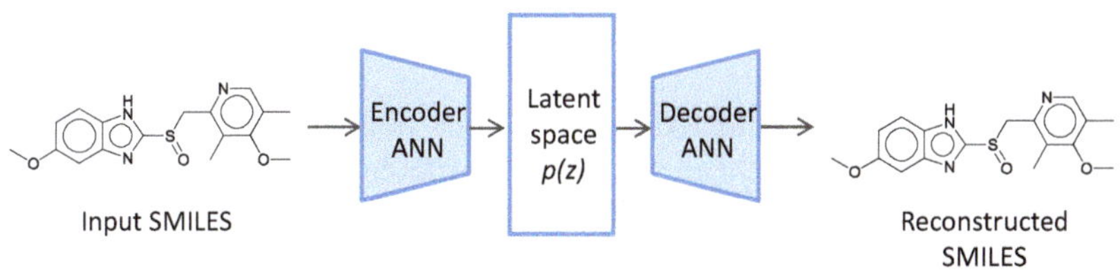

Figure 10.4 Structure generation using VAE. During training, the encoder ANN maps a molecule into latent space and the decoder ANN attempts to reconstruct the molecule. During structure generation, only the decoder is used to assemble molecules from the latent space.

10.3.4 Graph-based Neural Networks

While SMILES strings currently are the most popular encoding of a molecular graph for generative modeling, molecular graphs might also be used directly. For the reasons mentioned above, structure generation using graphs is more complex than using textural representations. However, graph-based generative modeling is another area of ongoing research where two different concepts have emerged. One methodology stores a molecular graph in multi-dimensional arrays and attempts the construction of graphs on the basis of this data structure.[26] However, this approach typically requires a computationally expensive graph matching procedure to evaluate the likelihood of a generated structure. The other approach attempts training of RNNs that directly operate on a molecular graph by adding new atoms and bonds from a list of pre-defined possible solutions during each operation.[27] To these ends, a graph convolutional policy network[27] or deep RL[28] is used to generate molecular structures. Because these approaches are rule-based, they typically produce formally correct structures. However, the ability to generate novel structures with desired properties remains to be further evaluated.

10.4 Property Prediction through Deep Learning

For property and activity prediction, descriptor-based linear regression models and machine learning methods such as RF and SVM have been widely used. In addition, reward- or optimization-based generative DL approaches such as RL or VAE rely on accurate property prediction to guide structure generation. For example, Dahl *et al.*[29] and Mayr *et al.*[30] reported predictions using DNNs on the Merck Kaggle Challenge and Tox21 Challenge data set, respectively. In both cases, large numbers of molecular descriptors were used for property prediction, without the need for feature selection. DNN overfitting was balanced by dropout modeling,[6] as discussed above. However, property predictions were substantially influenced by DNN hyper-parameter optimization including, among others, the number of layers, number of neurons per layer, or type of activation functions. Overall, multi-task DNNs, which simultaneously predict multiple properties at once, were found to further increase the predictive performance of conventional single-task models, which has also been observed by Ramsundar *et al.*[31] in an independent ANN investigation as well as for other machine learning approaches.[32]

Other studies have indicated that DL does not offer a significant advantage over alternative machine learning approaches in activity and property prediction when well-defined molecular representations are used.[33,34] This is consistent with observations made in other fields such as image processing where high performance of DL was largely attributable to the ability of DNN architectures to extract suitable representations from low-resolution data (such a pixels in image analysis). Similarly, the capability of DNNs to learn representations directly from molecular structures, without the use of pre-defined descriptors, distinguishes DL from other ML approaches. Representation learning is still in its infancy in chemical informatics. However, first attempts are being made. For example, graph convolutional (GCN) modeling using a "neural fingerprint" design has been introduced as a DL approach for chemical representation learning.[35] Following this approach, a molecular graph is represented as a matrix containing atom type and connectivity information for each atom. This matrix passes through a single-layer ANN, which is followed by a pooling operation to generate a fixed-length vector. The pooling operation is applicable at different levels taking atom environment information into account. Vectors resulting from pooling operations are transformed and combined to yield a final vector representation of a given compound. These neural fingerprints are then passed through another ANN to generate representations for training a QSPR model.[35] Furthermore, Bjerrum *et al.*[36] used redundant and canonical SMILES strings as an input for RNNs for property prediction and Goh *et al.*[37] used images of 2D drawings of molecules as input, achieving results comparable to those obtained applying DNNs trained on conventional fingerprints. Hence, currently it is too early to judge whether chemical representation learning will yield significant performance improvements in DL applied to *de novo* design and QSAR/QSPR.

10.5 Conclusions and Outlook

In this chapter, we have focused on the use of DNNs for *de novo* structure generation and property prediction, which currently are intensely investigated topics in chemical informatics and drug design. Different DNN architectures and application scenarios have been discussed. Despite the strong interest in DL, alternative approaches for structure generation that do not depend on DL are also being investigated. For example, Yoshikawa *et al.*[38] developed a genetic algorithm for editing SMILES strings and population-based evolutionary selection to generate molecules. This approach was found to meet or exceed the performance of different DL methods.

The ability to learn task-specific molecular representations sets DL apart from other ML approaches. Although this capability is at present only little explored in chemical informatics, it provides an alternative to conventional descriptor-based representations going forward. It will be interesting to see if such representations can be extracted from unstructured or fuzzy chemical data for ML that rival performance based on pre-defined molecular representations. For generating alternative molecular representations *via* DL, large compound data sets will generally be required. On the other hand, "one-shot learning" has been introduced for representation learning from small data sets augmented with additional information.[39] However, such learning strategies still await more extensive assessment in compound activity/property prediction. Another opportunity for future research is the exploration of alternative QSAR-type models and optimization functions to guide the generation of novel structures with desired biological activities.

For the drug design field, the increasing exploration of DL strategies is an exciting development, which will further expand the spectrum of relevant computational methods. The investigation of new learning strategies is a valuable exercise for the evolution of ML in drug design. Whether or not DL will put the field onto a new level, at least for specific applications, remains to be seen. The jury is still out, but exciting times are ahead.

References

1. C. M. Dobson, *Nature*, 2004, **432**, 824–828.
2. G. Schneider, T. Geppert, M. Hartenfeller, F. Reisen, A. Klenner, M. Reutlinger, V. Hähnke, J. A. Hiss, H. Zettl, S. Keppner, B. Spänkuch and P. Schneider, *Future Med. Chem.*, 2011, **3**, 415–424.
3. P. S. Kutchukian and E. I. Shakhnovich, *Expert Opin. Drug Discovery*, 2010, **5**, 789–812.
4. M. Hartenfeller and G. Schneider, *Methods Mol. Biol.*, 2011, **672**, 299–323.
5. G. Schneider and U. Fechner, *Nat. Rev. Drug Discovery*, 2005, **4**, 649–663.
6. N. Srivastava, G. Hinton, A. Krizhevsky, I. Sutskever and R. Salakhutdinov, *J. Mach. Learn. Res.*, 2014, **15**, 1929–1958.
7. S. Dreyfus, *IEEE Trans. Automat. Contr.*, 1973, **18**, 383–385.
8. C. Cortes and V. Vapnik, *Mach. Learn.*, 1995, **20**, 273–297.
9. I. Barandiaran, *IEEE Trans. Pattern Anal. Mach. Intell.*, 1998, 20.
10. L. Wan, M. Zeiler, S. Zhang, Y. L. Cun and R. Fergus, in *Proceedings of the 30th International Conference on Machine Learning*, ed. S. Dasgupta and D. McAllester, PMLR, Atlanta, Georgia, USA, **vol. 28**, 2013, pp. 1058–1066.
11. V. Nair and G. E. Hinton, in *Proceedings of the 27th International Conference on Machine Learning*, Omnipress, Haifa, Israel, 2010, pp. 807–814.
12. S. Fernández, A. Graves and J. Schmidhuber, in *Artificial Neural Networks – ICANN 2007*, ed. J. M. de Sá, L. A. Alexandre, W. Duch and D. Mandic, Springer, Berlin, Heidelberg, **vol. 4669**, 2007, pp. 220–229.
13. S. Hochreiter and J. Schmidhuber, *Neural Comput.*, 1997, **9**, 1735–1780.
14. K. Cho, B. van Merrienboer, C. Gulcehre, D. Bahdanau, F. Bougares, H. Schwenk and Y. Bengio, in *Proceedings of the 2014 Conference on Empirical Methods in Natural Language Processing (EMNLP)*, Association for Computational Linguistics, Stroudsburg, PA, USA, 2014, pp. 1724–1734.
15. Y. Bengio, *FNT Mach. Learn.*, 2009, **2**, 1–127.
16. D. Weininger, *J. Chem. Inf. Model.*, 1988, **28**, 31–36.
17. D. Weininger, A. Weininger and J. L. Weininger, *J. Chem. Inf. Model.*, 1989, **29**, 97–101.
18. R. J. Williams and D. Zipser, *Neural Comput.*, 1989, **1**, 270–280.
19. M. H. S. Segler, T. Kogej, C. Tyrchan and M. P. Waller, *ACS Cent. Sci.*, 2018, **4**, 120–131.
20. W. Yuan, D. Jiang, D. K. Nambiar, L. P. Liew, M. P. Hay, J. Bloomstein, P. Lu, B. Turner, Q.-T. Le, R. Tibshirani, P. Khatri, M. G. Moloney and A. C. Koong, *J. Chem. Inf. Model.*, 2017, **57**, 875–882.

21. N. Jaques, S. Gu, D. Bahdanau, J. M. Hernández-Lobato, R. E. Turner and D. Eck, in *Proceedings of the 34th International Conference on Machine Learning*, ed. D. Precup and Y. W. Teh, PMLR, International Convention Centre, Sydney, Australia, **vol. 70**, 2016, pp. 1645–1654.
22. M. Olivecrona, T. Blaschke, O. Engkvist and H. Chen, *J. Cheminform.*, 2017, **9**, e48.
23. D. P. Kingma and M. Welling, 2013, arXiv:stat-ML/1312.6114.
24. R. Gómez-Bombarelli, J. N. Wei, D. Duvenaud, J. M. Hernández-Lobato, B. Sánchez-Lengeling, D. Sheberla, J. Aguilera-Iparraguirre, T. D. Hirzel, R. P. Adams and A. Aspuru-Guzik, *ACS Cent. Sci.*, 2018, **4**, 268–276.
25. T. Blaschke, M. Olivecrona, O. Engkvist, J. Bajorath and H. Chen, *Mol. Inform.*, 2018, **37**, e1700123.
26. M. Simonovsky and N. Komodakis, in *Proceedings of the 27th International Conference on Artificial Neural Networks and Machine Learning – ICANN 2018*, ed. V. Kůrková, Y. Manolopoulos, B. Hammer, L. Iliadis and I. Maglogiannis, Springer International Publishing, Cham, **vol. 11139**, 2018, pp. 412–422.
27. J. You, B. Liu, R. Ying, V. Pande and J. Leskovec, 2018, arXiv:cs-LG/1806.02473.
28. Z. Zhou, S. Kearnes, L. Li, R. N. Zare and P. Riley, 2018, arXiv:cs-LG/1810.08678.
29. G. E. Dahl, N. Jaitly and R. Salakhutdinov, 2014, arXiv:stat-ML/1406.1231.
30. A. Mayr, G. Klambauer, T. Unterthiner and S. Hochreiter, *Front. Environ. Sci.*, 2016, **3**, e80.
31. B. Ramsundar, B. Liu, Z. Wu, A. Verras, M. Tudor, R. P. Sheridan and V. Pande, *J. Chem. Inf. Model.*, 2017, **57**, 2068–2076.
32. R. Rodríguez-Pérez and J. Bajorath, *ACS Omega*, 2019, **4**, 4367–4375.
33. G. B. Goh, N. O. Hodas and A. Vishnu, *J. Comput. Chem.*, 2017, **38**, 1291–1307.
34. R. Rodríguez-Pérez, T. Miyao, S. Jasial, M. Vogt and J. Bajorath, *ACS Omega*, 2018, **3**, 4713–4723.
35. D. K. Duvenaud, D. Maclaurin, J. Aguilera-Iparraguirre, R. Gómez-Bombarelli, T. Hirzel, A. Aspuru-Guzik and R. P. Adams, in *Advances in Neural Information Processing Systems 28*, ed. C. Cortes, N. D. Lawrence, D. D. Lee, M. Sugiyama and R. Garnett, Curran Associates, Inc., 2015, pp. 2224–2232.
36. E. J. Bjerrum, 2017, arXiv:cs-LG/1703.07076.
37. G. B. Goh, C. Siegel, A. Vishnu, N. O. Hodas and N. Baker, 2017, arXiv:stat-ML/1706.06689.
38. N. Yoshikawa, K. Terayama, M. Sumita, T. Homma, K. Oono and K. Tsuda, *Chem. Lett.*, 2018, **47**, 1431–1434.
39. H. Altae-Tran, B. Ramsundar, A. S. Pappu and V. Pande, *ACS Cent. Sci.*, 2017, **3**, 283–293.

CHAPTER 11

Junction Tree Variational Autoencoder for Molecular Graph Generation

WENGONG JIN,* REGINA BARZILAY AND TOMMI JAAKKOLA

MIT Computer Science & Artificial Intelligence Laboratory, 9A Academy St, Arlington, MA 02476, USA
*Email: wengong@csail.mit.edu

11.1 Introduction

The key challenge of drug discovery is to find target molecules with desired chemical properties. Currently, this task takes years of development and exploration by expert chemists and pharmacologists. Our goal is to accelerate this process using machine learning.[1] From a computational perspective, we decompose the challenge into two complementary subtasks: learning to represent molecules in a continuous manner (encoding); and learning to map a continuous representation back into a molecular graph (decoding). While deep learning has been extensively investigated for molecular graph encoding,[2–7] the harder combinatorial task of molecular graph generation from latent representation remains under-explored.

Prior work on drug design formulated the graph generation task as a string generation problem,[8–14] in an attempt to side-step direct generation of graphs. Specifically, these models generate SMILES,[15] a linear string notation used in chemistry to describe molecular structures. However, this design has two critical limitations. First, the SMILES representation is not designed to capture molecular similarity. For instance, two molecules with similar

Drug Discovery Series No. 75
Artificial Intelligence in Drug Discovery
Edited by Nathan Brown

Published by the Royal Society of Chemistry, www.rsc.org

chemical structures may be encoded into markedly different SMILES strings (*e.g.*, Figure 11.1(a)). This prevents generative models like variational autoencoders from learning smooth molecular embeddings. Second, essential chemical properties such as molecule validity are easier to express on graphs rather than linear SMILES representations. We hypothesize that operating directly on graphs improves generative modeling of valid chemical structures.

Our primary contribution is a novel neural architecture for generating molecular graphs.[16–18] While one could imagine solving the problem in a standard manner – generating graphs node by node[19,20] – the approach is not ideal for molecules. This is because creating molecules atom-by-atom would force the model to generate chemically invalid intermediaries (see, *e.g.*, Figure 11.1(b)), delaying validation until a complete graph is generated. Instead, we propose to generate molecular graphs in two phases by exploiting valid subgraphs as components. The overall generative approach, cast as a *junction tree variational autoencoder*, first generates a tree structured object (a junction tree) whose role is to represent the scaffold of subgraph components and their coarse relative arrangements. The components are valid chemical substructures automatically extracted from the training set using tree decomposition and are used as building blocks. In the second phase, the subgraphs (nodes in the tree) are assembled together into a coherent molecular graph.

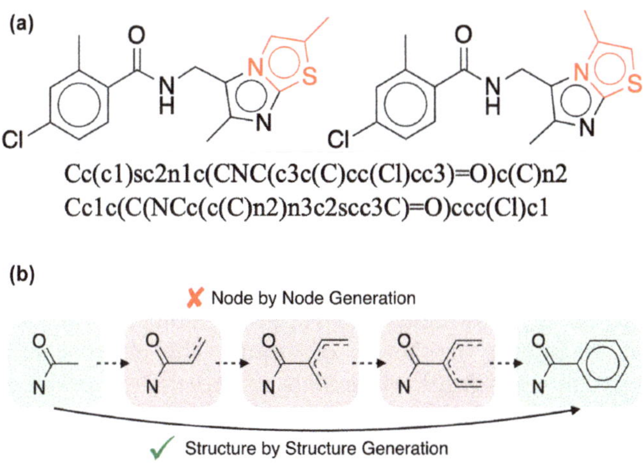

Figure 11.1 Key challenges of SMILES-based and atom-by-atom generation methods. (A) Two almost identical molecules with markedly different canonical SMILES in RD- Kit. The edit distance between two strings is 22 (50.5% of the whole sequence). (B) Comparison of two graph generation schemes: Structure-by-structure approach is preferred as it avoids invalid intermediate states (marked in red) encountered in node by node approach.
Reproduced from ref. 16, http://proceedings.mlr.press/v80/jin18a.html, under the terms of a CC BY 4.0 license, https://creativecommons.org/licenses/by/4.0/.

The proposed graph generation method is applied to *molecular generative modeling*[16] and *molecule-to-molecule translation*[17] for molecular optimization. In the generative modeling task, our model is trained on a large corpus of molecules and learns to propose new compounds with desired properties *via* Bayesian optimization in the latent space. For the molecular translation task, we train our model to learn to optimize input compounds according to desired properties, given a corpus of molecular pairs $\{(X, Y)\}$ where Y is a structural analog of X with improved properties. As baselines, we utilize previous SMILES-based[10,13] and graph-based generation approaches.[20–22] In addition, we compare against state-of-the-art matched molecular pair analysis (MMPA) method[23] on the molecular translation task. Our empirical results demonstrate that our model always produces valid molecules when sampled from the prior distribution of variational autoencoder, and it significantly outperforms other baselines in terms of the property score of generated compounds in molecular optimization tasks.

11.2 Neural Generation of Molecular Graphs

Deviating from previous work[8,9] that generates molecules *via* their SMILES strings, our model directly generates their graph representation. Specifically, we interpret each molecule as having been built from subgraphs chosen out of a vocabulary of valid fragments. These fragments are used as building blocks both when encoding a molecule into a vector representation as well as when decoding latent vectors back into valid molecular graphs. The key advantage of this view is that the decoder can realize a valid molecule piece by piece by utilizing the collection of valid fragments and how they interact, rather than trying to build the molecule atom-by-atom through chemically invalid intermediaries (Figure 11.1(b)). An aromatic bond, for example, is chemically invalid on its own unless the entire aromatic ring is present. It would be, therefore, challenging to learn to build rings atom-by-atom rather than by introducing rings as part of the basic fragment vocabulary.

Our vocabulary of fragments, such as rings, bonds and individual atoms, is chosen to be large enough so that a given molecule can be covered by overlapping fragments or *clusters* of atoms. The clusters serve the role analogous to cliques in graphical models, as they are expressive enough that a molecule can be covered by overlapping clusters without forming cluster cycles. In this sense, the clusters serve as cliques in a (non-optimal) triangulation of the molecular graph. We form a junction tree of such clusters and use it as the tree representation of the molecule, analogous to their scaffolds.

The original molecular graph and its associated junction tree offer two complementary representations of a molecule X. We therefore encode the molecule into a two-part latent representation $\boldsymbol{x}^T$,$\boldsymbol{x}^G$ where $\boldsymbol{x}^T$ encodes the tree structure and what the clusters are in the tree without fully capturing how exactly the clusters are mutually connected. $\boldsymbol{x}^G$ encodes the graph to capture the fine-grained connectivity. Both parts are created by tree and graph encoders based on neural message passing networks. To generate a molecule Y, we first generate its junction tree using a tree decoder, then we predict the fine grain connectivity between the clusters in the junction tree using a graph decoder to realize the full molecular graph. This coarse-to-fine approach allows us to maintain chemical feasibility during generation and provides an enriched representation that encodes molecules at different scales (Figure 11.2).

11.2.1 Junction Tree

Given a molecular graph G, we construct its junction tree through a tailored tree decomposition algorithm. First, we find its smallest set of smallest rings (SSSR) and bonds not belonging to any rings. Next, two rings are merged together if they have more than two overlapping atoms, so that any two clusters in the junction tree will have at most two atoms in common. Each of those fragments is considered as a cluster. Next, a cluster graph is constructed by adding edges between all intersecting clusters. We note that the junction tree of a molecule is not unique when its cluster graph contains cycles. This introduces additional uncertainty for our probabilistic modeling. To reduce such variation, for any of the three (or more) intersecting bonds, we add their intersecting atom as a cluster and remove the cycle connecting them in the cluster graph. Finally, we select one of its spanning trees as the junction tree of G (Figure 11.3).

By applying the above procedure over a set of molecules, we derive a cluster vocabulary including chemical structures such as bonds, rings and bi-cyclic structures (Figure 11.3). We found that the vocabulary size is usually limited enough that we can assign each fragment a distinct label and represent each fragment with a neural embedding vector. For instance, the vocabulary collected over 250 K molecules in the ZINC database[8,24] has no more than 800 unique structures.

11.2.2 Tree and Graph Encoder

Viewing trees as graphs, we encode both junction trees and graphs using graph message passing networks. Specifically, a graph is defined as $G=(V, E)$ where V is the vertex set and E the edge set. Each node v has a feature vector. For atoms, it includes the atom type, valence, and other atomic properties. For clusters in the junction tree, $\boldsymbol{f}_u$ is a one-hot vector indicating its cluster label. Similarly, each edge $(u,v) \in E$ has a feature vector $\boldsymbol{f}_{uv}$. Let $N(v)$ be the set of neighbor nodes of v. There are two hidden vectors $\boldsymbol{m}_{uv}^{(t)}$ and $\boldsymbol{m}_{vu}^{(t)}$ for each edge (u, v) representing the message from u to v and *vice versa*. These messages are updated iteratively with a Gated Recurrent Unit (GRU):

$$\boldsymbol{m}_{uv}^{(t)} = \mathrm{GRU}\left(\boldsymbol{f}_u, \boldsymbol{f}_{uv}, \sum_{w \in N(u)} \boldsymbol{m}_{wu}^{(t-1)}\right) \tag{11.1}$$

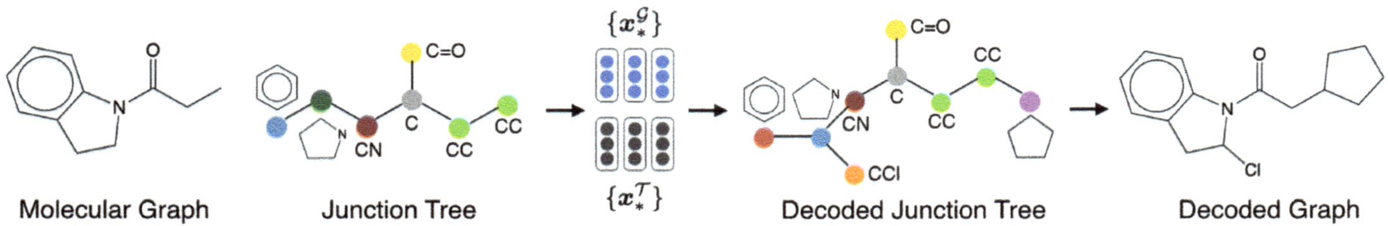

Figure 11.2 Illustration of our encoder–decoder model. Molecules are represented by their graph structures and junction trees encoding the *scaffold* of molecules. Nodes in the junction tree (which we call *clusters*) are valid chemical substructures such as rings and bonds. During decoding, the model first generates a junction tree and then combines clusters in the predicted tree into a molecule.
Reproduced from ref. 17 with permission.

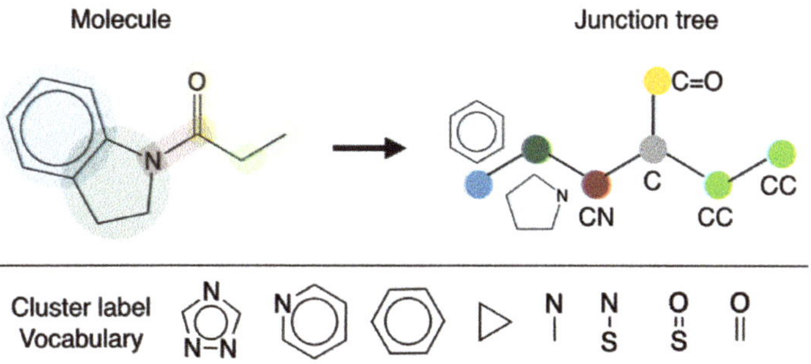

Figure 11.3 Illustration of tree decomposition and sample of cluster label vocabulary. Reproduced from ref. 16, http://proceedings.mlr.press/v80/jin18a.html, under the terms of a CC BY 4.0 license, https://creativecommons.org/licenses/by/4.0/.

where $\boldsymbol{m}_{uv}^{(t)}$ is the message computed in the *t*th iteration, initialized with $\boldsymbol{m}_{uv}^{(0)} = 0$. In each iteration, all messages are updated asynchronously, as there is no natural order among the nodes. After T steps of iteration, we aggregate messages *via* a one-layer neural network to derive the latent vector of each vertex, which captures its local graph (or tree) structure:

$$\boldsymbol{x}_u^G = \mathrm{ReLU}\left(\boldsymbol{U}_1^g \boldsymbol{f}_u + \sum_{v \in N(u)} \boldsymbol{U}_2^g \boldsymbol{m}_{vu}^{(T)}\right) \tag{11.2}$$

Applying the above message passing network to junction tree T and graph G yields two sets of vectors $\{\boldsymbol{x}_i^T\}$ and $\{\boldsymbol{x}_j^G\}$. The *tree vector* $\boldsymbol{x}_i^T$ is the embedding of tree node i, and the *graph vector* $\boldsymbol{x}_j^G$ is the embedding of graph node j.

11.2.3 Junction Tree Decoder

We generate a junction tree T with a tree recurrent neural network. The tree is constructed in a top-down fashion by expanding the tree one node at a time. Formally, let E be the edges traversed in a depth first traversal over tree T, where $m = 2|E|$ as each edge is traversed in both directions. Let E_t be the first t edges in E. At the *t*th decoding step, the model visits node i_t and receives message vectors $\boldsymbol{h}_{ij}$ from its neighbors. The message $\boldsymbol{h}_{i_t j_t}$ is updated through GRU function:

$$\boldsymbol{h}_{i_t,j_t} = \mathrm{GRU}\left(\boldsymbol{f}_{i_t}, \boldsymbol{f}_{i_t,j_t}, \sum_{(k,i_t) \in E_t} \boldsymbol{h}_{k,i_t}\right) \tag{11.3}$$

Topological Prediction When the model visits node i_t, it first computes a predictive hidden state $\boldsymbol{h}_t$ by combining node features $\boldsymbol{f}_{i_t}$ and inward messages $\{\boldsymbol{h}_{(k,i_t)}\}$ *via* a one hidden layer network. The model then makes a binary prediction on whether to expand a new node or backtrack to the parent of i_t. This probability is computed based on the current hidden state as well as the latent representations $\boldsymbol{x}^T,\boldsymbol{x}^G$ of input molecule X, with $\sigma()$ denoting a sigmoid function:

$$\boldsymbol{h}_t = \text{ReLU}\left(\boldsymbol{W}_1^d \boldsymbol{f}_{i_t} + \boldsymbol{W}_2^d \sum_{(k,i_t)\in E_t} \boldsymbol{h}_{k,i_t}\right) \tag{11.4}$$

$$\boldsymbol{c}_t^d = \text{aggregate}(\boldsymbol{h}_t, \boldsymbol{x}^T) \tag{11.5}$$

$$\boldsymbol{p}_t = \sigma(\boldsymbol{u}^d \cdot \text{ReLU}(\boldsymbol{W}_3^d \boldsymbol{h}_t + \boldsymbol{W}_4^d \boldsymbol{c}_t^d)) \tag{11.6}$$

The aggregate () function is used to retrieve information from latent representation of molecule X. The aggregation method can either be neural attention mechanism[25] or a concatenation with input representations, depending on the application (molecular translation or molecular generative modeling, see Section 3).

Label Prediction If node j_t is a new child to be generated from parent i_t, we predict its label by another neural network which outputs a distribution $\boldsymbol{q}_t$ over the label vocabulary:

$$\boldsymbol{q}_t = \text{softmax}(\boldsymbol{U}^l \text{ReLU}(\boldsymbol{W}_1^l \boldsymbol{h}_{i_t,j_t} + \boldsymbol{W}_2^l \boldsymbol{c}_t^l)) \tag{11.7}$$

$$\boldsymbol{c}_t^l = \text{aggregate}(\boldsymbol{h}_{i_t,j_t}, \boldsymbol{x}^{\text{T}}) \tag{11.8}$$

Learning The tree decoder aims to maximize the likelihood of the ground truth tree structure. Similar to sequence generation, during training we perform *teacher forcing*: after topological and label prediction at each step, we replace them with their ground truth so that the model makes predictions given correct histories.

Decoding and Validity During decoding, the tree is constructed recursively guided by topological predictions without any external supervision used in training. To ensure the decoded tree could be realized into a valid molecule in the next step, we define set V_i to be cluster labels that are chemically compatible with node i and its current neighbors. When a child node j is generated from node i, we predict its label from V_i to avoid invalid options. We note that this approach cannot be applied when the graph is generated atom-by-atom, as it must go through chemically invalid intermediate steps.

11.2.4 Graph Decoder

The second step in the decoding process is to construct a molecular graph G from a predicted junction tree T. This step is not deterministic since multiple molecules may correspond to the same junction tree. For instance, the junction tree in Figure 11.4 can be assembled in three different ways. The

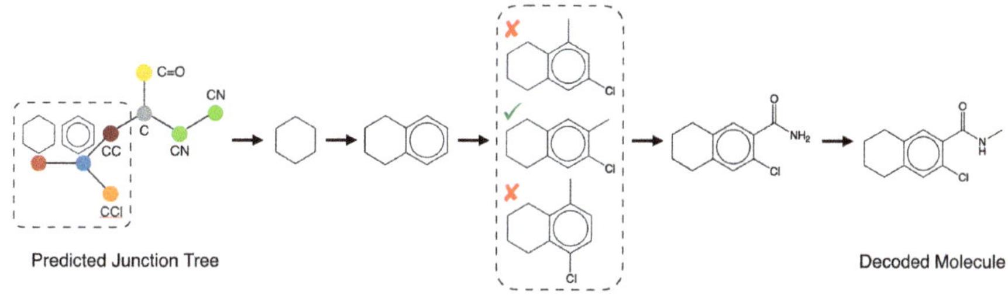

Figure 11.4 Illustration of graph decoding. We assemble the molecular graph one neighborhood at a time, starting from the root of the junction tree. Notably, the neighborhood highlighted in the dashed box can be assembled into three different ways. The correct configuration is predicted by scoring function *f*( ) parameterized by a graph message passing network.

underlying degree of freedom pertains to how neighboring clusters are attached to each other. Let Γ_i be the set of possible candidate attachments at tree node i. Each graph G_i is a particular realization of how cluster C_i is attached to its neighboring clusters C_j. The goal of the graph decoder is to predict the correct attachment between the clusters.

To this end, we design the following scoring function $f()$ for ranking candidate attachments within the set Γ_i. We first apply a graph message passing network over graph G_i to compute atom representations $\{G_{i,v}\}$. Then we derive a vector representation of G_i through sum-pooling:

$$\gamma_{G_i} = \sum_{v} \gamma_{G_{i,v}} \tag{11.9}$$

Finally, we score candidate G_i by computing dot products between γ_{G_i} and the graph representation of input molecule X:

$$f(G_i) = \gamma G_i \cdot \mathbf{x}^{\mathrm{G}} \tag{11.10}$$

The graph decoder is trained to maximize the log-likelihood of ground truth subgraphs at all tree nodes. During training, we apply teacher forcing by feeding the graph decoder with ground truth junction tree as input. During testing, we assemble the graph one neighborhood at a time, following the order in which the junction tree was decoded. During decoding, we assemble the molecular graph one neighborhood at a time, following the order in which the junction tree was decoded. Specifically, we start by predict the assembly of the root and its neighbors according to their scores. Then we proceed to assemble the neighbors and their associated clusters (removing the degrees of freedom set by the root assembly) and so on.

11.3 Application to Molecular Design

In this section, we apply the proposed graph generation method to molecular *de novo* design. We simulate the drug design process under two scenarios. In the first scenario, we train our model on a corpus of molecules $\{(X, y)\}$ with additional attributes y quantifying the property score of compound X. The goal is to discover some compounds having the desired properties. For this purpose, we build a molecular generative model and apply Bayesian optimization to discover the desired compounds. Our second scenario focus on conditional generation: given a starting compound X, how can we modify its structure to improve its chemical properties? To this end, we train our model to learn to modify an input compound to improve its properties, based on a corpus of molecular pairs $\{(X, Y)\}$ where Y is a structural analog of X with improved properties.

11.3.1 Molecular Generative Model

We demonstrate how to use the junction tree encoder and decoder to learn a variational autoencoder for molecules.[16] The variational autoencoder (VAE) contains an encoder that learns to represent the input as a continuous vector, and a decoder that maps the continuous representation back to the molecular graph. In particular, given a compound X, we embed its molecular graph and junction tree into continuous vectors using the tree and graph encoder, whose output contains two sets of vectors $\{\mathbf{x}_i^T\}$ and $\{\mathbf{x}_j^G\}$. Its tree and graph representations $\mathbf{x}^T$,$\mathbf{x}^G$ are sampled from a normal distribution whose mean and log variance is computed by two separate affine layers $\boldsymbol{\mu}()$ and $\Sigma()$:

$$\mathbf{x}^{\mathrm{T}} \sim N\left(\mu_T\left(\sum_i \mathbf{x}_i^T\right), \Sigma_T\left(\sum_i \mathbf{x}_i^T\right)\right)\mathbf{x}^G \sim N\left(\mu_G\left(\sum_i \mathbf{x}_i^G\right), \Sigma_G\left(\sum_i \mathbf{x}_i^G\right)\right) \tag{11.11}$$

The final representation of molecule X is the concatenation of the two: $E(X) = [\mathbf{x}^T, \mathbf{x}^G]$.

Next, we train the corresponding tree and graph decoder to reconstruct compound X from its continuous embedding. The aggregation function in the tree decoder is a concatenation:

$$\text{aggregate}\ (\boldsymbol{h}, \mathbf{x}^T, \mathbf{x}^{\mathrm{G}}) = [\boldsymbol{h}, \mathbf{x}^T, \mathbf{x}^G] \tag{11.12}$$

To facilitate the optimization of chemical properties, we additionally learn a property predictor F on top of the latent space. Suppose a molecule X has a property with value y, we compute its continuous representation $E(X)$ and feed it to the property predictor which is parameterized as a Gaussian process. We choose a Gaussian process because it provides an uncertainty estimate of each prediction. The property predictor is learned to minimize the mean square error between the predicted property value and its ground truth (parameters of the VAE are fixed in this phase):

$$\text{loss}^{\text{prop}}\ (X,y) = (F(E(X)) - y)^2 \tag{11.13}$$

By coupling the generative model with the property predictor, we can perform Bayesian optimization in the latent space to find molecules with optimal properties.[8,10] As illustrated in Figure 11.5, Bayesian optimization is an iterative procedure whose goal is to find a latent vector $\mathbf{z}$ having optimal predicted property score $F(\mathbf{z})$ with high confidence (*i.e.*, the acquisition function). The discovered vector $\mathbf{z}$ is then decoded into its corresponding molecule using our junction tree and graph decoder.

11.3.2 Molecule-to-Molecule Translation

Our encoder and decoder model can also be applied to optimize one molecule into another with better properties based on an available corpus of paired molecules. This task, coined as *molecule-to-molecule translation*,[17] is closely related to *matched molecular pair analysis* (MMPA)[26,27] in drug *de novo* design. During training, we are given a dataset of paired molecules $\{(X, Y)\}$ where Y is a structural analog of X with better chemical properties. It is important to note that this joint distribution is a many-to-many mapping. For instance, there exist many ways to modify molecule X to increase its solubility. Given a new molecule X, the model must generate a diverse set of outputs.

We model the molecule translation task using a conditional variational autoencoder.[28] In particular, we propose to learn a function $F(X, \mathbf{z})$ that transforms the input molecule X into Y. The latent code $\mathbf{z}$ is drawn from a prior distribution $P(\mathbf{z}) = N(\mathbf{0}, \mathbf{I})$, which allows us to explicitly model the multimodal aspect of the output distribution. As illustrated in Figure 11.5, the model learns to translate a compound into its different structural analogs with different latent code $\mathbf{z}$. In other words, each $\mathbf{z}$ represents a meaningful "direction" to optimize the given structure towards desired properties.

During training, the latent code $\mathbf{z}$ is computed as follows. Given a molecular pair (X, Y) in the training set, we embed them into their tree and graph vectors $\{\mathbf{x}_i^T\},\{\mathbf{x}_i^G\}$; $\{\mathbf{y}_i^T\},\{\mathbf{y}_i^G\}$ using the same encoder with shared parameters (Section 2.2). Then we compute the difference vector $\boldsymbol{\delta}_{(X,Y)}$ between molecules X and Y as in eqn (11.14). Since each tree and graph vector represents local substructure in the junction tree and molecular graph, the difference vector encodes the structural changes occurred from molecule X to Y:

$$\delta_{X,Y}^{\mathrm{T}} = \sum_i \mathbf{y}_i^{\mathrm{T}} - \sum_i \mathbf{x}_i^{\mathrm{T}} \quad \delta_{X,Y}^{\mathrm{G}} = \sum_i \mathbf{y}_i^{\mathrm{G}} - \sum_i \mathbf{x}_i^{\mathrm{G}} \tag{11.14}$$

Following Kingma and Welling,[29] the approximate posterior $Q(\cdot|X, Y)$ is modeled as a normal distribution, allowing us to sample latent codes $\mathbf{z}^T$ and $\mathbf{z}^G$ *via* a reparameterization trick. The mean and log variance of $Q(\cdot|X,Y)$ is computed from $\boldsymbol{\delta}_{(X,Y)}$ with two separate affine layers $\boldsymbol{\mu}(\cdot)$ and $\Sigma(\cdot)$:

$$z^T \sim N\left(\mu_1\left(\delta_{X,Y}^T\right), \Sigma_1\left(\delta_{X,Y}^T\right)\right) \quad z^G \sim N(\mu_2(\delta_{X,Y}^G), \Sigma_2(\delta_{X,Y}^G)) \tag{11.15}$$

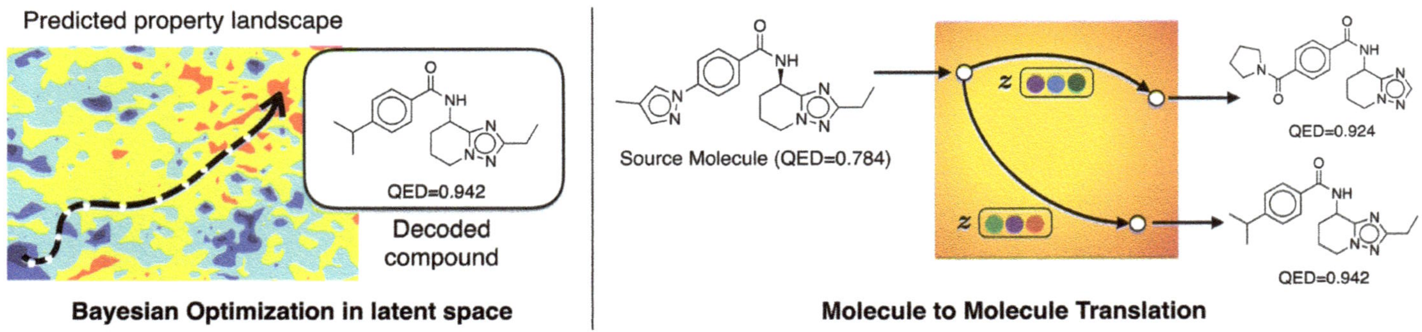

Figure 11.5 Left: Illustration of Bayesian optimization. The VAE latent space and its predicted property landscape is visualized with a contour map. The Bayesian optimization procedure traverse through the latent space to find compounds with desired properties (red region). Right: Illustration of molecule-to-molecule translation, whose goal is to transform a given compound into another with better property (*e.g.*, improving QED score in this example). With different latent code z, the model translates the input compound into different structural analogs.

Finally, we combine the latent code z^T and z^G with source tree and graph vectors:

$$\tilde{x}_i^T = ReLU(W_1^e x_i^T + W_2^e z^T)\ \tilde{x}_i^G = ReLU(W_3^e x_i^G + W_4^e z^G) \tag{11.16}$$

where $\tilde{x}_i^T$ and $\tilde{x}_i^G$ are "perturbed" tree and graph vectors of molecule X. The perturbed vectors are then fed into the decoder to synthesize the target molecule Y. The aggregation function in the tree decoder is a neural attention layer[25] that computes attention scores $\{\alpha_i^T\}$ over the encoded tree vectors. The attention score $\alpha_i{}^T$ indicates how relevant the context $\mathbf{x}_i{}^T$ is to the current hidden state $\mathbf{h}$ and its following prediction.

$$\text{aggregate}(\boldsymbol{h}, \boldsymbol{x}^T) = \sum_i \alpha_i^T \boldsymbol{x}_i^T\ \alpha_i^T = \frac{\exp(\boldsymbol{h}\boldsymbol{A}_T \boldsymbol{x}_i^T)}{\sum_k \exp(\boldsymbol{h}\boldsymbol{A}_T \boldsymbol{x}_k^T)} \tag{11.17}$$

The overall training objective follows a conditional variational autoencoder, including a reconstruction loss and a KL regularization term:

$$\text{loss}^{\text{VAE}}\ (X,Y) = -E_{(z\sim Q)}\left[\log P(Y|z,X)\right] + \lambda_{KL}\ D_{KL}\left[Q(z|X,Y)\,||\,P(z)\right] \tag{11.18}$$

The KL regularization forces the empirical distribution of z over the training set to be similar to the prior distribution $p(z)$. During testing, multiple z are sampled from the prior to transform compound X unseen during training.

11.4 Experiments

11.4.1 Molecular Variational Autoencoder

Our evaluation efforts measure various aspects of molecular variational autoencoders based on previously proposed tasks in Kusner *et al.*:[10]

- **Molecule reconstruction and validity** We test the VAE models on the task of reconstructing input molecules from their latent representations, and decoding valid molecules when sampling from prior distribution. (Section 4.1.1)
- **Bayesian optimization** Moving beyond generating valid molecules, we test how the model can produce novel molecules with the desired properties. To this end, we perform Bayesian optimization in the latent space to search molecules with specified properties. (Section 4.1.2)

Below we describe the data, baselines and model configuration that are shared across the tasks. Additional setup details are provided in the task-specific sections.

Data We use the ZINC molecule dataset from Kusner *et al.*[10] for our experiments, with the same training and testing split. It contains about 250 K drug molecules extracted from the ZINC database.[24]

Baselines We compare our approach with previous SMILES and graph based baselines:

- **Character VAE** (CVAE)[8] which generates SMILES strings character by character
- **Grammar VAE** (GVAE)[10] that generates SMILES following syntactic constraints given by a context-free grammar
- **Syntax-directed VAE** (SD-VAE)[13] that incorporates both syntactic and semantic constraints of SMILES *via* attribute grammar.
- **GraphVAE**[21] that directly generates atom labels and adjacency matrices of graphs.
- **Atom-by-Atom LSTM:**[20] An LSTM-based autoregressive model that generates molecular graphs atom-by-atom.

Model Configuration To be comparable with the above baselines, we set the latent space dimension as 56, *i.e.*, the tree and graph representation $\mathbf{x}^T,\mathbf{x}^G$ have 28 dimensions each. The hidden state dimension is 450 for both encoder and decoder. We represent each cluster with a neural embedding vector, similar to word embedding in natural language processing. The graph encoder and decoder runs three iterations of neural message passing.

11.4.1.1 Molecule Reconstruction and Validity

Setup The first task is to reconstruct and sample molecules from latent space. Since both encoding and decoding processes are stochastic, we estimate reconstruction accuracy by Monte Carlo method used in Kusner *et al.*:[10] Each molecule is encoded 10 times and each encoding is decoded 10 times. We report the portion of the 100 decoded molecules that are identical to the input molecule. To compute validity, we sample 1000 latent vectors from the prior distribution $N(\mathbf{0}, \mathbf{I})$, and decode each of these vectors 100 times. We report the percentage of decoded molecules that are chemically valid (checked by RDKit). For ablation study, we also report the validity of our model without validity check in the decoding phase.

Results Table 11.1 shows that JT-VAE outperforms previous models in molecule reconstruction, and always produces valid molecules when sampled from prior distribution. In contrast, the atom-by-atom based generation only achieves 89.2% validity as it needs to go through invalid intermediate states (Figure 11.1(b)). Our model bypasses this issue by utilizing valid substructures as building blocks. As shown in Figure 11.6, the sampled molecules have non-trivial structures such as simple chains.

We further sampled 5000 molecules from the prior distribution and found they are *all distinct* from the training set. Thus our model is not a simple memorization.

Table 11.1 Reconstruction accuracy and prior validity results.

Method	Reconstruction	Validity
CVAE	44.6%	0.7%
GVAE	53.7%	7.2%
SD-VAE	76.2%	43.5%
GraphVAE	—	13.5%
Atom-by-Atom LSTM	—	89.2%
JT-VAE	**76.7%**	**100.0%**

We qualitatively examine the latent space of JT-VAE by visualizing the neighborhood of molecules. Given a molecule, we follow the method in Kusner *et al.*[10] to construct a grid visualization of its neighborhood. Figure 11.6 shows the local neighborhood of the same molecule visualized in Dai *et al.*[13] In comparison, our neighborhood does not contain molecules with huge rings (with more than seven atoms), which rarely occur in the dataset.

11.4.1.2 Bayesian Optimization

Setup The second task is to produce novel molecules with desired properties. Following Kusner *et al.*,[10] our target chemical property $y(m)$ is an octanol–water partition coefficient (logP) penalized by the synthetic accessibility (SA) score and number of long cycles.[†] To perform Bayesian optimization (BO), we first train a VAE and associate each molecule with a latent vector, given by the mean of the variational encoding distribution. After the VAE is learned, we train a sparse Gaussian process (SGP) to predict $y(m)$ given its latent representation. Then we perform five iterations of batched BO using the expected improvement heuristic. For comparison, we report (1) the predictive performance of SGP trained on latent encodings learned by different VAEs, measured by log-likelihood (LL) and root mean square error (RMSE) with 10-fold cross validation. (2) The top-3 molecules found by BO under different models.

Results As shown in Table 11.2, JT-VAE finds molecules with significantly better scores than previous methods. As shown in Figure 11.7, JT-VAE finds over 20 molecules with scores over 4.02 (roughly the best molecule proposed by SD-VAE). Moreover, the SGP yields better predictive performance when trained on JT-VAE embeddings.

11.4.2 Molecular Translation

Data Our molecular translation models are evaluated on three molecular optimization tasks. Following standard practice in MMPA, we construct training sets by sampling molecular pairs (X, Y) with significant property improvement and molecular similarity $\text{sim}(X, Y) \geq \delta$. The similarity

[†] $y(m) = \text{log}P(m) - \text{SA}(m) - \text{cycle}(m)$ where $\text{cycle}(m)$ counts the number of rings that have more than six atoms.

Figure 11.6 Visualization of the local neighborhood of a molecule in the center.
Reproduced from ref. 16, http://proceedings.mlr.press/v80/jin18a.html, under the terms of a CC BY 4.0 license, https://creativecommons.org/licenses/by/4.0/.

Table 11.2 The best molecule property scores discovered by each method and the predictive performance of sparse Gaussian processes using different VAEs. Baseline results are from previous work.

Method	1st	2nd	3rd	Log Likelihood	RMSE
CVAE	1.98	1.42	1.19	-1.812 ± 0.004	1.504 ± 0.006
GVAE	2.94	2.89	2.80	-1.739 ± 0.004	1.404 ± 0.006
SD-VAE	4.04	3.50	2.96	-1.697 ± 0.015	1.366 ± 0.023
JT-VAE	**5.30**	**4.93**	**4.49**	$\mathbf{-1.658 \pm 0.023}$	$\mathbf{1.290 \pm 0.026}$

constraint is also enforced at evaluation time to exclude arbitrary mappings that completely ignore the input X. We measure the molecular similarity by computing Tanimoto similarity over Morgan fingerprints.[30] Next we describe how these tasks are constructed.

- **Penalized logP** We first evaluate our methods on the constrained optimization task proposed by Jin *et al.*[16] The goal is to improve the penalized logP score of molecules under the similarity constraint. Following their setup, we experiment with two similarity constraints ($\delta = 0.4$ and 0.6), and we extracted 99 K and 79 K translation pairs respectively from the ZINC dataset[16,24] for training. We use their validation and test sets for evaluation.
- **Drug likeness (QED)** Our second task is to improve the drug likeness of compounds. Specifically, the model needs to translate molecules with QED scores[31] within the range [0.7, 0.8] into the higher range [0.9, 1.0]. This task is challenging as the target range contains only the top 6.6% of molecules in the ZINC dataset. We extracted a training set of 88 K molecule pairs with similarity constraint $\delta = 0.4$. The test set contains 800 molecules.
- **Dopamine Receptor (DRD2)** The third task is to improve a molecule's biological activity against a biological target named the dopamine type 2 receptor (DRD2). We use a trained model from Olivecrona *et al.*[12] to assess the probability that a compound is active. We ask the model to translate molecules with predicted probability $p < 0.05$ into active compounds with $p > 0.5$. The active compounds represent only 1.9% of the dataset. With similarity constraint $\delta = 0.4$, we derived a training set of 34 K molecular pairs from ZINC and the dataset collected by Olivecrona *et al.*[12] The test set contains 1000 molecules.

Baselines We compare our molecule-to-molecule translation model, called Variational Junction Tree Neural Network (VJTNN), with the following baselines:

- **MMPA:** We utilized the implentation of Dalke *et al.*[23] to perform molecular pair analysis.[‡] Molecular transformation rules are extracted

[‡]https://github.com/rdkit/mmpdb

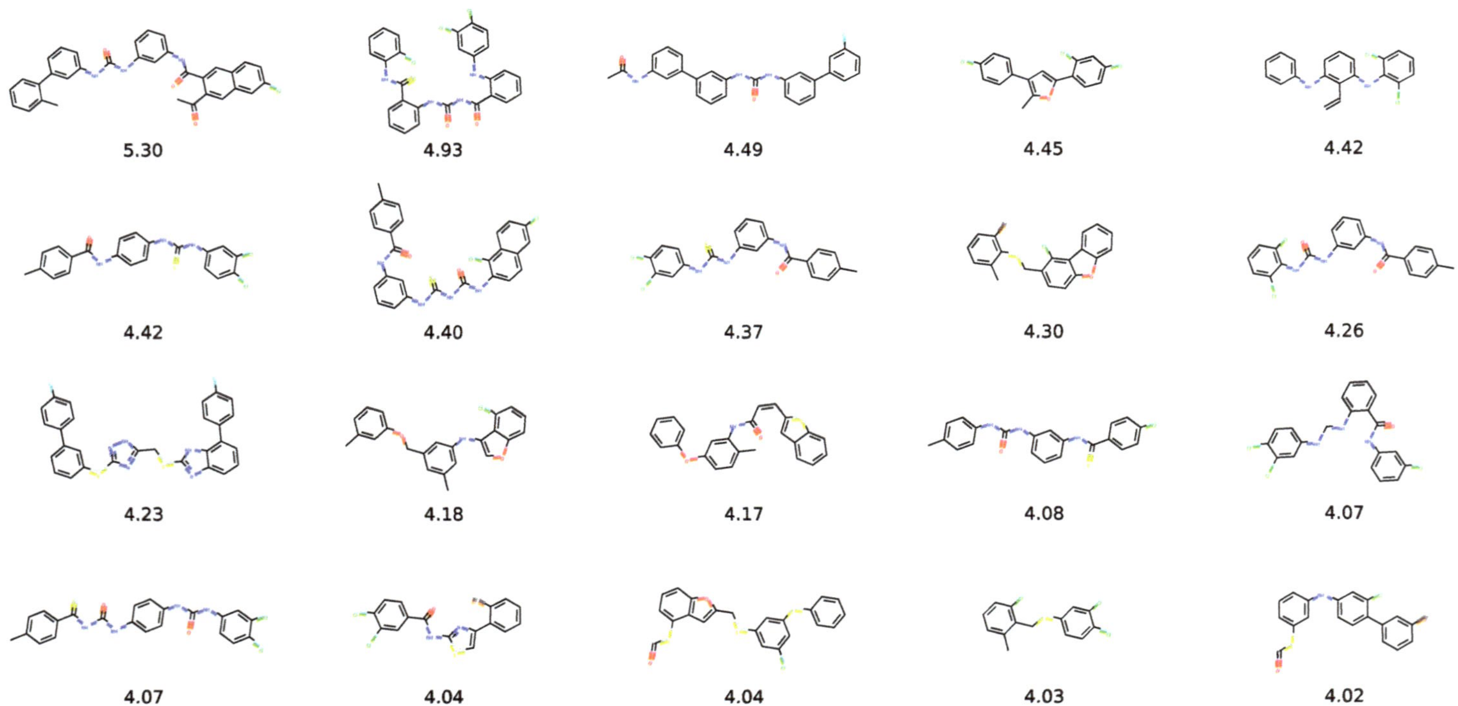

Figure 11.7 Molecules with best penalized logP scores found by Bayesian optimization. Reproduced from ref. 16, http://proceedings.mlr.press/v80/jin18a.html, under the terms of a CC BY 4.0 license, https://creativecommons.org/licenses/by/4.0/.

from the ZINC and Olivecrona *et al.*[12] dataset for corresponding tasks. During testing, we translate a molecule multiple times using $K=20$ matching transformation rules that have the highest average property improvements in the database.

- **VSeq2Seq**: Our second baseline is a variational sequence-to-sequence translation model that uses SMILES strings to represent molecules and has been successfully applied to other molecule generation tasks[8]. Specifically, we augment the architecture of with stochastic latent codes learned in the same way as our VJTNN model.
- **GCPN**: GCPN[22] is a reinforcement learning based model that modifies a molecule by iteratively adding or deleting atoms and bonds. In addition, they adopt adversarial training to enforce naturalness of the generated molecules.
- **MolDQN**: MolDQN[32] is a reinforcement learning based model similar to GCPN but trained through Q-learning instead of policy gradient.

Model Configuration Both VSeq2Seq and our models use latent codes of dimension $|\mathbf{z}|=8$, and we set the KL regularization weight $\lambda_{KL}=1/|\mathbf{z}|$. For the VSeq2Seq model, the encoder is a one-layer bidirectional LSTM and the decoder is a one-layer LSTM with hidden state dimension 600. For fair comparison, we control the size of both VSeq2Seq and our models to be around 4M parameters.

11.4.2.1 Results

We quantitatively analyze the translation accuracy, diversity, and novelty of different methods.

Translation Accuracy We measure the translation accuracy as follows. On the penalized logP task, we follow the same evaluation protocol as JT-VAE. That is, for each source molecule, we decode K times with different latent codes $\mathbf{z} \sim \mathcal{N}(\mathbf{0}, \mathbf{I})$, and report the molecule having the highest property improvement under the similarity constraint. We set $K=20$ so that it is comparable with the baselines. On the QED and DRD2 datasets, we report the success rate of the learned translations. We define a translation as successful if one of the K translation candidates satisfies the similarity constraint and its property score falls in the target range (QED $\in [0.9, 1.0]$ and DRD2 > 0.5).

Tables 11.3 and 11.4 give the performance of all models across the three datasets. Our models outperform the MMPA baseline with a large margin across all the tasks, clearly showing the advantage of the molecular translation approach over rule-based methods. Compared to JT-VAE and GCPN baselines, our models perform significantly better because they are trained on parallel data that provides direct supervision, and are therefore more sample efficient. Overall, our graph-to-graph approach performs better than the VSeq2Seq baseline, indicating the benefit of graph-based representation. The proposed adversarial training method also provides a slight improvement over VJTNN model.

Table 11.3 Translation performance on the penalized logP task. The GCPN and MolDQN results are copied from previous work.[22]

	$\delta = 0.6$		$\delta = 0.4$	
Method	Improvement	Diversity	Improvement	Diversity
MMPA	1.65 ± 1.44	0.329	3.29 ± 1.12	**0.496**
GCPN	0.79 ± 0.63	—	2.49 ± 1.30	—
MolDQN	1.86 ± 1.21	—	3.37 ± 1.62	—
VSeq2Seq	$\mathbf{2.33 \pm 1.17}$	0.331	3.37 ± 1.75	0.471
VJTNN	$\mathbf{2.33 \pm 1.24}$	**0.333**	$\mathbf{3.55 \pm 1.67}$	0.480

Table 11.4 Translation performance on the QED and DRD2 task. Baseline results are copied from ref. 10.

	QED			DRD2		
Method	Success	Diversity	Novelty	Success	Diversity	Novelty
MMPA	32.9%	0.236	99.9%	46.4%	**0.275**	99.9%
VSeq2Seq	58.5%	0.331	99.6%	75.9%	0.176	79.7%
VJTNN	**59.9%**	**0.373**	98.3%	**77.8%**	0.156	83.4%

Diversity We define the diversity of a set of molecules as the average pairwise Tanimoto distance between them, where Tanimoto distance $\text{dist}(X, Y) = 1 - \text{sim}(X, Y)$. For each source molecule, we translate it K times (each with different latent codes) and compute the diversity over the set of validly translated molecules. As we require valid translated molecules to be similar to a given compound, the diversity score is upper-bounded by the maximum allowed distance (*e.g.* the maximum diversity score is around 0.6 on the QED and DRD2 tasks). As shown in Tables 11.3 and 11.4, our methods achieve higher diversity score than MMPA and VSeq2Seq on two and three tasks respectively. Figure 11.8 shows some examples of diverse translation over the QED and DRD2 tasks.

Novelty Finally, we report how often our model discovers new molecules in the target domain that are unseen during training. This is an important metric as the ultimate goal of drug discovery is to design new molecules. Let M be the set of molecules generated by the model and S be the molecules given during training. We define novelty as $1 - |M \cap S|/|S|$. On the QED and DRD2 datasets, our models discover new compounds most of the time, but less frequently than MMPA and GCPN. Nevertheless, these methods have much lower translation success rate.

11.5 Conclusion

In this paper we present a junction tree variational autoencoder for generating molecular graphs. We apply our model on molecular generative modeling and molecule-to-molecule translation. The empirical results demonstrate that the proposed method significantly outperforms previous work in molecule generation and optimization.

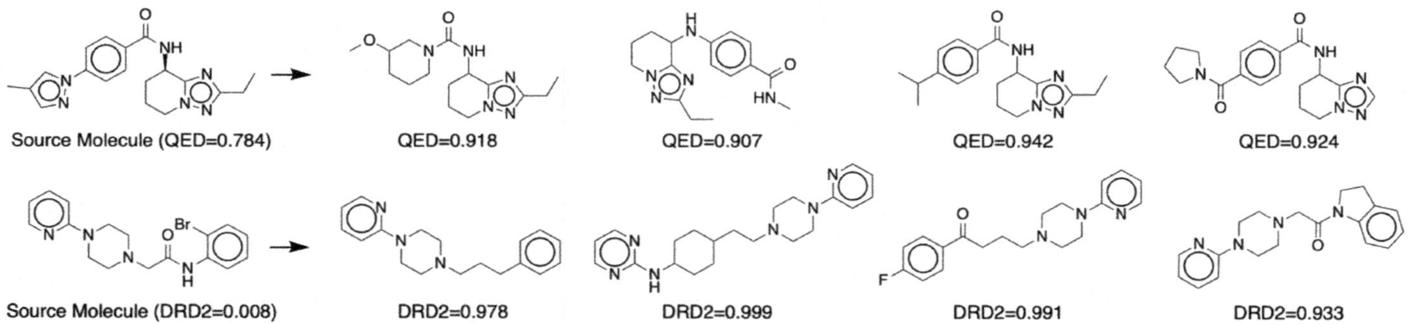

Figure 11.8 Examples of diverse translations learned by VJTNN on QED and DRD2 dataset. Reproduced from ref. 17 with permission.

References

1. B. Sanchez-Lengeling and A. Aspuru-Guzik, *Science*, 2018, **361**, 360–365.
2. D. K. Duvenaud, D. Maclaurin, J. Iparraguirre, R. Bombarell, T. Hirzel, A. Aspuru-Guzik and R. P. Adams, *Advances in Neural Information processing Systems*, 2015, pp. 2224–2232.
3. S. Kearnes, K. McCloskey, M. Berndl, V. Pande and P. Riley, *J. Comput. Aided Mol. Des.*, 2016, **30**, 595–608.
4. H. Dai, B. Dai and L. Song, *International Conference on Machine Learning*, 2016, pp. 2702–2711.
5. J. Gilmer, S. S. Schoenholz, P. F. Riley, O. Vinyals and G. E. Dahl, *International Conference on Machine Learning*, 2017, pp. 1263–1272.
6. T. Lei, W. Jin, R. Barzilay and T. Jaakkola, *International Conference on Machine Learning*, 2017, pp. 2328–2337.
7. W. Jin, C. Coley, R. Barzilay and T. Jaakkola, *Advances in Neural Information Processing Systems*, 2017, pp. 2607–2616.
8. R. Gomez-Bombarelli, J. N. Wei, D. Duvenaud, J. M. Hernandez-Lobato, B. Sanchez-Lengeling, D. Sheberla, J. Aguilera-Iparraguirre, T. D. Hirzel, R. P. Adams and A. Aspuru-Guzik, *ACS Cent. Sci.*, 2018, **4**(2), 268.
9. M. H. Segler, T. Kogej, C. Tyrchan and M. P. Waller, *ACS Cent. Sci.*, 2017, **4**(1), 120.
10. M. J. Kusner, B. Paige and J. M. Hernandez-Lobato, *International Conference on Machine Learning*, 2017, pp. 1945–1954.
11. G. L. Guimaraes, B. Sanchez-Lengeling, P. L. C. Farias and A. Aspuru-Guzik, 2017, arXiv preprint arXiv:1705.10843.
12. M. Olivecrona, T. Blaschke, O. Engkvist and H. Chen, *J. Cheminf.*, 2017, **9**, 48.
13. H. Dai, Y. Tian, B. Dai, S. Skiena and L. Song, *International Conference on Learning Representations*, 2018.
14. M. Popova, O. Isayev and A. Tropsha, *Sci. Adv.*, 2018, **4**, eaap7885.
15. D. Weininger, *J. Chem. Inf. Comput. Sci.*, 1988, **28**, 31–36.
16. W. Jin, R. Barzilay and T. Jaakkola, *International Conference on Machine Learning*, 2018, pp. 2328–2337.
17. W. Jin, K. Yang, R. Barzilay and T. Jaakkola, *International Conference on Learning Representations*, 2019.
18. W. Jin, R. Barzilay and T. Jaakkola, 2019, arXiv preprint arXiv:1907.11223.
19. Y. Li, L. Zhang and Z. Liu, *J. Cheminf.*, 2018, **10**(1), 33.
20. Y. Li, O. Vinyals, C. Dyer, R. Pascanu and P. Battaglia, 2018, arXiv preprint arXiv:1803.03324.
21. M. Simonovsky and N. Komodakis, 2018, arXiv preprint arXiv:1802.03480.
22. J. You, B. Liu, R. Ying, V. Pande and J. Leskovec, *Advances in Neural Information Processing Systems*, 2018, pp. 6410–6421.
23. J. Dalke, Hert and C. Kramer, *J. Chem. Inf. Model.*, 2018, **58**(5), 902.
24. T. Sterling and J. J. Irwin, *J. Chem. Inf. Model.*, 2015, **55**, 2324–2337.

25. D. Bahdanau, K. Cho and Y. Bengio, 2014, arXiv preprint arXiv:1409. 0473.
26. E. Griffen, A. G. Leach, G. R. Robb and D. J. Warner, *J. Med. Chem.*, 2011, **54**, 7739–7750.
27. A. G. Dossetter, E. J. Griffen and A. G. Leach, *Drug Discovery Today*, 2013, **18**, 724–731.
28. K. Sohn, H. Lee and X. Yan, *Advances in Neural Information processing Systems*, 2015, pp. 3483–3491.
29. D. P. Kingma and M. Welling, 2013, arXiv preprint arXiv:1312.6114.
30. D. Rogers and M. Hahn, *J. Chem. Inf. Model.*, 2010, **50**, 742–754.
31. G. R. Bickerton, G. V. Paolini, J. Besnard, S. Muresan and A. L. Hopkins, *Nat. Chem.*, 2012, **4**, 90.
32. Z. Zhou, S. Kearnes, L. Li, R. N. Zare and P. Riley, 2018, arXiv preprint arXiv:1810.08678.

CHAPTER 12

AI via *Matched Molecular Pair Analysis*

ED GRIFFEN,* ALEXANDER DOSSETTER AND ANDREW G. LEACH

MedChemica Ltd, Alderley Park, Macclesfield, Cheshire SK10 4TG, UK
*Email: ed.griffen@medchemica.com

12.1 Introduction

Matched molecular pair analysis (MMPA) is an SAR analysis technique that has a particular attraction for medicinal chemists. As most medicinal chemists are initially trained as synthetic chemists, the concept of linking two molecules by an 'alchemical reaction' is quite intuitive to them. We also borrow from synthetic chemistry and use the term 'transformation' to denote the chemical change from one molecule in a matched pair to another. Although MMPA has been part of the 'folk practice' of medicinal chemistry for decades it is only in the last 20 years that it has been more formalised.[1] This formalisation has led to opportunities for automation, application beyond the initial visions of the early users, and as so frequently happens when a domain is more formalised, identification of issues, 'edge cases', assumptions and areas of imprecise thinking. It is a technique ripe for exploitation within an artificial intelligence framework, as it has the benefit of being already easily understood and adopted by many users. A key concept in MMPA is that it is both analytical and generative. Discovering that 'generally addition of a methyl to a secondary amide increases solubility'[2] is imperative knowledge – it tells the user what can be done. To rephrase the information: 'if you have a secondary amide in your molecule, you can

Drug Discovery Series No. 75
Artificial Intelligence in Drug Discovery
Edited by Nathan Brown

Published by the Royal Society of Chemistry, www.rsc.org

improve solubility by methylating the amide'. This means if the information is captured appropriately then it can be applied directly to a set of molecules and all those possessing a secondary amide will have a new methylated analogue generated. There are a number of practical steps necessary for MMPA: identifying matched pairs, analysing the biological or physicochemical data, storing the results and exploiting the knowledge. Beyond being able to both understand SAR and propose new molecules, it has the critical advantage that it is a highly auditable and transparent method. The importance of the latter feature will be explored in later sections as it is an advantage over other more 'black box' AI methods.

12.2 Essential Features of Artificial Intelligence

There are six critical features of intelligence. The first three are: the synthesis of information, pattern recognition and decision making. The second three are all related to autonomy: intelligence reacts to information and based on some goals, generates a decision to increase the likelihood of achieving that goal, it makes mistakes, but then learns from them and at the most sophisticated level exhibits strategic thinking and situational awareness to avoid being trapped in local minima. Within a medicinal chemistry context, examples of the first three features are: the compilation of chemical structures and biological data, the generation of a (Q)SAR model and the ability to prioritise suggested compounds based on a model. The second set of features are more strategic activities, for example: a system that monitors *in vitro* clearance data being added to a database and proposes new compounds for synthesis to decrease metabolism understanding that low metabolism is a project goal. By also updating the local model for metabolism if the previous round of compounds synthesised have not achieved their goal, a new round of compounds can be proposed based on the updated model. Finally, in addressing a goal, the system could have a hierarchy of global and local models and understand that the current local model of clearance is insufficiently powered such that compounds it supports are less likely to achieve the goal than those advocated by a global model based on a larger data set.

12.3 Matched Molecular Pair Analysis

Matched molecular pair analysis is the process of finding and looking at the properties of 'pairs of molecules within a set that are related to one another by a defined structural difference.'[2] Within this deceptively simple definition lies a range of subtleties. Many of the generic issues in cheminformatics are rapidly found in matched molecular pair analysis. This is another consequence of its transparency as a method: assumptions cannot be hidden within an encoded descriptor, everything is clear for good or ill. Seven archetypal examples are shown for illustration (Figure 12.1). Some of these issues are generic, irrespective of how matched pairs are identified, others are consequences of the algorithms used to identify matched pairs

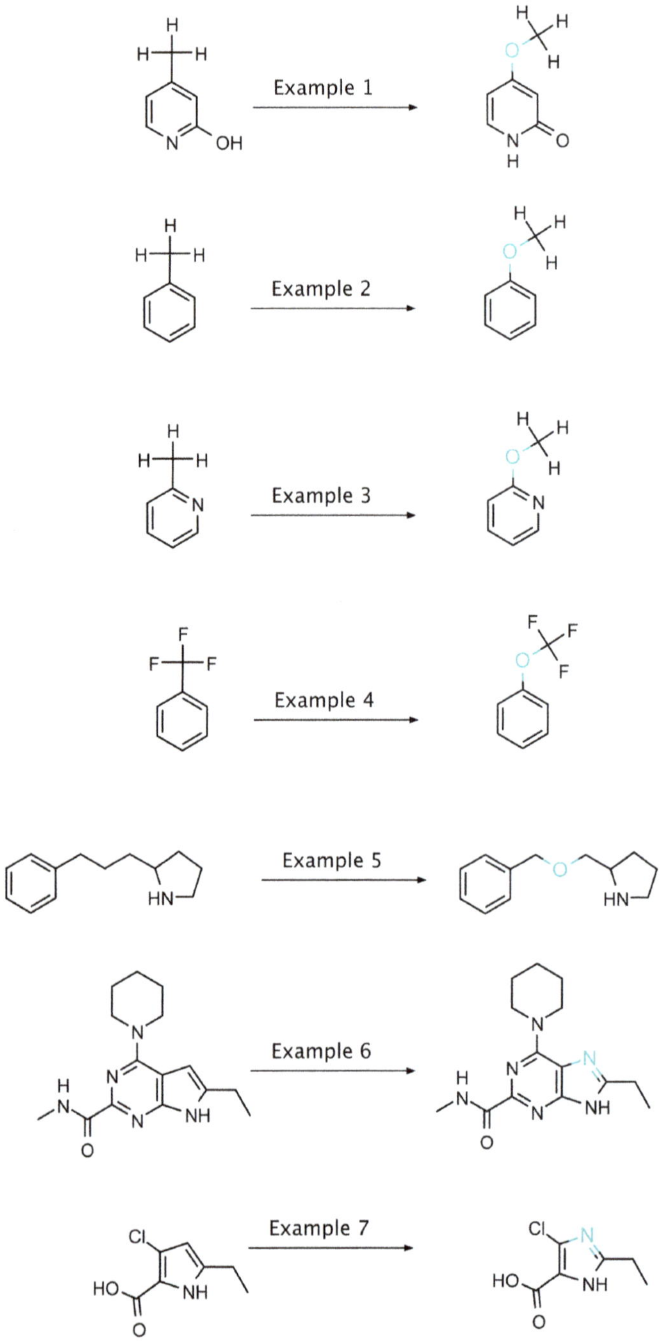

Figure 12.1 Archetypal matched molecular pairs.

automatically. Although the area has been reviewed extensively,[3,4] it is worth focussing on a few points that are particularly relevant when the process is automated in an AI system. When dealing with the identification and exploitation of MMPs, more heuristics have to be applied than are initially obvious.

12.3.1 Generic Issues in Identifying Matched Molecular Pairs

We will first address issues that are generic before examining algorithms used to identify matched pars and the particular additional challenges that process incurs. The first and most obvious issue is that the compounds to be analysed must be in a consistent salt, charge and tautomeric state. The recorded structural format does not need to be the most physiologically relevant (because changing substituents might alter this) but must be identical for all molecules of a class otherwise the matched pair definition will be missed – the compounds will not be a pair connected by a *single* structural change. Obviously, an ammonium or carboxylate salt needs to be both desalted and stored in the neutral form with the counter ion removed. Tautomers are more complex, a simple example is demonstrated in Figure 12.1: example 1, where the methyl hydroxy pyridine will not be paired with the methoxy-pyridone, as the hydroxy pyridine and pyridine do not match. Examples 2, 5 and 6 are what chemists commonly describe as a substituent change, a linker change and a core change (or 'scaffold hop'). However, looking at Example 2 compared to Example 5, both consist of inserting an oxygen between two carbons, so when is a substituent change actually a linker exchange? This may seem a trivial point, however if a system is implemented that understands the concepts of these three classes, then a chemist may apply a linker exchange and be surprised to observe methyl groups being mutated.

The next issue is that of capturing the local chemical environment around a point of chemical change. To any chemist this is essential – a substituent ortho to a nitrogen in a pyridine ring would be expected to behave differently to a substituent in a phenyl ring, this is shown in Figure 12.1 examples 2 and 3. Should the examples of aryl methyl → aryl methoxy only contain phenyl groups, should the pyridines be separated out, what about other heterocyclic pairs, how are they encoded and aggregated? The significant effect of environment has been elegantly demonstrated in an extended study by Papadatos and co-workers at GSK who encoded the local environment (described as 'context' in their paper),[5] in multiple different ways: Daylight fingerprints, Murcko frameworks, reduced graphs, atom environments and localised reduced graph nodes. These ranged from least specific Murko frameworks to most specific circular atom environments. The circular fingerprints have also been captured as the hash of the concatenation of the circular fingerprints rooted at each attachment atom, ordered canonically in a study from Roche.[6] Warner *et al.*[7] at AstraZeneca adopted an approach which explicitly captured the local atomic environment in multiple shells round the point of change and recorded this as part of the transformations so they could be exactly analysed. This approach has been applied on a very large scale to multiple pharmaceutical company datasets.[8] Whichever method is preferred, recording the chemical environment of a transformation is essential.

A variant on the 'is it a substituent change or is it a linker change?' issue is shown in Figure 12.1 example 5, here it appears to be an exchange of a $CH_2 \rightarrow O$ as a linker, but equally it could be described as $CH_2CH_2CH_2 \rightarrow CH_2OCH_2$. This may seem like a minor detail, however if the transformation is recorded as $CH_2 \rightarrow O$, then the examples would be aggregated with examples 2, 3 and 4, whereas if it is recorded as $CH_2CH_2CH_2 \rightarrow CH_2OCH_2$, then it would not. When analysing linkers this can become more extreme with longer chains, so the homologous change $CH_2CH_2CH_2CH_2CH_2 \rightarrow CH_2CH_2OCH_2CH_2$, could also be recorded as $CH_2CH_2CH_2 \rightarrow CH_2OCH_2$, and $CH_2 \rightarrow O$. A heuristic needs to be adopted as to how to record these matched pair relationships – as all the options, the largest transformation, or the smallest?

The equivalent issue for 'core exchanges' or 'scaffold hops', is the challenge of how much core is swapped. Figure 12.1 examples 6 and 7 show this. Do we record and aggregate the aromatic C–H exchange for N (which would also include examples of phenyl to pyridine and the exchange of pyrrole to imidazole) which would encompass both the examples shown, or the full ring system exchange (pyrollopyrimidine → imidazopyrimidine and pyrrole → imidazole) in which case these two examples would not be aggregated?

Many of these issues reflect attempting to analyse pairs of molecules in different ways, the more specific the transformation, the fewer examples there will be to compare, and although the signal may be stronger, the statistical support may be less. If the definitions are instead left very broad, although many example pairs will contribute to a transformation, the different environments and definitions will blur the signal and no significant effect may be discerned.

12.3.1.1 Methods for Automatically Identifying Matched Molecular Pairs

There are two classes of methods for identifying MMPs: supervised and unsupervised. If we assume that molecules have been normalised with regards to salt form, charge and tautomeric form they can be described as below.

12.3.1.1.1 Supervised Analysis. A transformation is encoded as a reaction and the following algorithm executed. For each molecule in the set:

- apply the transformation,
- search the product against the initial set,

- if there is a match – the molecule and the transformation product are a pair,
- else – move to next compound.

This method is highly robust and will scale with the number of compounds in the set. It was the original approach used to computationally identify matched pairs.[1] It has the advantage that complex and precise transformations for substituents, linker swaps and core exchanges can all be defined. The two principle drawbacks of the method are: (1) it is supervised – only relationships that are applied will be found, it won't discover new relationships; (2) it requires the creation of a mapped transformation which is a non-trivial manual task, an example is provided in Figure 12.2. Here a mapped reaction is specified with a pair of compounds that it will identify. In practice, defining all but the simplest transformations is an iterative process where a set of pairs are found, analysed, and in examining outliers in the set of pairs, refinement to the transformation is required to achieve the correct level of specificity. For a single transformation this may prove acceptable, but if multiple transformations are being created, ensuring that the same heuristics are applied for the capture of the environment and level of substitution may be difficult. Chemists who are experienced at reaction searching are familiar with the problem of iteratively refining a query, however in this case given each step requires analysis rather than just inspection, it is significantly slower. Nevertheless, significant ADMET analyses have been conducted using this approach.[9–13]

12.3.1.1.2 Unsupervised Analysis. The three central issues of unsupervised matched molecular pair analysis are problem specification, speed and automation. Finding and tuning algorithms that generate all the pairs of compounds chemists consider to be connected by a 'single well defined transformation' while both identifying pairs they had not thought of and not swamping them with irrelevancies is a significant challenge. Being unsupervised, the second challenge is speed; even with sophisticated software engineering, unsupervised MMPA is essentially an all-against-all problem. Finally there is the issue of automating the entire process.

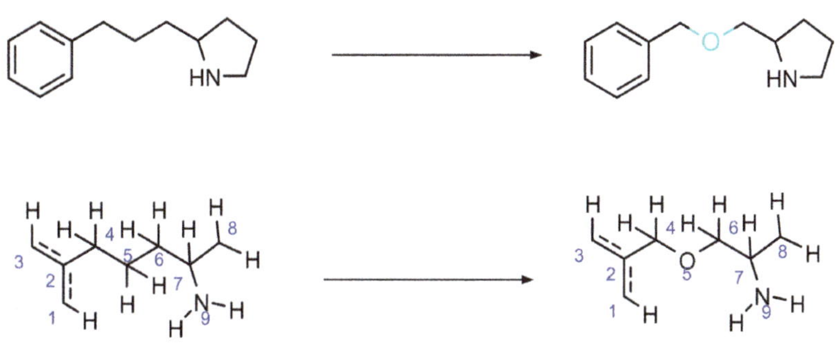

Figure 12.2 Example of a mapped transformation.

There are two families of algorithms that can be applied, the first initially published by Sheridan *et al.* at Merck[14] and Warner *et al.* at AstraZeneca[7] uses identification of the maximum common substructure (MCSS) as a key component in the algorithm.

For each molecule in the set:

- compare to each other molecule in turn:
- generate the maximum common substructure,
- identify the remainder components excluding the common substructure,
- if the remainders are each only one component the pair are a matched pair
    - identify the point of change in the maximum common substructure,
    - delete atoms distant from the point of change to capture the local environment and to generate a reduced common substructure,
    - generate a transformation by adding the changing components to the reduced common substructure.

This algorithm scales with $\frac{n^2}{2} - n$ as every molecule is compared to every other molecule except itself. The second speed issue is that the process of identifying the maximum common substructure is intrinsically slow, it is the chemistry embodiment of the graph theory maximum common subgraph problem which is classed as an NP-hard problem. The time taken to identify the maximum common substructure of two molecules scales with k^x where k is the number of atoms in the smaller molecule, x cannot be guaranteed to be less than 2, and may be significantly higher. A further issue is what constitutes an acceptable size for the minimum common substructure – in the limit most compounds would have a carbon as a single common substructure – but specification of the heuristics of what is a reasonable minimum size has further implications which we will discuss below. A further issue of MCSS algorithms is that usually they are maximum common *connected* substructure algorithms, the implication of which is shown in Figure 12.3. For example, the maximum common connected substructures A→B is chlorobenzene, but for A→C is the full two ring compound with the ether, likewise C→D has just the chlorobenzene as the MCSS, but likewise B→D has the full amine. A chemist would view A and B and B and D as pairs of compounds with different linkers between the aryl rings. Searching thoroughly for disconnected maximum common substructures is significantly more complex.[15]

In an attempt to avoid the MCSS issues, Hussein and Rea at GSK[16] pioneered an approach based on fragmenting molecules into subcomponents and then identifying molecules that have all components bar one in common.

For each molecule in the set:

- split the molecule at specified bonds creating a table of cores and fragments,

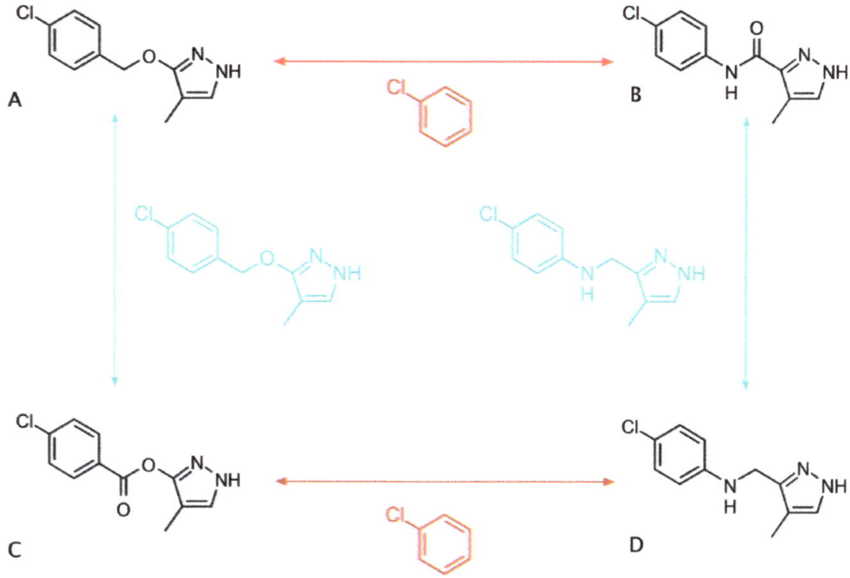

Figure 12.3 Size of MCSS determines designation of pairing.

- repeat the process cutting two and three bonds at a time again generating sets of fragments and cores – store the results to a table,
- identify all the molecules that have the same core but different fragments,
  - those coming from a single cut are substituent exchanges, those coming from double cuts are linker exchanges and from triple cuts core swaps.

This algorithm scales much more closely to being linear with the number of molecules being analysed, however this is done by storing the fragment table. The size of the stored core and fragment table rapidly becomes extremely large. As it does so, even with efficient indexing the algorithm begins to slow down. Further issues with the algorithm are the definition of exactly which bonds are cut. For efficiency, only exocyclic bonds are usually considered for cutting to generate fragments, which creates consequences we will discuss below. In addition some limits are usually placed on the relative sizes of the core and fragment (as for the MCSS approach). However, if cores are allowed to be small enough to include methylene then again everything can be paired with everything. The normal rule of only cutting exocyclic bonds and avoiding cutting amidic or sulphonamide N–C(=O) or N–S bonds can cause matched pairs to be missed such as in the case of changes within a macrocyclic ring (Figure 12.4).

There are in addition some pairs and transformations that both algorithms find difficult to identify. For example (Figure 12.5) the exchange of an

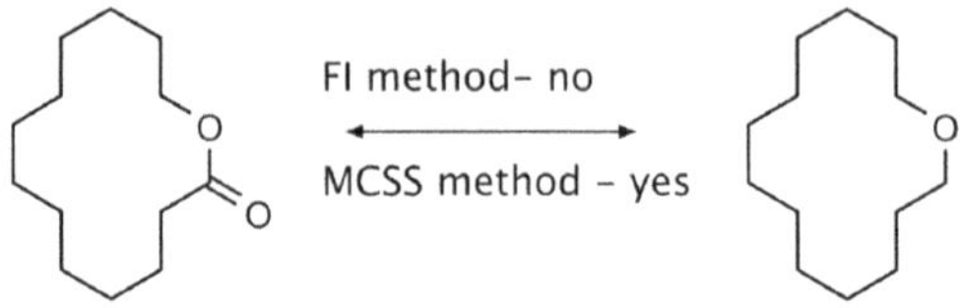

Figure 12.4 Identifying pairs in macrocyclic ring changes.

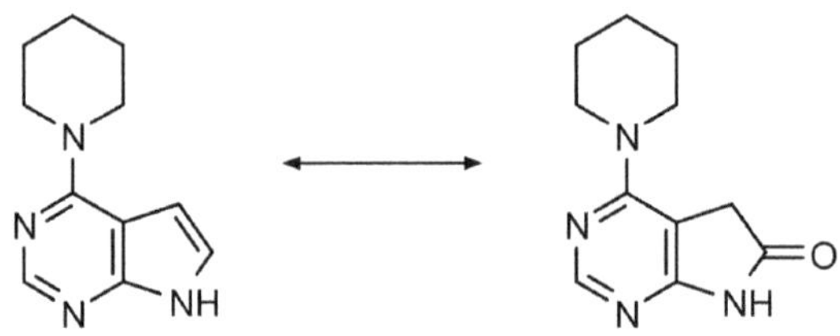

Figure 12.5 A hard matched pair to identify.

indole aromatic atom for a sp^2 aliphatic amidic atom, though intuitive to a medicinal chemist is harder to identify automatically as it is a change in atom type that the algorithms routinely miss.

Interestingly, the two automated methods often cover each other's limitations, and using both together is a highly effective strategy for finding the vast majority of pairs of compounds. This has been extensively explored by converting the outputs of both algorithms into the same format and comparing their performance on a number of different datasets.[17] Comparison of the speeds of the algorithm is a distraction since once two compounds have been identified as a pair and the transformations that connect them generated, that is a static feature of their structures and can be stored in a database. Further biological or physicochemical data may be added, but as long as the structures don't change, the compound pair relationship is fixed.

Despite these special cases, it should be emphasised that above very small sets (*e.g.* tens of molecules), automated MMPA systems are the only reliable way of identifying the vast majority of MMP relationships. For 10 molecules, 40 comparisons ($10^2/2–10$) need to be made which is tiresome and error prone manually but possible. For 30 compounds: 420 comparisons ($30^2/2–30$) are needed which is already impractical. This also indicates why, analysing all the matched pairs in ADMET assay sets (in large companies this could be 50 000 compounds) can result in billions to trillions of comparisons, this necessitates the use of highly parallelised calculations on large scale high performance clusters with the results stored in well-designed databases. Moving to compound collection scale MMPA in the $10^6–10^7$ scale compounds is a further step in engineering which has been delivered but is a serious undertaking.

12.3.1.2 Aggregating Biological Data

Once matched pairs have been identified, the data that is associated with the compounds need to be aggregated in some way. There are two key issues in aggregating any biological data that are a feature of the data type: to use statistical language it is usually non-linear and it is censored. The non-linearity is due to the relationship between ΔG and Ki (or IC50), a 1.4 kcal change in binding energy will give a 10-fold change in potency. So the same change in binding energy can decrease potency from 1 to 10 nM or from 10 to 100 μM. Using the arithmetic difference in potency will be 9 nM in the first case but 90 000 nM in the second. Logging the data to give pIC50's of 9 and 8 or 5 and 4, shows in both cases that the ΔpIC50 is 1 log unit. If the Ki, IC50, or EC50 is to be related to other properties such as logP or pK_a which are linearly related to ΔG regression analysis will be valid as both the independent variable and dependant have been logged. The second issue is censored data (this is often referred to by scientists as 'out-of-range' or 'qualified' data): biological data usually has some limits of detection both high and low, so an IC50 may be quoted as <1 nM or >100 depending on the dose range used in testing. This causes a further, more difficult to deal with, issue in that there will be bunching at the top and bottom of a data set.

Understanding these features of the data then allows us to rationally decide on the appropriate form of analysis. For matched pairs, the simplest approach is to use the difference between the logged values for a pair of compounds, the ΔpIC50, these values can then be aggregated over multiple pairs and the summary statistics (such as mean or median) can be reported to the user as is or anti-logged to give a ratio. It is generally poor practice to use the mean of low numbers of examples, and it is better to assume that a data set is not normally distributed unless that can be demonstrated, therefore the median is a better representative of the centre of a distribution than the mean. Censoring presents a different problem. For matched molecular pair analysis some of the most useful insights will be where a structural change causes the biological data to significantly increase or decrease, therefore excluding censored data from the analysis causes a loss of useful understanding. Clearly if both members of a pair are out of range in the same direction then nothing can be said about the effect of the change, however if one member is in range and the other out of range, the difference between the effects can still be used. Deciding if a set of changes are significant requires some form of statistical test. Being consistent in avoiding the assumption of normality means that parametric statistical tests should be avoided. Using the count of the number of times a structural change has led to an increase or decrease in the biological effect enables the non-parametric binomial test to be used which has well understood statistical pedigree and is straightforward to explain.[17]

12.3.1.3 Auditability and Transparency

There is a simple choice in creating an AI system for use in compound design. Either the system works with chemists or it is part of a closed loop design–make–test–analyse system that can operate without human intervention. The latter, though appealing to many computational chemists, requires a very high degree of capability in synthesis optimisation and route design as well as a high degree of robotic integration of both synthesis and purification. The former option, acting to augment the working chemist requires the interaction between the chemist and the system to be highly effective. Like any other collaboration, to be effective one would expect the human AI interface to 'require good communication, trust, clarity and understanding'.[18] Much of contemporary predictive chemistry research has focussed on the quality of prediction rather than the interpretability of the models generated. Although some models are more interpretable, such as linear regression, principle component analysis and random forest models, support vector machines and in particular neural networks and deep neural networks can be particularly opaque. This presents a problem: whereas in the consumer field the user may not care about how a book or song recommendation is made, or how a route is recommended for navigation because the effect of a poor choice is of relatively low impact; in drug discovery where the cost of making and testing a compound until it is found to be inadequate may incur costs in the thousands to tens of thousands of dollars. The user as a skilled professional has much more invested in the decision to make a compound and therefore requires more understanding to trust a recommendation. There is considerable research underway in attempting to explain black box models but this has proven to be difficult,[19] however it has been proposed that attempting to explain black box models is perpetuating bad practice and a better option is to use interpretable models from the start.[20] Matched molecular pair based methods obviously fulfil this criterion.

12.3.1.4 *From Matched Molecular Pair Analysis to Creating an AI System*

Matched molecular pair analysis is just a method of identifying precise relationships between pairs of molecules. The development of an AI system is far more than this.

The obvious way to integrate MMPA into an AI system is to generate an expert system which uses the MMPs as medicinal chemistry rules. The concept of creating a database of rules by hand originated with Fujita[21] and was extended in the implementation of Drug Guru by Stewart.[22] The automated generation of rules using matched molecular pair analysis[1,2] on measured compounds was developed by workers at Merck and AstraZeneca.[7,14] The expert system enumerator and automated rule extraction methods were then combined to generate a learning machine[23] which mined rules for ADMET data and then applies them to seed molecules to generate new 'better' structures. This architecture has even been used to compare and share rules between companies.[8,24] This system is obviously more transparent than the neural network approaches and because it uses only real and measured compounds it is far less likely to generate invalid or undesirable structures. It is also very straightforward to 'drill back' to the underlying data supporting the rules, enabling auditing of suggestions. There is a parallel with the approaches in computer aided synthesis planning (CASP) between the 'database of rules', 'Good Old Fashioned AI' (GOFAI) approaches and the neural network methods. The GOFAI approaches can be highly effective and fall into the class of 'explainable AI'[18] which is considered a strategic priority[25] both for checking biases and to support uptake by a usually cautious and sceptical professional user community.

12.3.2 Automation

Automation is often seen as a binary property: a system is 'automated ' or it is not. This is however insufficiently nuanced for developing automated systems for medicinal chemistry. A useful metaphor is that of managing a colleague: the relationship can pass through instruction, coaching, process monitoring and finally delegating. With an AI system that is insufficiently knowledgeable it will have to be 'told' exactly what to do – the interface will be complex and significant supervision will be required. As the system becomes more reliable and independent, then it may be sufficient to have it operating with just monitoring of the process, finally the interface becomes very simple, merely an on/off switch. This represents a useful metaphor as automation and interfaces both require significant engineering to deliver them. Automating an insufficiently trustworthy system runs the risk of creating a 'sorcerer's apprentice' system running out of hand but producing little of value. Conversely, giving the operator too much control over an unbiased and well automated system may result in them 'fiddling' and reducing the possible benefits.[26] This metaphor also presents a path to plan system development.

A further aspect of implementing an automated system is the level of robustness required for software to run autonomously compared to under human direction. This is particularly relevant in the research domain where novel structures may not be correctly processed by the system leading to errors. This issue and the steps needed to address it has been reviewed in the context of developing QSAR models from publicly available data.[27] Human generated data may be mistyped leading to alphabetic characters where a system is expecting a number, or the out of range qualifier (</>) may be included with numeric data. Whereas a human user can spot these common errors and remove them, a system running autonomously needs to catch the error, log it and move onto the next task without dropping out of the cycle.

One factor that has become particularly challenging within the last five years is the quality of interfaces expected by chemists. As large technology companies such as Apple and Google have delivered apparently seamless systems to the general consumer, in some cases with no charge to the user, the expectation of highly engineered and intuitive interfaces has risen. This is in conflict with the global user base of medicinal chemistry software being tiny in comparison to consumer software. The business case for large development teams delivering intuitive software to a small user base is hard to make, and it may be more efficient to embed data scientists within project teams to work with chemists until the automation is performing at a sufficient level to simplify the interfaces.

Automating an MMPA system consists of creating several modules: components automating structure and data clean up and loading (known as ETL for Extract-Transform-Load), components finding the matched pairs and then aggregating the biological or physical chemistry data and generating rules, modules that can apply the rules to a seed compound or compounds and apply any preset filters to the output. The whole process can be further connected to produce a learning machine that ingests structures and data, generates rules, suggests compounds for synthesis and testing, and then as the data is generated the cycle begins again.

If the data sets include ADME properties and automated route design can be added with a prediction model available for potency, finally the option becomes available to close the design–make–test cycle and fully automate the entire process.[23] (Figure 12.6) At this point, it would be expected for the system to become learning, as data is fed back into the system the models are re-built and better predictions made on the next generation of compounds. Assuming that the correct optimisation criteria have been specified, the system should optimise compounds to the point of prioritising compounds for *in vivo* efficacy and toxicity studies. Active learning is the process whereby the system seeds into the suggestions not just compounds targeted

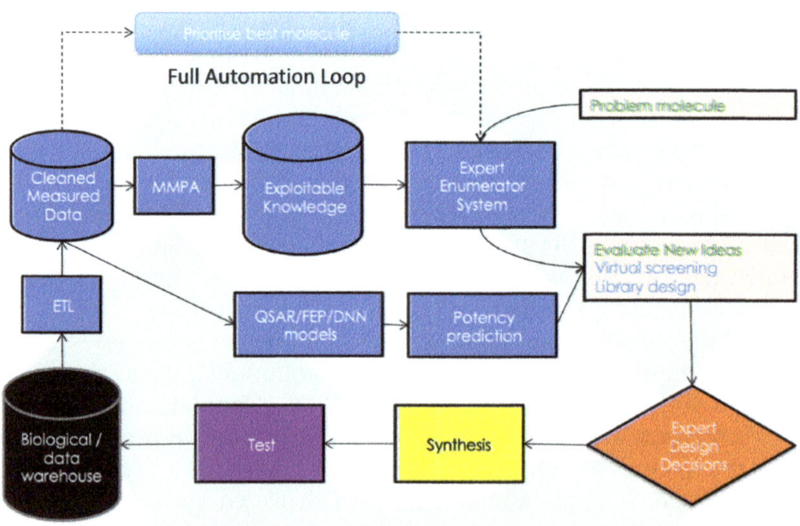

Figure 12.6 Fully automated and integrated MMPA system.

at improving against the required goal, but also those that will extend and improve the model.[28] Here we see the tension in an 'explore *versus* exploit' balance. Initially, it would be more productive for the system to explore chemical space until the models are of sufficient quality to focus on exploiting the knowledge. Once the models are of sufficient quality then switching to the exploit phase should allow identification of better compounds. This is an algorithmic equivalent of the medicinal chemist's task to balance in their strategy between 'looking for new chemistry' and optimising the area they are in.

It can be envisaged that eventually a system could consider the current properties of a lead compound and then set the goal of optimising specific properties where the compound is deficient. As the data are obtained that suggest the goal is sufficiently delivered, the optimisation could switch to a different goal. An example would be the shift from optimising for potency to optimising metabolic clearance. Although chemists often claim both are being optimised in parallel, the reality is that one is frequently optimised first, and a decision to choose to stop attempting to make even more potent compounds can be a difficult one. The two 'flavours' of project: trajectories starting from higher blood levels, and moderate intrinsic potency or lower blood levels and high potency represent two approaches to achieving the same *in vivo* effect. An algorithmic analysis of this problem may be challenging, but it might be useful to have some unbiased view. Attempting to optimise several properties at the same time is extremely difficult in any field, one of the more accessible approaches is to use Pareto optimisation where the objective is to identify compounds that are the best in each property in turn for each weighting of all the multiple properties.[29,30]

12.3.3 Other Matched Pair Technologies

Beyond the simple comparison of pairs of compounds there have been other developments in pair based analysis. As these are newer, their use has been less widespread and therefore both the cases studies for success and their maturity for integration into AI systems less evolved, however they represent areas for development.

12.3.4 Fuzzy Matched Pairs

An alternative approach to increasing the aggregation of matched pair descriptions to enhance the signal, is to combine examples into classes. The assignment of different chemical structures into pharmacophore families is well established in the QSAR domain and Boehringer–Ingelheim workers described how they could create 'Fuzzy Matched Pairs'.[31] Here both the environment round a point of change and the change itself are described in terms of classical pharmacophores: hydrogen bond donor and acceptor, donor/acceptor, aromatic ring, halogen. This is an effective method for capturing knowledge and describing the SAR of a target. This approach does introduce the issue of how pharmacophores are defined, either as a coarse-grained description as done by Gobbi and Poppinger,[32] or to define hydrogen bond donors and acceptors in a more complex, but probably more realistic way such as described by Taylor and Cosgrove?[33] Capturing the information as pharmacophore pairs however, does not tell the user specifically what compounds to make next rather recommendations are at the pharmacophore level. Although a two step enumeration process can be envisaged, where first pharmacophore suggestions are made and then preferred examples of given pharmacophores generated, reduction of this concept to practice has not been published to date. As yet this gap precludes integrating this method into a closed loop AI process.

12.3.5 Matched Molecular Series

MMPA has proved to be effective in delivering insights for the small set of biological targets that have very large data sets such as ADMET assays, however it is less applicable in the analysis of target potency. This is unsurprising, as the same substituent might be binding in completely different receptor sub-pockets in different protein targets and therefore the same transformation of that substituent would be expected to have radically different effects on the binding affinity. This was robustly demonstrated by Hadjuk and Sauer[34] who analysed 50 127 compound pairs across 30 targets and showed that 'a nearly normal distribution of potency changes is observed for all substituents'. One approach to splitting out the different situations is to use the compounds in chemical series. This method described as 'Matched Molecular Series' (MMS)[35] was introduced as a graphical way of understanding the SAR of sets of compounds and is a rigorous approach to the medicinal chemistry practice of describing a congeneric chemical series. A matched molecular series 'contains all compounds that differ only by a single modification at a specific site.' These could be a set of compounds all identical except for at one position where they are substituted by different halogen groups. This concept was formalised in the development of the Matsy algorithm by workers at NextMove and AstraZeneca.[36] A key insight in this algorithm is realising that for a given matched molecular series, for example the hydrogen + halogen series: Br, Cl, F, H, the order of the potencies of the compounds do not occur at random – because sub-pockets that bind a halogen have an order of affinity based on the characteristics of that pocket. So if all the incidences of this MMS are observed, the frequency of a given potency sequence can be compared to what would be expected at random and an enrichment factor calculated. For example: Br > Cl > F > H would be expected to be observed 1/4! of the time in a random distribution, but frequency analysis by the Matsy authors reveals that Br > Cl > F > H actually occurs 5.6× more frequently than this. This fits with a medicinal chemist's instinctive 'correlation with size or lipophilicity' mental model. Once large data sets of MMS have been compiled, it is then possible to compare a series for the target under study, with all the series' that it correlates with and predict what the most likely better substituent would be. This method suffers a version of the same issue that environment partitioned MMPA suffers: the quality of prediction is correlated with the number of members in a series, but the larger the size of a series, the fewer examples of that series exist. In the NextMove study, they show that, for practical use, matched molecular series should have at least five members, this will obviously be a minority of cases and may limit the method to those with extremely large data sets available. A very detailed analysis of the best method of comparing MMS has been carried out by Kramer *et al.* at Roche[37] where they demonstrate that the centred RMSD represents the best approach for comparing the similarity of series. Implementation of the MMS methodology onto an industrial scale has been carried out by Keefer and Chang at Pfizer, where they analysed over 5.3 million data points from 7759 assays.[38] The Pfizer workers used the squared Pearson's correlation coefficient (R^2) as a metric for correlation between series, and to cope with the scale of the problem used a graph database to store and query the inter-series relationships. On this scale, the Pfizer workers encountered significant levels of spurious correlations, but addressed this by exploiting the network behaviour of series essentially clustering sets of series – if a series belongs to a cluster of series that all show correlated series orderings, then a prediction is more likely to be correct.

The primary uses of MMS are parallel to MMPA to either suggest a better substituent (linker or core) or to predict the probable potency of a proposed compound. In addition, in highlighting biological targets that show parallel SAR in a quantitative manner, MMS may be used as an early warning of secondary pharmacology. In particular, MMS may highlight if the primary target potency and an off target potency are displaying parallel SAR and so it may be more fruitful to modify a different area of the molecule to gain selectivity. MMS is of clear interest in pharmaceutical companies and early examples of its use in academic drug hunting are starting to appear.[39]

12.3.6 MMPA Enhanced by Protein Structural Data

An alternative solution to the problem of noise introduced by different substituents binding in different protein environments is to directly use protein–ligand structural data. Several different methods have been published based on this idea and the topic has been reviewed recently.[40] A common challenge is that only rarely do both members of a matched pair have crystal structures determined, and that due to the low number of examples, statistical significance of rules is hard to establish. A brief summary of the approaches with their strengths and weaknesses is given below.

The first: VAMMPIRE[41] generates pairs of structures by using the maximum common substructure from a ligand with a crystal structure to generate an overlay for its pair partner without a crystal structure and then minimises the structure within the protein cavity. The interaction is then characterised in three ways: by identifying the three nearest amino acids to the changing fragment, just the side chain amino acids interacting with the fragment, and finally the SYBYL atom types of the atoms closest to the fragment. This allows a query to be made consisting of the fragment to be changed and the environment surrounding it. From public data the authors were able to identify 8972 matched molecular pairs (MMPs) from 142 biological targets. In one-third of cases, however, only single atom fragments were identified which draws out the key issue, the sparsity of x-ray crystallographic data compared to measured IC_{50} or K_i data. One strength of this approach is the simplicity of the coding of the environment which is easy to communicate to the user. This approach was then further developed by the same group with the VAMMPIRE-LORD[42] algorithm where the characterisation of the protein–ligand interaction was replaced by the use of a fingerprint constructed from the atom–atom distances in the ligand and receptor–ligand distances using Estate[43] atom types. This approach improves the specificity of receptor sub-pocket characterisation, but at the expense of clarity of communication.

In parallel with the development of VAMMPIRE, a group at BMS described their approach to enhancing matched molecular pair analysis.[44] Their 3DMP is a single target focussed tool where the key step is generating high quality overlays within an active site from multiple crystal structures and inferring probable binding modes by similarity. The attachment points of the fragment of multiple matched pairs are then mapped as points within the active site. A query can then be generated that filters possible substituents to those that are likely to bind in the same region as the query. Binding pockets can be annotated with regions where there is positive and negative effect of fragments on binding. Unlike VAMMPIRE, no explicit use of the protein sequence is used, the ligand–protein structures are just used to generate a biologically meaningful overlay.

OOMMPPAA[45] from GSK is another single target focussed technique, here the critical difference is to use pharmacophore features similar to those used by Geppert and Beck in their fuzzy matched pairs approach.[31] Again matched pairs of compounds are identified, and the compounds without known crystal structures are modelled by common substructure template overlay, conformation enumeration and minimisation. The protein structure is deliberately omitted in the minimisation, as the objective is to understand how pairs of compounds are binding differently, therefore features that generate a reduction in activity need to be maintained in position. The pharmacophore points of each compound in a pair are then identified and those points that are different identified. This enables the change in pharmacophore disposition to be associated with a change in potency. Significant care is taken to generate 2D and 3D visualisations that can summarise the data to the user.

To date, reports of the use of 3D MMP methods have been fewer than the 2D methods, this may be due to their more recent development, or may be that the lack of structural data hampers the generation of the statistical support that 2D MMPA enjoys. Inevitably their use will be more limited as fewer biological targets currently have protein–ligand structural data, however this may change as high-resolution cryo electron microscopy becomes both industrialised and more widespread. They do however enable the summary of activity and protein–ligand structural data in a format that is more readily digestible to the user. In groups that focus on structure based drug design, 3D methods may be more influential, however how they can be automated and industrialised for use within a closed loop AI driven design–make–test–analyse cycle is an unaddressed challenge.

12.4 Future Developments

Matched molecular pair based methods are one of the most commonly used approaches in drug hunting. As a data mining methodology they have the advantages that they are open to coarse parallelisation and the results are both understandable, and exploitable in both descriptive and imperative (or generative) modes. However, the scale of the calculations and size of the databases generated makes many of the challenges in the software engineering and integration phases rather than development of the basic science. The integration of MMS for potency and MMPA for ADMET appears well suited to an AI platform and 3D methodologies appear to be an area where there is an opportunity to integrate the description of binding sub-pockets with binding data in a matched pairs format. However, making the descriptions and visualisation of 3D cavities clearer and quantifying the differences between two 3D fields still requires development.

12.5 Summary

Matched molecular pair analysis is a subset of QSAR analysis almost uniquely well accepted by medicinal chemists because of its interpretability. MMPA can also generate useful, credible new molecular structures. Though requiring large data sets to be statistically well supported, the development of unsupervised algorithms has made very large scale MMPA possible. The latter is now more of a software engineering than an algorithm development challenge. The extensions of fuzzy MMPA, matched molecular series analysis and 3D MMPA shows the field is still developing and has applicability in areas of both potency analysis and understanding and solving ADMET issues in drug discovery.

References

1. P. W. Kenny and J. Sadowski, Structure Modification in Chemical Databases, in *Methods and Principles in Medicinal Chemistry [Internet]*, ed. T. I. Oprea, Wiley-VCH Verlag GmbH & Co. KGaA, Weinheim, FRG, [cited 2016 Feb 25], 2005, pp. 271–285, Available from: http://doi.wiley.com/10.1002/3527603743.ch11.

2. A. G. Leach, Matched molecular pairs as a guide in the optimization of pharmaceutical properties; a study of aqueous solubility, plasma protein binding and oral exposure, *J. Med. Chem.*, 2006, **49**, 6672–6682.
3. E. Griffen, A. G. Leach, G. R. Robb and D. J. Warner, Matched Molecular Pairs as a Medicinal Chemistry Tool: Miniperspective, *J. Med. Chem.*, 2011, **54**(22), 7739–7750.
4. A. G. Dossetter, E. J. Griffen and A. G. Leach, Matched Molecular Pair Analysis in drug discovery, *Drug Discovery Today*, 2013, **18**(1513), 724–731.
5. G. Papadatos, Lead optimization using matched molecular pairs: inclusion of contextual information for enhanced prediction of hERG inhibition, solubility, and lipophilicity, *J. Chem. Inf. Model.*, 2010, **50**, 1872–1886.
6. A. Dalke, J. Hert and C. Kramer, mmpdb: An Open-Source Matched Molecular Pair Platform for Large Multiproperty Data Sets, *J. Chem. Inf. Model.*, 2018, DOI: 10.1021/acs.jcim.8b00173.
7. D. J. Warner, E. J. Griffen and S. A. St-Gallay, WizePairZ: A Novel Algorithm to Identify, Encode, and Exploit Matched Molecular Pairs with Unspecified Cores in Medicinal Chemistry, *J. Chem. Inf. Model.*, 2010, **50**(8), 1350–1357.
8. C. Kramer, A. Ting, H. Zheng, J. Hert, T. Schindler and M. Stahl, *et al.*, Learning Medicinal Chemistry Absorption, Distribution, Metabolism, Excretion, and Toxicity (ADMET) Rules from Cross-Company Matched Molecular Pairs Analysis (MMPA): Miniperspective, *J. Med. Chem.*, 2017, DOI: 10.1021/acs.jmedchem.7b00935.
9. M. L. Lewis and L. Cucurull-Sanchez, Structural pairwise comparisons of HLM stability of phenyl derivatives: introduction of the Pfizer metabolism index (PMI) and metabolism-lipophilicity efficiency (MLE), *J. Comput.-Aided Mol. Des.*, 2009, **23**, 97–103.
10. A. G. Dossetter, A statistical analysis of in vitro human microsomal metabolic stability of small phenyl group substituents, leading to improved design sets for parallel SAR exploration of a chemical series, *Bioorg. Med. Chem.*, 2010, **18**, 4405–4414.
11. A. G. Dossetter, A matched molecular pair analysis of in vitro human microsomal metabolic stability measurements for methylene substitution or replacements for parallel SARof those transforms more likely to have beneficial effects, *MedChemComm*, 2012, **3**(12), 1518.
12. P. Gleeson, G. Bravi, S. Modi and D. Lowe, ADMET rules of thumb II: a comparison of the effects of common substituents on a range of ADMET parameters, *Bioorg. Med. Chem.*, 2009, **17**, 5906–5919.
13. T. J. Ritchie and S. J. F. Macdonald, Heterocyclic replacements for benzene: Maximising ADME benefits by considering individual ring isomers, *Eur. J. Med. Chem.*, 2016 Nov, **124**, 1057–1068.
14. R. P. Sheridan, P. Hunt and J. C. Culberson, Molecular Transformations as a Way of Finding and Exploiting Consistent Local QSAR, *J. Chem. Inf. Model.*, 2006, **46**(1), 180–192.

15. R. Hariharan, A. Janakiraman, R. Nilakantan, B. Singh, S. Varghese and G. Landrum, *et al.*, MultiMCS: A Fast Algorithm for the Maximum Common Substructure Problem on Multiple Molecules, *J. Chem. Inf. Model.*, 2011, **51**(4), 788–806.
16. J. Hussain and C. Rea, Computationally Efficient Algorithm to Identify Matched Molecular Pairs (MMPs) in Large Data Sets, *J. Chem. Inf. Model.*, 2010, **50**(3), 339–348.
17. I. Lukac, J. Zarnecka, E. J. Griffen, A. G. Dossetter, S. A. St-Gallay and S. J. Enoch, *et al.*, Turbocharging Matched Molecular Pair Analysis: Optimizing the Identification and Analysis of Pairs, *J. Chem. Inf. Model.*, 2017, **57**(10), 2424–2436.
18. Accenturelabs, Understanding-Machines-Explainable-AI, https://www.accenture.com/_acnmedia/PDF-85/Accenture-Understanding-Machines-Explainable-AI.pdf [Internet], Accenture-Understanding-Machines-Explainable-AI, Available from: https://www.accenture.com/_acnmedia/PDF-85/Accenture-Understanding-Machines-Explainable-AI.pdf.
19. R. P. Sheridan, Interpretation of QSAR Models by Coloring Atoms According to Changes in Predicted Activity: How Robust Is It?, *J. Chem. Inf. Model.*, 2019, DOI: 10.1021/acs.jcim.8b00825.
20. C. Rudin, Stop explaining black box machine learning models for high stakes decisions and use interpretable models instead, *Nat. Mach. Intell.*, 2019, **1**(5), 206–215.
21. T. Fujita, M. Adachi, M. Akamatsu, M. Asao, H. Fukami and Y. Inoue, *et al.*, Background and features of emil, a system for database-aided bioanalogous structural transformation of bioactive compounds, in *Pharmacochemistry Library [Internet]*, Elsevier, [cited 2016 Mar 23], 1995, pp. 235–273. Available from: http://linkinghub.elsevier.com/retrieve/pii/S0165720806800528.
22. K. D. Stewart, M. Shiroda and C. A. James, Drug Guru: A computer software program for drug design using medicinal chemistry rules, *Bioorg. Med. Chem.*, 2006, **14**(20), 7011–7022.
23. E. Griffen, The rise of the intelligent machines in drug hunting?, *Future Med. Chem.*, 2009, **1**, 405–408.
24. E. J. Griffen, A. G. Dossetter, A. G. Leach and S. Montague, Can we accelerate medicinal chemistry by augmenting the chemist with Big Data and artificial intelligence?, *Drug Discovery Today*, 2018, **23**(7), 1373–1384.
25. National Science and Technology Council, *The National Artificial Intelligence Research and Development Strategic Plan [Internet]*, CreateSpace Independent Publishing Platform, 2016, p. 48, Available from: https://catalog.data.gov/dataset/the-national-artificial-intelligence-research-and-development-strategic-plan.
26. J. Delaney, Modelling iterative compound optimisation using a self-avoiding walk, *Drug Discovery Today*, 2009, **14**(340), 198–207.
27. D. Fourches, E. Muratov and A. Tropsha, Trust, But Verify: On the Importance of Chemical Structure Curation in Cheminformatics and QSAR Modeling Research, *J. Chem. Inf. Model.*, 2010, **50**(7), 1189–1204.

28. D. Reker and G. Schneider, Active-learning strategies in computer-assisted drug discovery, *Drug Discovery Today*, 2015, **20**(4), 458–465.
29. V. J. Gillet, M. J. Bodkin and D. Hristozov, Multiobjective *De Novo* Design of Synthetically Accessible Compounds, in *De Novo Molecular Design [Internet]*, ed. G. Schneider, Wiley-VCH Verlag GmbH & Co. KgaA, Weinheim, Germany, [cited 2019 Jul 3], 2013, pp. 267–285. Available from: http://doi.wiley.com/10.1002/9783527677016.ch11.
30. V. J. Gillet, P. Willett, P. J. Fleming and D. V. S. Green, Designing focused libraries using MoSELECT, *J. Mol. Graphics Modell.*, 2002, **20**(6), 491–498.
31. T. Geppert and B. Beck, Fuzzy Matched Pairs: A Means To Determine the Pharmacophore Impact on Molecular Interaction, *J. Chem. Inf. Model.*, 2014, **54**(4), 1093–1102.
32. A. Gobbi and D. Poppinger, Genetic optimization of combinatorial libraries, *Biotechnol. Bioeng.*, 1998, **61**(1), 47–54.
33. R. Taylor, J. C. Cole, D. A. Cosgrove, E. J. Gardiner, V. J. Gillet and O. Korb, Development and validation of an improved algorithm for overlaying flexible molecules, *J. Comput.-Aided Mol. Des.*, 2012, **26**(4), 451–472.
34. P. J. Hajduk and D. R. Sauer, Statistical analysis of the effects of common chemical substituents on ligand potency, *J. Med. Chem.*, 2008, **51**, 553–564.
35. M. Wawer and J. Bajorath, Local Structural Changes, Global Data Views: Graphical Substructure on ligand potencyactionQSAR Mo, *J. Med. Chem.*, 2011, **54**(8), 2944–2951.
36. N. M. O'Boyle, J. BostrAp, R. A. Sayle and A. Gill, Using Matched Molecular Series as a Predictive Tool To Optimize Biological Activity, *J. Med. Chem.*, 2014, **57**(6), 2704–2713.
37. E. S. R. Ehmki and C. Kramer, Matched Molecular Series: Measuring SAR Similarity, *J. Chem. Inf. Model.*, 2017, **57**(5), 1187–1196.
38. C. E. Keefer and G. Chang, The use of matched molecular series networks for cross target structure activity relationship translation and potency prediction, *MedChemComm*, 2017, **8**(11), 2067–2078.
39. A. Crusco, H. Whiteland, R. Baptista, J. E. Forde-Thomas, M. Beckmann and L. A. J. Mur, *et al.*, Antischistosomal Properties of Sclareol and Its Heck-Coupled Derivatives: Design, Synthesis, Biological Evaluation, and Untargeted Metabolomics, *ACS Infect. Dis.*, 2019, 5(7), 1188–1199.
40. E. S. R. Ehmki and M. Rarey, Exploring structure-activity relationships with three-dimensional matched molecular pairs – A review, *ChemMedChem*, , DOI: 10.1002/cmdc.201700628, 2017.
41. J. Weber, J. Achenbach, D. Moser and E. Proschak, VAMMPIRE: A Matched Molecular Pairs Database for Structure-Based Drug Design and Optimization, *J. Med. Chem.*, 2013, **56**(12), 5203–5207.
42. J. Weber, J. Achenbach, D. Moser and E. Proschak, VAMMPIRE-LORD: A Web Server for Straightforward Lead Optimization Using Matched Molecular Pairs, *J. Chem. Inf. Model.*, 2015, **55**(2), 207–213.

43. L. H. Hall and L. B. Kier, Electrotopological State Indices for Atom Types: A Novel Combination of Electronic, Topological, and Valence State Information, *J. Chem. Inf. Model.*, 1995, **35**(6), 1039–1045.
44. S. L. Posy, B. L. Claus, M. E. Pokross and S. R. Johnson, 3D Matched Pairs: Integrating Ligand- and Structure-Based Knowledge for Ligand Design and Receptor Annotation, *J. Chem. Inf. Model.*, 2013, **53**(7), 1576–1588.
45. A. R. Bradley, I. D. Wall, D. V. S. Green, C. M. Deane and B. D. Marsden, OOMMPPAA: A Tool To Aid Directed Synthesis by the Combined Analysis of Activity and Structural Data, *J. Chem. Inf. Model.*, 2014, **54**(10), 2636–2646.

CHAPTER 13

Molecular De Novo *Design Through Deep Generative Models*

OLA ENGKVIST,* JOSEP ARÚS-POUS, ESBEN JANNIK BJERRUM AND HONGMING CHEN

Hit Discovery, Discovery Sciences, R&D Biopharmaceuticals, AstraZeneca Gothenburg, Sweden
*Email: ola.engkvist@astrazeneca.com

13.1 Introduction

Machine learning (ML) and artificial intelligence (AI) have had a renaissance during the last few years and have become a hot topic not only in drug discovery but in the whole of society. There are many reasons for the comeback: access to a larger volume of data through automation, faster computers (*i.e.* GPUs) and methodological progress within deep learning. Drug discovery has also benefited from these trends and, as shown in this book, ML and AI are becoming much more prominent.[1] Besides impacting areas that have been using ML for many years such as QSAR modelling, completely new areas have opened up with deep learning (DL). One is synthesis prediction, where rule-based methods have been replaced by ML methods.[2,3] But most importantly, an area that has been transformed by DL is molecular *de novo* generation, which will be discussed in this chapter.

The goal with deep-learning-based *de novo* molecular design is to be able to sample the whole chemical space. Estimations of the size of the chemical

Drug Discovery Series No. 75
Artificial Intelligence in Drug Discovery
Edited by Nathan Brown

Published by the Royal Society of Chemistry, www.rsc.org

space vary wildly, but the most common estimate is that it consists of 10^{60} molecules.[4] Irrespective of how large the chemical space is everyone agrees that it is too large to be explicitly enumerated.

Historically, molecular *de novo* design has been done in several ways.[5] Most commonly, when structure or ligand-based constraints are given, molecules can be generated *in silico* to fulfil them. This can be done using brute-force: fully enumerating a virtual library and scoring each of the compounds on how well they fulfil the constraints. The best scoring molecules are then prioritized for synthesis. These virtual libraries are mainly constructed from in-house or commercially available building blocks and reactions that are assumed to be robust. There have also been efforts to search libraries that are non-enumerable with various search techniques such as genetic algorithms. While many successes using these techniques have been reported in the literature, it is also possible to point out drawbacks.[6] The main difference with deep learning approaches that will be described in this chapter is the lack of prior knowledge of what a drug-like molecule should look like. This concept is neither present in combinatorial enumeration of libraries nor in genetic algorithm types of approach.

DL-based molecular *de novo* generation has recently been reviewed extensively.[7] The prospective user needs to make several choices on how to generate molecules. The first one is to decide if the generation will be string- or graph-based. Another is to decide which architecture to use. The main advantage of DL is that a huge array of architectures can be used. For example, recurrent neural networks (RNN), variational auto-encoders (VAE) or generative adversarial networks (GAN). In this chapter both string-based and graph-based methods are reviewed, and the different DL architectures discussed. With the explosion of articles describing DL-based molecular *de novo* generation in the last 2–3 years there has been an increased awareness that it is necessary to create benchmarks to measure the diversity and the coverage of the chemical space of generated molecules. That is why a large part of the chapter will focus on discussing the latest developments in benchmarking. It is important that benchmarks cover both explorative aspects, which corresponds to identifying a new chemical series and exploitative aspects which corresponds to optimizing a chemical series.

13.2 Sequence-based Methods for *De Novo* Generation of Small Molecules

Sequence-based methods for *in silico* generation of molecules generally uses the simple molecular line entry system (SMILES) format.[8] The format was originally proposed as a way to describe molecules as strings and uses a sequence of letters to specify elements combined with special characters that enables branching "(" and ")", ring closures ("1", "2", "3", ...) and different bond orders ("–", "=" and "#"). Special properties such as charges, isotopes and explicit hydrogens are handled by adding modifiers (" + ", " − ", "H", *etc.*)

to atoms enclosed in square brackets. The character case can be used to denote aromaticity of a molecule and the format can handle stereochemistry. Together these characters fully describe the molecular structure. Figure 13.1 A and B illustrate how a molecule is encoded as a SMILES string by choosing a path through the molecule and using branching and ring-closures for non-serial parts.

A molecule can have several different SMILES strings, but one SMILES will always be parsed into the same molecule. The derivation of the SMILES will be dependent on how the path through the molecular graph is taken. As this is a problem for identification based on the SMILES string, algorithms were quickly developed that canonicalize how it should be derived and ensure a one to one relationship between SMILES and molecule.[9] However, most cheminformatics toolkits do not use the same canonicalization algorithm.

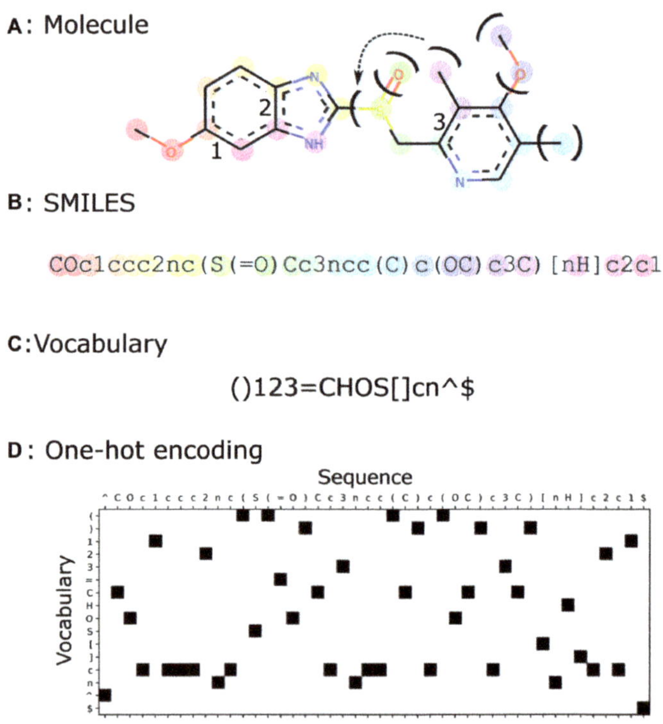

Figure 13.1 From molecule to one-hot encoded "piano-roll". A: The molecule is parsed to a SMILES string by taking a path through the molecule, inserting branches (brackets) and ring-closures (numbers) as needed. A long-range branch is indicated with a dashed arrow. B: SMILES representation of the molecule with the same color highlighting as the molecule above. C: The vocabulary is here shown simplified as the characters of the SMILES added ^ and $ as start- and end-tokens respectively. It should be derived from all the training data. D: In one-hot encoding a bit is set for each character corresponding to the index in the vocabulary.

This one-to-many relationship between molecules and SMILES strings has been exploited as data augmentation for SMILES-based neural network modelling.[10–14]

De novo generation is not a new field and various methods to generate novel chemical matter *in silico* have been proposed such as fragment combinations in virtual library enumerations and genetic algorithm-based methods.[5] However, the realization that the SMILES syntax could be utilized together with recurrent neural network to generate novel SMILES strings after training the networks token by token have led to a renewed interest in the field.[15–18] The flexibility of the neural network architectures alongside the simplicity of the SMILES syntax have led to many different architectures and solutions.

13.2.1 Embeddings and Tokenization

In all the SMILES-based generative architectures, the SMILES string are represented as one-hot encoded vectors, in which each character or token is represented by setting a bit in a vector. The first step is to determine which characters (or tokens) are used as a vocabulary. It is also possible to deconstruct the SMILES into tokens with multiple characters. This enables each atom-type to get its own representation. As an example, the chlorine atom can be represented with the code for "C" followed by the code for "l" or be merged into the code for "Cl", better representing the actual chemistry. The encoded SMILES are usually prepended with a "<BEGIN>" token and the end of the sequence appended with an "<END>" token. These characters can be anything, given that they are not part of the regular SMILES syntax. Traversing through the SMILES from one end to the other, the one-hot encoded token vectors are concatenated into a two-dimensional array having the tokens along one side and the position in the original SMILES along the other in a piano-roll-like format. An example of the encoded format and how it relates to the SMILES string is shown in Figure 13.1 part D. To enable training on SMILES of different length in the same batch multiple "<END>" tokens can be appended or use SMILES of the same length in each mini-batch.

13.2.2 Recurrent Neural Networks

Recurrent neural networks (RNNs) are a type of neural network that feed their output back to themselves with a recurrent connection spanning a timestep. The influence of previous computations makes it possible that an output at a given timestep could be different even though the input is completely the same as in another timestep. This makes these networks ideally suited to work with sequences of arbitrary length and to handle the case where the same input is dependent on the context and should lead to different output. In the context of a SMILES string, an oxygen atom encoded as an "O" could mean different things depending on the surrounding characters, such as a double bond "=" or implicit single bond. In the first case, it is much more likely to be terminated with a branch closure ")" or maybe the end token than in the second.

The RNNs performance when handling long range relations are significantly improved with the use of long short-term memory cells (LSTMs)[19] or gated recurrent units (GRUs).[20] These special neural network mini-architectures contain gates that control the flow of information, and allow the network to store computations for more steps through their hidden state. Thus, the modified RNNs can better fit long-range correlations in the sequences analyzed, which is important when handling SMILES ring closures and closing brackets. As an example, in the SMILES string in Figure 13.1, the ring closure pair "1" and "1" spans nearly the whole SMILES and setting a premature "<END>" token during sampling before the last "1" would lead to an invalid SMILES strings. Likewise, the model must keep account of opening and closing brackets. Both LSTM and GRU units have been used with success for *de novo* generation,[15–18,21,22] and most deep learning frameworks have the LSTM and GRU cell layers implemented.[23,24] There is a limit to the sequence length that can be handled with LSTM and GRU cells, and it has been shown that temporal convolutional networks (TCNs) can handle certain types of long-range tasks better LSTM.[25] Fortunately, the maximum SMILES length of most drug-like small molecules is well within the length correctly handled by GRU or LSTM. Molecular validity of the generated SMILES is generally higher than 90%, yet a modified SMILES grammar, DeepSMILES,[26] has been proposed to improve these rates. This grammar changes how ring numbering and branching works and claims that it can perform better in deep learning tasks, but a more careful assessment is needed.

RNNs are trained in a sequential fashion, so that the network at each timestep is fed a token and must predict the next one from a probability distribution with the same size as the available characters or tokens (Figure 13.2). The result of the training is that the neural network fits a posterior probability distribution of the next character that is sampled. After a start token there are certain tokens or characters that are more common, such as C, N and O, and some that are never seen (*e.g.* ")", "]" or bonds "=", "#", "–"). These probabilities change depending on the previous sequence and current input. A good way to illustrate this is through heatmaps of the output probabilities as shown in Figure 13.3, where it is possible to follow how the model varies the probabilities of the next token or character depending on the sequence of inputs so far.

13.2.3 Sampling SMILES from RNNs

After fitting to a training set, these stepwise output probabilities provide a way to sample new molecules. It is possible to get the next character/token in the sequence by using multinomial sampling where the chance of sampling a token is related to the probability distribution (Figure 13.2). After the start

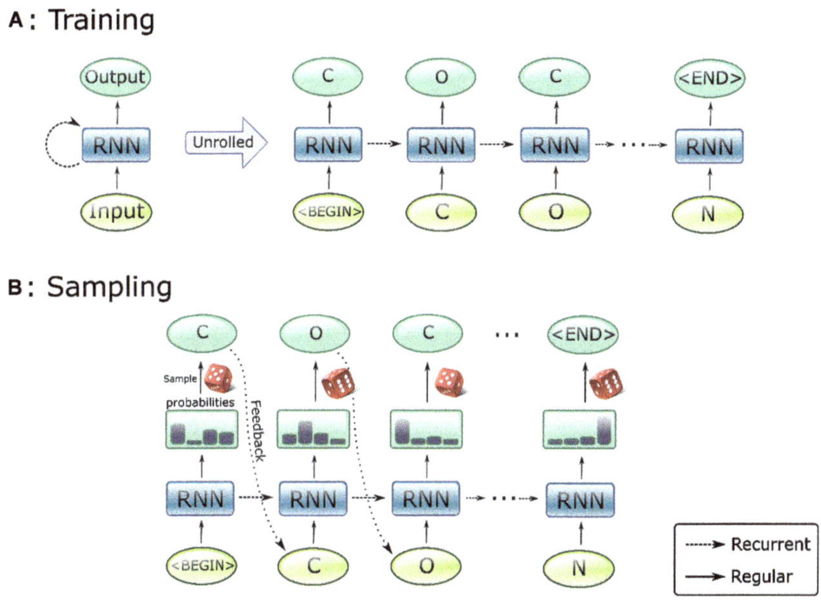

Figure 13.2 Training and sampling of recurrent neural networks. A: Training phase. The recurrent neural network is shown simplified on the left and unrolled on the right. The network is trained to predict the next token or character of existing molecules. The target output is simply the input SMILES with an offset of one step (without the "<BEGIN>" token). The RNN is depicted as a single cell but contains multiple cells and multiple connections to all gates. The recurrent connections (dashed) allows information to flow from previous inputs and computations to the next prediction. B: Sampling phase. It is done one character or token at a time. The learned output probabilities are sampled randomly with multinomial sampling. The choice is recorded and fed back to the network as the next input. Each time the sampling is performed a different SMILES is produced due to the stochasticity of the multinomial sampling.

token, the chance is biggest for sampling a C (Figure 13.3, leftmost column). The sampled character is then fed back as input to the network, giving a probability output for the next character that is then sampled and the result fed back as new input. This process is repeated until the end token is sampled. Due to the stochasticity of the multinomial sampling, the SMILES string produced will not be the same, but will follow the rules for sequence construction extracted from the training set. This works surprisingly well and gives valid SMILES easily. The molecules sampled are similar to, but not the same as in the training set. For metrics and ways to measure the performance of the sampling, see section below on benchmarks and metrics.

Temperature scaling can be used to alter the raw, non-scaled probability values (logits). With *temperature* > 1, the lower probabilities are increased and allow the networks to sample more uncommon combinations of tokens with a greater chance. This gives more diversity to the sampled SMILES.

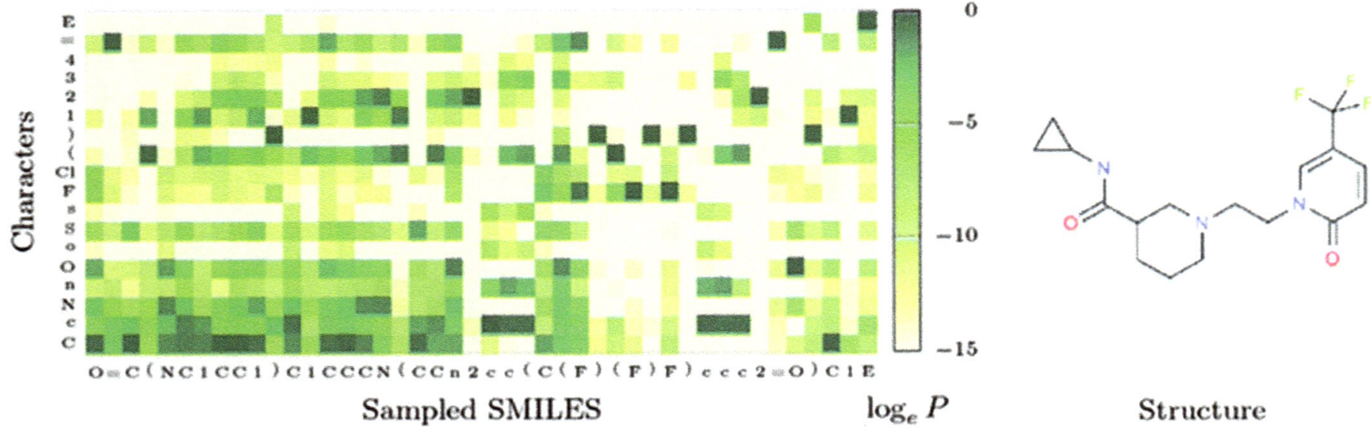

Figure 13.3 Sampling a molecule one character at a time. The heatmap shows the output probabilities given the sequence up to the point of prediction. The multinomial sampling does not always sample the highest probability, and this will give different molecules for each sampling run, as an example, the token "O" is sampled for the first character although "C" has higher probability (darker green). Areas with fewer possibilities are found in between areas of larger possible variation. Decoding the CF_3 group seems to be very limited, possibly a reflection of the fact that CF_3 groups are much more prevalent in training database (ChEMBL) than CF and CF_2 groups.

However, higher temperatures also lead to an increased rate of invalid SMILES strings that cannot be parsed as molecules.[15] Formally, given the vocabulary V and a temperature Y:

$$\forall v \in V\,;\ \mathrm{logit}'_v = \frac{\mathrm{logit}_v}{\gamma}$$

$$p_v = \mathrm{softmax}(\mathrm{logit}'_v) = \frac{e^{\mathrm{logit}'_v}}{\sum\limits_{w \in V} e^{\mathrm{logit}'_w}}$$

13.2.4 Properties and Synthesizability

In general, the distribution of properties follows the prior training set. If fragment-based molecules are used for training, fragment-based molecules will be generated and similarly, training on drug-like molecules will lead to generation of drug-like molecules.[15] Distributions of measured descriptors also follow the training sets closely. A similar effect happens with synthesizability. The generated molecules are of limited use if the ligand candidates suggested only exists as virtual molecules and cannot be made in the laboratory in a stable form. To solve this problem, surrogate scores for synthetic accessibility (*e.g.* SAscore[27]) can be employed to asses this aspect and sometimes retrosynthetic evaluation using in-silico tools.[15] The distribution of the SAscores follows the training sets, although the retrosynthetic analysis showed issues for the compounds with the worst SAscore.[15] The synthesis of generated compounds has also been evaluated experimentally,[28,29] albeit with a significant amount of work-up and filtering after the *de novo* generation.

13.2.5 Advanced Neural Architectures

Gaining better control over the molecular generation has led to a variety of proposed architectures and protocols for controlling and tuning a recurrent neural network. A simple solution is to seed the RNN with a SMILES fragment and let the RNN continue the fragment by sampling.[21,28] This is useful if there is a known core structure crucial for binding, but is limited as it builds up the molecule from only one point of attachment because of the serial nature of SMILES strings. The RNN output "module" described above is either changed post training (Figure 13.4 top) or the RNN is steered by setting the initial states of the RNN in encoder style networks (Figure 13.4 bottom). Combinations of the approaches and various modifications to the same basic architectures have led to a large number of proposed solutions. The lack of agreed standards and metrics make it difficult to compare the approaches based on the publications themselves.

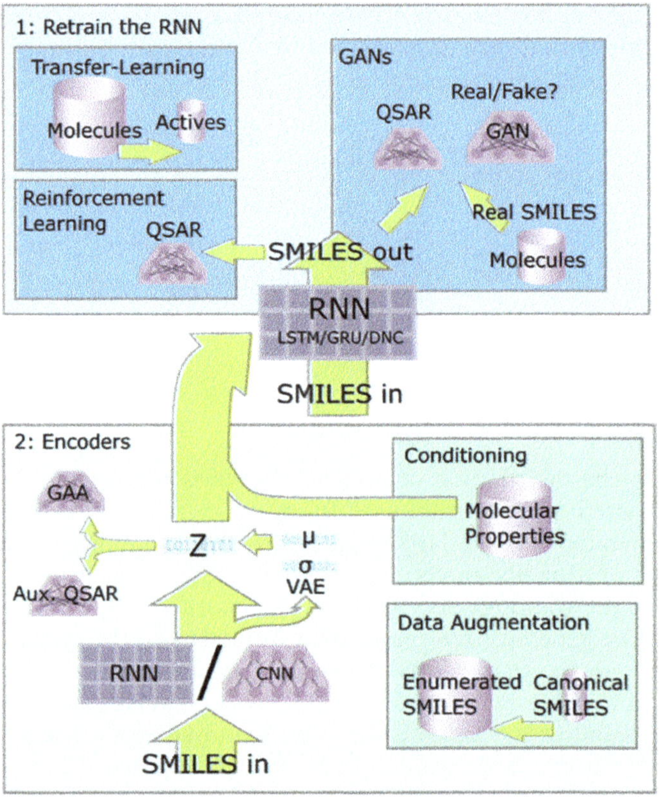

Figure 13.4 Approaches to adjust and tune the molecular generation. 1: Manipulate the RNN by retraining on smaller, specific datasets (transfer learning) or directly tune it to output the desired molecules (reinforcement learning). 2: Train the RNN to output molecules indistinguishable from molecules with known desirable properties using the RNN in a GAN. 3: Manipulate the initial state or generation process itself. Encoder type networks encode the SMILES into a latent code (z) and use this to start the molecular generation. Z can be sampled from known vectors (μ and σ) in variational encoders (VAE) compared to a wanted distribution (GAA) or coupled to QSAR tasks to disentangle the latent space. 4: Manipulation of the training data itself with data augmentation or different representations also influences the properties of the RNN.

13.2.5.1 Retuning the Generator by Transfer Learning

The simplest approach to finely-tune the RNN is transfer learning (Figure 13.4: 1, A), which uses a large prior dataset of general molecules (*e.g.* the whole ChEMBL database[30]) to train the RNN the SMILES syntax and chemical rules. After training on the prior, the networks are refocused by training on a smaller, focused dataset. This has shown that an RNN generates increased amounts of actives similar to the focused dataset.[17,21] Learning the general SMILES syntax on a large dataset first gives much better results than training only on the small dataset itself. The approach was experimentally validated with synthesis of five selected molecules after transfer learning on and fragment growing,[28] in which four molecules were found to be active in the target assay. The prior does not necessarily need to be from the same distribution as the set used for transfer learning, considering the results obtained from training a prior from a database of small drug-like fragments (FDB-17[31]) with drug-like molecules.[32]

13.2.5.2 Reinforcement Learning

In reinforcement learning an agent takes actions that lead to a reward in an iterative process aimed to maximize the rewards. In the *de novo* generation, the RNN is the agent and the generation of molecules the action. The reward is then calculated from a designed score and the output from the RNN is modified through several iterations to optimize this score. The reinforcement learning re-training of the network enables the definition of reward functions that focus on molecular properties, predicted ADME properties, existence of certain motifs and predicted activity in QSAR models, *etc.* This makes it possible to get the RNN to produce molecules with very specific and desirable properties. Reinforcement learning has been used with success to refocus the character-based RNN to generate molecules that better fulfill the criteria of the reward function.[18,33]

13.2.5.3 Generative Adversarial Networks

A generative adversarial network (GAN) has two components, a generator and a discriminator, that compete against each other during training. The generator tries to generate new SMILES that are indistinguishable by the discriminator from a pool of SMILES from known compounds. With each round of training the generator becomes better at generating SMILES that follow the properties of the dataset, while at the same time the discriminator learns to discriminate between real or generated SMILES. To get the generator to sample molecules with specific properties, the GAN training is coupled to an auxiliary task where the generator produces SMILES with the desirable properties and the generator is jointly optimized with reinforcement learning during the GAN training. Examples of such implementations are ORGAN[34] and ORGANIC.[35] The former was tested with both molecular generation as well as musical scores, whereas the latter was targeted directly at an inverse design of molecules. The latter had trouble optimizing towards the discrete values from Lipinski's rule of five[36] heuristic score, but showed some success in optimizing the QED[37] score. GAN + RL-based training was also used in RANC[38] and ATNC[39] where the central RNN was substituted by a differential neural computer (DNC)[40] without and with an adversarial threshold block. The differential neural computer is a more advanced recurrent architecture, where the units have an associated memory block that can be written, accessed and deleted in contrast to the single hidden state of regular GRUs and LSTMs. The authors demonstrated that the DNC-based architectures can handle longer SMILES and have more diversity in the output than the ORGANIC implementation.

13.2.5.4 Encoder–Decoder Architectures

An encoder–decoder network tries to obtain control of the generative process directly by feeding information to the generator module either by setting the initial states or at each timestep. One of these architectures, the autoencoder, consists of an encoder – a code layer – and a decoder that are trained together to give the same output as that fed into the network. During training the autoencoder will have to produce a representation in the code layer that makes it possible to reconstruct the input with the decoder part. For molecular generation the decoder is usually similar to the character-based RNNs described above where the initial states are set using information from the code layer before sampling begins. The task of the encoder is to convert the SMILES string into a latent vector representation that the RNN can decode to the same SMILES. After training on large molecular databases, the code layer can be used to navigate the "molecular space" of the training set with optimization algorithms, such as Bayesian optimization.[16,22,41,42] With this optimization process, the molecular generation can be directed towards given properties such as cLogP, QED and TPSA. Moreover, the latent space representations of molecules have also been used as descriptors in QSAR models[11,12,43] and represent a sort of automatically engineered fingerprint.

In regular autoencoders, the code layer is restricted in its fitting capacity, which means that one-to-one copies are not possible. This can be achieved through compression to lower number of dimensions (bottlenecks). Alternatively, the code layer can be regularized with L2, dropouts, noise layers, *etc.*

In variational autoencoders the fitting capacity is lowered by inserting a sampling from learned means and variances. The mean and variance of the code is set by the decoder and used to sample the latent representation before passing the information to the decoder. Further restrictions are imposed on the code-layer. Specifically, the distribution on average should look similar to a uniform Gaussian. To achieve that, the distributions are compared using the Kullback–Leibler divergence (KLD). This is important as the encoder could otherwise set the variance of the latent representation to be very low and thus transfer it unaltered to the decoder. This adds an uncertainty to the latent information and prevents the decoder from overfitting. The assumption is that meaningful (and normal distributed) variances are extracted from the data. Additionally, when the latent space dimensionality is too large, there is a risk that the network transfers the information with a selected number of bits with a low variance, but still fulfill the KLD by having bits not used by the decoder output a Gaussian.

To further smoothen and disentangle the latent code and introduce *smoothness* and relevance in the code space, the encoder network is often co-trained to predict calculated properties of the molecules from the latent code. In this approach the information from the code layer is not just passed to the decoder, but also to an additional neural network that tries to build a QSAR model of computed properties of the molecules in the training set. The assumption is that the code space itself will be changed to be relevant to both the QSAR task and to the decoding of molecules. This architecture has been used to sample and optimize molecules in the ChemVAE,[16] GrammarVAE[22] and SD-VAE.[42] The two latter add extra processing on both input and output to pre-process and correct the syntax of the SMILES before training and during sampling.

Adversarial autoencoders (AAEs) also enforce a prior distribution to the latent vector. They do this by coupling the code layer to a discriminator (similar to the GANs) that tries to distinguish the resulting code layers from a defined prior distribution. The random prior distribution is usually uniform or normal. Knowing the distribution that the latent layer resembles enables an easier random sampling of novel points but can include artificial biases in it. After comparing several options, an AAE trained to have a uniform distributed latent space and an external support vector machine (SVM) model to optimize the molecular generation using an AAE of compounds active against DRD2.[41]

Instead of optimizing the latent space that is fed to the decoder, the calculated or predicted molecular properties can be fed to the autoencoder during training to further condition the decoder. This approach has been used both for an AAE, the entangled conditional AAE architecture (ECAAE)[29] and for a conditional VAE.[44] After training a network with the provided conditions together with the compressed code layer, it can be used to direct the generator to produce compounds with specific properties.

A third interesting option to manipulate the properties of the latent space is to use format translation or heteroencoders. Instead of using autoencoders, where the same SMILES is encoded and decoded to itself, the encoders are trained to translate between different formats (*e.g.* INCHI to SMILES, enumerated SMILES to SMILES)[12] or from one example of a non-canonical SMILES to another (enumerated to enumerated SMILES).[11] The fundamental assumption is that by presenting variations of the same underlying latent information (the molecule), the code layer is forced to represent the molecular information in the latent space rather than one representation or another. This training approach was shown to lead to code layers much more relevant in QSAR and QSPR tasks.[11,12]

Using enumerated SMILES in the input shows that the sequence-based encoders handle information in the many-to-one cases. Different non-canonical SMILES of the same molecule encode to similar latent spaces. Moreover, heteroencoders can handle one-to-many cases where the same input information needs to match several outputs. Feeding the decoder from a heteroencoder the same latent space vector several times with subsequent sampling, makes the decoder produce different SMILES strings where the majority was non-canonical versions from the same molecule.[11] On the other hand, using the enumeration approach in the output also introduces an uncertainty and fuzziness in the decoding that can be beneficial or disturbing depending on the intended use-case. This one-to-many capability was also used in a case where reduced graph fingerprints were decoded back to probable molecules using LSTM networks. The reduced graphs identified as useful scaffolds, could be translated back to collections of SMILES for further filtering and processing.[45]

Remapping the latent space with dimensionality reduction has also been demonstrated as an effective way to steer the molecular generation into areas of interest. After mapping active molecules to the reduced space, generative topographic mapping in this case, the coordinates of the active areas are identified. The coordinates of the mapping can afterwards be sampled back to probable latent space vectors and then to molecules using the decoder.[46]

A problem seen with encoder architectures is that points in the latent space may not lead to any molecule. Specifically, the decoder will not be able to decode to a SMILES string that can be parsed into a molecule. The latent space points derived from real molecules in the test-set can have very high SMILES reconstruction accuracy and validity, but suggested vectors during interpolation or optimization can lead to production of a high percentage of invalid SMILES. There is the possibility that trying to impose a continuous representation on fundamentally discrete molecules is not feasible. Adding a penalty term to the optimization if the points suggested do not give valid SMILES string can help mitigating the problem.

13.3 Graph-based *De Novo* Structure Generation

The first approaches to deep learning based *de novo* molecule design focused on structure generation using SMILES strings. However, when generating molecules using SMILES intermediate smiles strings cannot be converted to valid molecules, which make it difficult to estimate the intermediate reward for reinforcement learning-based structure generation.[18] To address this issue, methods of generating molecules based on molecular graphs have been proposed and, using this type of method molecules can be directly generated step-by-step as molecular graphs. One of the first proposed approaches was the GGT-NN (gated graph transformation neural network), which modifies a graph based on a sequence of input sentences.[47] This idea was further refined by considering molecule structure generation as a sequence of graph transformations (Figure 13.5).[48] Specifically, a generative model was used to predict if an atom should be added or a bond should be formed among existing atoms.

The algorithm is described as such: For a graph $G = (V, E)$, for each node an embedding vector is generated. These embedding vectors are computed initially from node features and edge features, and then propagated on the graph to aggregate information from the local neighborhood. A GRU NN was

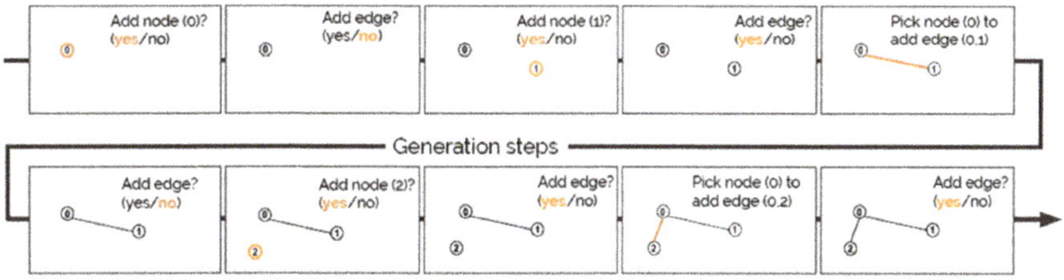

Figure 13.5 The demonstration of the action sequences in the graph-based structure generation.
Reproduced from ref. 48 with permission.

used to generate the node embeddings. The graph generative model defines a policy distribution over the sequence of graph generating decisions (Figure 13.5) and the action for each step is taken based on the NN predicted probability distribution over all possible actions. Several types of action are considered in the algorithm, which includes adding new atoms, forming new bonds among the new atoms and existing atoms and the structure termination. The probability distribution of these types of actions are predicted by individual NNs with the given molecular graph as input and these NNs are trained together to learn the generative model for molecular generation.

Another graph-based structure creation method based on the previous architecture is MolMP.[49] Three types of action are defined for structure generation (Figure 13.6): (1) Append, adding a specific new atom to the graph and forming certain bond types; (2) Connect, forming intra-graph bond between a pair of existing atoms in the graph; (3) Stop, the generation process stopped. The molecule generation process can be treated as a Markov decision process (MDP) or have a recurrence in the generative model. For the MDP approach, a model is trained to predict the policy of making one action at each step by stacking together several graph convolution NNs and feed-forward NNs. Introducing recurrence in the network is done by adding a RNN that keeps the information of the previous steps. Results show that using a RNN improves the physicochemical properties of generated structures.

Graph convolutions[50] can also be used successfully to generate molecules.[51] A graph convolutional policy network is firstly used to calculate the node embedding vectors of the intermediate graph. Then, four different link prediction-based actions are decided using a generative model: (1) Selection of first atom; (2) Selection of second atom; (3) Bond connect between two selected atoms; (4) Structure termination. After a structure generative model is trained, the adversarial-based reinforcement learning is performed to generate structures satisfying certain objectives. Results show that GCPN can achieve a 61% improvement on chemical property optimization over state-of-the-art baselines while resembling known molecules and achieve 184% improvement on the constrained property optimization tasks.

The SMILES-based variational autoencoder methods in general suffer from low validity of generated structures due to the inability of learning

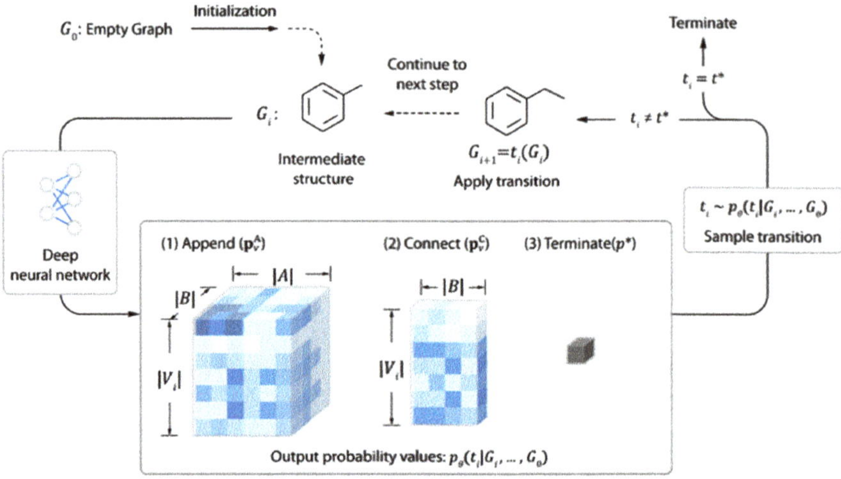

Figure 13.6 The representation of the structure generation process proposed by Li *et al.*[49] The intermediate graph is used as the input to the graph convolution neural networks to predict the probability distribution (*i.e.* policy prediction) among all possible actions. Once the probability distribution is generated, next action is taken by sampling the policy. Reproduced from ref. 49, https://doi.org/10.1186/s13321-018-0287-6, under the terms of a CC BY 4.0 license, https://creativecommons.org/licenses/by/4.0/.

smooth molecular embedding. A graph-based autoencoder method, the junction tree neural network (JTNN), was proposed to address this issue.[52] Specifically, the pharmacophore definition used in the feature tree method[53] can be employed to first convert molecule structures to junction trees. In these, each node represents a cluster of atoms and there are no cycles. As can be seen in Figure 13.7, two different message passing neural networks (MPNN)[54] are used to encode the molecular graph (graph embedding vector Z) and its respective junction trees (tree embedding vector Z) separately. The tree embedding vector is then decoded to recover the junction tree and, in the end, using together graph embedding Z_G and the decoded junction tree, the molecule graph is decoded to reproduce the original input structure. Then, the tree decoder and graph decoder are trained to achieve maximal likelihood of tree structure given Z and predicting correct subgraphs G_i of the ground true graph G at each tree node. In the experiment, their method is able to generate 100% valid molecule structures which significantly improves the results of SMILES-based VAE methods. This is largely due to the usage of pre-defined chemical substructures in the junction tree which helps to focus the neural network on learning the connectivity among chemical substructures instead of the substructures themselves.

Generative adversarial networks (GANs)[55] have been used in graph-based molecular generation. One of the first approaches to appear was MolGAN,[56] which behaves in a similar way as the sequence-based models. In detail, the generator is a feed-forward neural network that takes a multi-dimensional

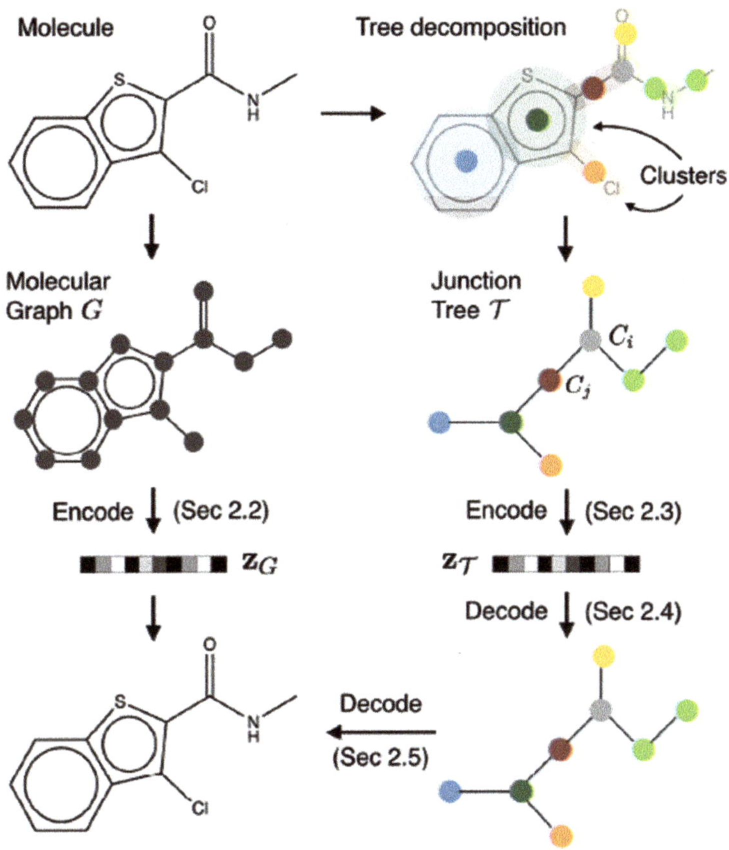

Figure 13.7 The scheme of the junction tree autoencoder method. Reproduced from ref. 52 with permission.

vector sampled from a standard normal distribution as input and has two outputs: an atom feature matrix and a 3D adjacency matrix representing the connectivity among the atoms across the different atom types. A softmax function is applied on the output values to form the probability distribution for the atom type and connectivity at each node of the graph. A categorical sampling is then carried out to obtain a graph. The generated graphs are then sent to the discriminator to see if the discriminator can distinguish them from the real graph data. Besides the normal GAN loss, a reinforcement learning related reward is fed back to the generator to optimize the generated structures toward pre-defined molecular properties. This method does not rely on making a sequence of action to generate the graph and performs faster than other approaches, but it can only generate graphs with fixed molecular size.

Graph-based generative models are a new frontier in *de novo* molecular design. Comparing them with SMILES-based methods, they enable of injecting chemistry directly into the structure generation process. On the other hand, they generally perform slower than the SMILES based methods. This is because molecules need to be broken down into many subgraphs, thus increasing the size of training samples. We believe that graph-based *de novo* design method is still at its infancy, yet it would be interesting to have some research comparing the structures generated by SMILES and graph methods in the future.

13.4 Benchmarking Generative Molecular *De Novo* Design Models

One of the main challenges when working with generative models is trying to assess whether a model has been trained correctly for the task it has been created for. A benchmark, comprised of many complementary metrics, helps assessing the properties of the chemical space sampled and the suitability of the generated molecules given the challenges being addressed. Moreover, it can be used to optimize the training of a model or to compare different generative architectures. Unfortunately, the metrics used in benchmarks for generative models of different content types (*i.e.* image, audio, video) generally cannot be directly applied to molecule generation for two reasons. Firstly, each content type has completely different ways of assessing its quality and secondly, the content generated by any given model affects its domain in a unique way. For example, evaluating correctness in continuous formats such as images or audio is difficult: an image can be mostly correct, but this is not the case for molecules: they either have correct valency in all atoms or not. In addition, a careful choice of metrics must be made to prevent ambiguity: if we compare a metric that evaluates the correctness of molecules generated with another that checks its diversity (*i.e.* how different sampled molecules are from those in the training set), both metrics are independent of each other, because non-diverse sets can have a perfectly high percentage of valid molecules, such as the case of a model generating mostly valid molecules from the training set. Furthermore, all metrics can be labelled depending on the type of model that they are benchmarking. Models range between *explorative* and *exploitative*. Explorative models focus on generating as much chemical space as possible within the limitations of the training set a typical example is scaffold hopping, where a new chemical series needs to be identified. Exploitative models focus on obtaining novel molecules with specific structural or physicochemical properties, which correspond to optimization of a chemical series. Therefore, any benchmark must include a careful selection of metrics that is able to holistically characterize the generative model and account for all possible shortcomings.

13.4.1 Benchmarking Explorative Models

The chemical space explored by a generative model, or domain, is comprised of a set of molecules, each of which having a probability of being sampled. It has been shown that generative models whose generated molecule probability distribution tends to be uniform (*i.e.* all molecules have the same probability of being sampled) maximizes the sampleable chemical space.[57] This is important for explorative models since their main objective is to maximize the chemical space generated. That is why metrics that measure the breadth and the diversity of the chemical space generated are of great importance. Until late 2018, publications describing new explorative generative model architectures used a diverse set of metrics to characterize the models. They can be summarized in the following list:

1. **Validity** (fraction):[11,16–18,21,33,38,39,41,48–52,56,58–60] Fraction of sampled molecules that are valid according to a given chemistry software library. Note that some toolkits (*i.e.* RDKit) fix and alter some otherwise invalid SMILES.
2. **Uniqueness** (fraction):[18,33,38,39,49,56,59,60] Fraction of sampled molecules that are valid and different to each other. This uniqueness can be at a SMILES level (for models that train with SMILES) or at a molecular level, given the multiple SMILES representations for each molecule.
3. **Diversity** (fraction):[29,33,38,41,48,49,56,59–61] Fraction of valid sampled molecules that are not in the training set. This metric can compare to the exact molecules in the training set (*i.e.* the canonical SMILES) or it can otherwise use some similarity threshold.
4. **Drug-likeness** (QED) (scalar):[16,34,35,38,49] Fraction of molecules that resemble drug-like molecules. It is obtained by aggregating different descriptors (*e.g.* logP, HBD, HBA, MW, *etc.*) in a single score. We would advise caution when using this metric, as the applicability domain may be limited. Moreover, the metric contains no stability assessment, so unstable molecules with good QED scores may be generated.
5. **Synthetic availability score** (SAS) (scalar):[15,16,33,34,58,62] Score that predicts the potential synthesizability of molecules. It is obtained by enumerating the fragments of a molecule and checking if these fragments are common in PubChem molecules. After that a complexity score is added to penalize difficult features such as macrocycles or bridge atoms. This score has the same problems as QED, given that its applicability domain may be limited.
6. **Loss** (scalar):[15,21,61] The training process of a generative model is based on minimizing the loss function. The lower the loss function is, the more the sampled molecules will resemble those in the training set. This can evolve into overfitting problems.
7. **Descriptors** (scalar):[15,16,18,21,34,38,39,49,56,58,62] Descriptors such as clogP, molecular weight, number of rotatable bonds, *etc.* are calculated and the distribution is compared to the training set distribution either by mean $\pm$ standard deviation or using more sophisticated methods like the KLD.
8. **Plot representations:**[11,16,17,21] Some descriptors (or fingerprints) are calculated from sampled molecules and from molecules from the

training set. Optionally, a dimensionality reduction method such as principal component analysis (PCA) or t-distributed stochastic neighbor embedding (t-SNE) is used and the first pair of dimensions are plotted. This gives a visual representation of how the generated chemical space differs from the training set.

There is not a general rule on which metrics are used in each publication and it is problematic, as some of the published architectures do not give a holistic overview on the domain of the model. As can be seen in Table 13.1, publications that describe new explorative model architectures generally use validity, descriptors, diversity and uniqueness. Except for descriptors, these metrics are generally the best ones to assess the quality of any explorative model, as they can measure the size of the domain. The other metrics are more specific to each architecture and they must be used carefully because they can add substantial bias to the benchmarking. This must be specially noted for QED and SAS scores, given that their applicability domain is not known, and they may output invalid scores for some molecules.

Recently, a novel metric appeared that includes information from many of the basic metrics in a standardized way. The Fréchet ChemNet distance (FCD)[63] requires two molecule sets: a set of real-world molecules and a sample from the model. These sets are discretized by forward passing all the molecules in these sets up to the penultimate activation layer of ChemNet,[64] a bioactivity prediction model. Then, assuming that the distribution obtained for each set is a multi-dimensional Gaussian, the Fréchet distance is calculated between them. If the trained generative model can create molecules with similar chemical properties to the real-world molecules, the distance will be low. The researchers also showed that this distance is sensitive to changes on many of the widely known metrics, such as descriptors, diversity and drug-likeness. Nonetheless, there is no way of assessing which are the differences between both sets or what is the magnitude of that difference. Additionally, because the ChemNet was trained with a set of known drug-like molecules the domain of applicability of the FCD may be restricted to the known chemical space. This could imply that the metric is not suited for benchmarking models that explore novel regions of the chemical space.

Table 13.1 From a selected group of 23 publications of molecular generative models grouped by the architecture, the number of publications that use each explorative metric in benchmarks. Note that no metric is used throughout all published architectures.

	Total	Val.	Uniq.	Div.	QED	SAS	Loss	Desc.	Plots
SMILES-based RNN	6	4	2	1	0	3	2	4	2
VAEs	8	7	2	4	1	2	0	2	2
Graph-based	3	3	1	2	1	0	0	1	0
GANs	6	3	3	3	3	1	1	4	0
Total	23	17	8	10	5	5	3	11	4

13.4.2 Benchmarking Exploitative Models

Exploitative models focus on exploring a specific region of the chemical space. This is done by either including in the training process a score formed by a set of descriptors (reinforcement learning, loss function, *etc.*) or by using focused training sets (transfer learning, *etc.*). To benchmark these models some of the metrics used in explorative models, such as validity, drug-likeness, descriptors, *etc.*, can be used. Although more specific metrics that better fit the optimization problem are generally preferred. A list with some of the most common used in literature follows:

1. **Similarity** (scalar):[17,18,21,41,49,51,52,58] Fingerprint (*e.g.* ECFP4, MACCS, *etc.*) similarity of all generated molecules to a known molecule or to each other. This metric can be used to assess whether the generated focused chemical space is dense or sparse. Depending on the optimization problem this metric is important, as it is able to show the amount of novelty.
2. **Discovery** (scalar):[16,18,33,41,58,60] Measuring the time (training epochs, distance in latent space, number of iteration in Bayesian optimization, *etc.*) that takes a model to be able to sample a molecule not present in the training set. This metric is especially used to benchmark VAEs and models that use reinforcement learning.
3. **Plot representations:**[17,21,33,49] Using the same method as in the explorative method but this time plotting the sampled molecules with real molecules of interest. For example, plotting known inhibitors of a target with the molecules generated.
4. **Descriptors** (scalar):[16–18,33–35,49,51,52,56,58] Calculate descriptors on the sampled molecules. This metric is mainly used when a model is optimized to obtain molecules with given structural or physicochemical properties.
5. **Enrichment over Random** (EOR) (scalar):[17,49] This metric allows quantificaton of the improvement of a focused model compared to a non-focused one. It is calculated as the ratio of molecules in a training set T on the focused set of sampled molecules compared to the molecules found on a non-focused set. It is exclusively used when comparing explorative with exploitative models, such as those trained with transfer or reinforcement learning.
6. **Target prediction models** (TPM) (fraction):[17,18,29,33,49] Using TPMs or similar to predict whether generated molecules are active on a given set of targets. This metric should be used with care, as the TPM should be trained with molecules outside of the training set, to minimize bias.

As it can be seen in Table 13.2, the most common metrics used are descriptor-based statistics. This makes sense because most of the optimization problems in the published literature are about obtaining molecules with optimized properties. The second most common metric is similarity,

Table 13.2 From a selected group of 23 publications of molecular generative models grouped by the architecture, the number of publications that use each exploitative metric in the benchmarking section. Note, that as in the previous table, no metric is used throughout.

	Total	Sim.	Disc.	Plots	Desc.	EOR	TPM
SMILES-based RNN	6	3	2	3	3	1	3
VAEs	8	3	4	0	3	0	1
Graph-based	3	2	0	1	2	1	1
GANs	6	0	0	0	3	0	0
Total	23	8	6	4	11	2	5

which assesses if the generated chemical space is diverse enough. It is surprising that no published GAN model uses similarity metrics, and it may be to the specificities of the architecture.

13.4.3 Benchmarking Models During Training

One of the main uses of benchmarking is to optimize the training process of a given model architecture. Models with different hyperparameter configurations are trained and after each training epoch some metrics can be calculated and used to assess the quality of the models. Any of the metrics of the previous sections can be used to assess the quality of the half-trained models, although the metrics should not be computationally expensive to calculate. Most published models do not go into details on which metrics were used when training. It is important to note that in generative models we aim to maximize the likelihood of sampling molecules from the training set, but without controlling the diversity it may yield to models that output mostly molecules similar to the training set.

Details of a new benchmarking method have been published[57] where models are trained that sample as much as possible the GDB-13 database and describes an approach to training RNN/Graph/GAN-based models which can be applied to any other chemical space. Two sets of molecules are obtained beforehand: a training set and a validation set, which include molecules from GDB-13 but are not part of the training set. After each epoch, the negative log-likelihood (NLL) distributions of the molecules are calculated from a sample, the training set and the validation set. If the generated molecules from the model have the same probability of being generated to those from the training set and the validation set, it means that the domain of the model is bigger. To assess that, the three Jensen–Shannon divergences (JSD) between each pair of distributions are obtained and plotted for all epochs. Figure 13.1(a) is an example of a model trained with the 1 million molecules obtained randomly from GDB-13. Note that the three distances are minimal around epoch 70–100, meaning that generating a molecule from the training set has the same probability as generating a molecule outside of it. Afterwards, the likelihood of training molecules in the training set is higher than from outside, meaning that the model is overfitting. See that in Figure 13.1(b) the fraction of valid molecules is impervious to this change, as it is always increasing. This happens because when the model is overfit, it becomes more conservative and creates more molecules that are similar to those in the training set (Figure 13.8).

The same happens when the variance of the NLL distributions of the three sets is plotted. A low variance of the validation set NLL distribution implies that the generated molecule probability distribution is more uniform throughout the entire chemical space generated by the model. This means that every time the model is sampled, there will be more diversity. In Figure 13.1(c) the variance of the validation set NLL distribution is also minimal around epochs 60–90. Also note that in Figure 13.1(d), the loss function (NLL average) is also plotted and see that, even though the model is overfitting, the sampled set loss function continues to decrease. With this approach three different training stages were assessed for each model: a learning phase, followed by an optimally trained phase to then move to overfitting. The only downside of this method is that it is sensitive to noise, since it uses small samples to generate the plots for each iteration.

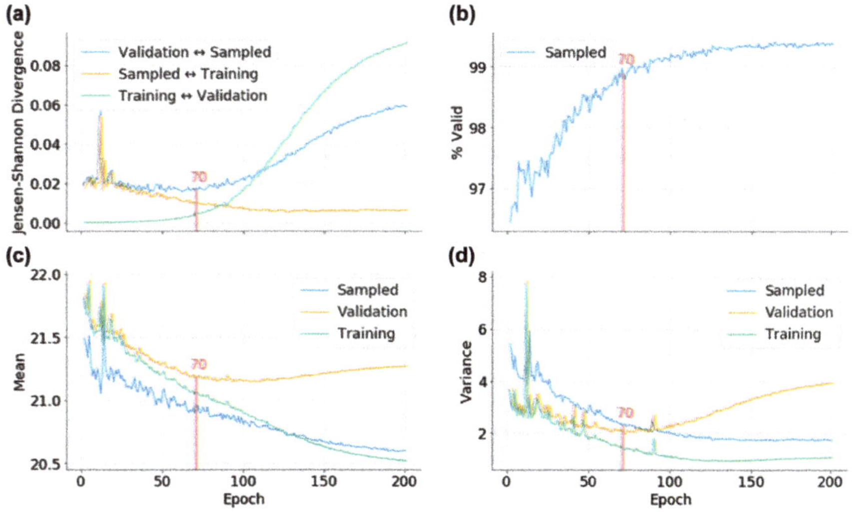

Figure 13.8 Metrics used to evaluate the training process of an RNN + SMILES molecular generative model trained with 1 million molecules from GDB-13.[57] The red line at epoch 70 represents the chosen epoch in the end. (a) JSD plot between the three NLL distributions for each of the 200 epochs. (b) Percentage of valid molecules in each epoch. Notice that the plot already starts at around 96,5%. Mean (c) and variance (d) of the three distributions. Note that spikes around epochs 1–20 are statistical fluctuations common in the beginning of the training process of a RNN, when the learning rate is high.

13.4.4 Comparing Model Architectures

In the recent years a wealth of molecular generative model architectures has been published. Unfortunately, the lack of a consistent and standardized set of metrics has made the process of comparing the improvements between them unreliable. One of the main issues is that there are neither datasets nor case problems to use when performing benchmarks. Moreover, some metrics have different meanings, or even are not applicable, depending on the generative model architecture benchmarked. Three works have been published[57,65,66] that offer a set of standardized methods that try to alleviate some of the current limitations.

A first approach, more akin to how image predictive models are benchmarked,[67] was suggested.[57] The objective of the benchmark is to train a model with a small subset of GDB-13 and use it to generate as much of it as possible sampling the model a specific number of times. The work by Arús-Pous *et al.* shows that a uniform and complete model, which samples uniformly molecules from and only from GDB-13, is the best possible. From this an upper bound can be obtained for any sample size: for example, sampling 2 billion molecules from the uniform model would yield on average 87.12% of the complete database. Different generative models can be trained with any combination of hyperparameters and the unique percentage of molecules generated from GDB-13 obtained. This gives a measure of the learning capability of a molecular generative model.

The second one, GuacaMol,[65] focuses on describing a range of metrics and tests that evaluate different characteristics of generative models. Some of the metrics have been also described in this section, and include validity, uniqueness, diversity, similarity, FCD, discovery, QED and descriptors. It also includes other interesting metrics, such as generating all possible isomers from a given molecular formula and a multi-objective optimization problem. A software implementation was also released alongside the published manuscript.

The last work, MOSES,[66] was made public and describes a more constrained benchmark that measures fragment, scaffold and nearest neighbor similarity, diversity, FCD and descriptors. Another important difference with GuacaMol is that it recommends using a specific subset of molecules to train from obtained from the ZINC Clean Leads database.[68] Also, a software implementation is available.

13.5 Conclusions

DL-based molecular *de novo* generation has emerged during the last three years as one of the most interesting and fast-moving fields in cheminformatics. In this chapter different molecular representations (strings and graphs) and different architectures such as RNNs, VAEs and GANs have been described. The openness in terms of using open source (*e.g.* RDKit,[69] TensorFlow[23] and PyTorch[24]), public databases (*e.g.* ChEMBL[30]) and readily preprint publication on preprint servers such as *arXiv* (https://arxiv.org/) and *ChemRxiv* (https://chemrxiv.org/) have contributed to the remarkable progress that has been seen in the field during the last 2–3 years. However, this *Cambrian explosion* of methods has made it necessary to start focusing on benchmarks to assess how well different methods work. The benchmark criteria used so far have been extensively reviewed in this chapter. Several different architectures can generate drug-like molecules and so far, no architecture has been shown to be superior to others in all metrics. It might very well be that different architectures will be superior under different circumstances. It can also be expected that new architectures might be used in the future.

Although there has been considerable progress in generating molecules covering the whole chemical space, much less progress has been made in scoring the ideas. While experimental validation of DL-based molecular *de novo* generation has been published,[28] it is still too early to fully assess the potential impact on drug design. However, we think that the full potential of DL-based molecular *de novo* design will most likely be in conjunction with synthesis prediction and automation.[70] The potential speed-up of the *design-make-test* cycle through applying AI together with automation has the potential to enhance the drug design process. Creating much faster high-quality tool compounds to validate targets *in vivo* will surely have a transformative impact on the drug discovery process.

Acknowledgements

The project leading to this article has received funding from the European Union's Horizon 2020 research and innovation program under the Marie Skłodowska-Curie grant agreement No 676434, "Big Data in Chemistry" ("BIGCHEM", http://bigchem.eu). The article reflects only the authors' view and neither the European Commission nor the Research Executive Agency are responsible for any use that may be made of the information it contains.

References

1. H. Chen, O. Engkvist, Y. Wang, M. Olivecrona and T. Blaschke, The Rise of Deep Learning in Drug Discovery, *Drug Discovery Today*, 2018, **23**(6), 1241–1250.
2. O. Engkvist, P.-O. Norrby, N. Selmi, Y. Lam, Z. Peng, E. C. Sherer, W. Amberg, T. Erhard and L. A. Smyth, Computational Prediction of Chemical Reactions: Current Status and Outlook, *Drug Discovery Today*, 2018, **23**(6), 1203–1218.
3. M. H. S. Segler, M. Preuss and M. P. Waller, Planning Chemical Syntheses with Deep Neural Networks and Symbolic AI, *Nature*, 2018, **555**(7698), 604–610.
4. R. S. Bohacek, C. McMartin and W. C. Guida, The Art and Practice of Structure-Based Drug Design: A Molecular Modeling Perspective, *Med. Res. Rev.*, 1996, **16**(1), 3–50.

5. G. Schneider and U. Fechner, Computer-Based de Novo Design of Drug-like Molecules, *Nat. Rev. Drug Discovery*, 2005, **4**(8), 649–663.
6. G. Schneider, M. L. Lee, M. Stahl and P. Schneider, De Novo Design of Molecular Architectures by Evolutionary Assembly of Drug-Derived Building Blocks, *J. Comput.-Aided Mol. Des.*, 2000, **14**(5), 487–494.
7. Y. Xu, K. Lin, S. Wang, L. Wang, C. Cai, C. Song, L. Lai and J. Pei, Deep Learning for Molecular Generation, *Future Med. Chem.*, 2019, **11**(6), 567–597.
8. D. Weininger, SMILES, a Chemical Language and Information System: 1: Introduction to Methodology and Encoding Rules, *J. Chem. Inf. Comput. Sci.*, 1988, **28**(1), 31–36.
9. D. Weininger, A. Weininger and J. L. Weininger, SMILES. 2. Algorithm for Generation of Unique SMILES Notation, *J. Chem. Inf. Comput. Sci.*, 1989, **29**(2), 97–101.
10. E. J. Bjerrum, SMILES Enumeration as Data Augmentation for Neural Network Modeling of Molecules, arXiv:1703.07076, 2017.
11. E. J. Bjerrum and B. Sattarov, Improving Chemical Autoencoder Latent Space and Molecular De Novo Generation Diversity with Hetero-encoders, *Biomolecules*, 2018, **8**(4), 1–17.
12. R. Winter, F. Montanari, F. Noé and D.-A. Clevert, Learning Continuous and Data-Driven Molecular Descriptors by Translating Equivalent Chemical Representations, *Chem. Sci.*, 2018, **10**(6), 1692–1701.
13. P. Schwaller, T. Gaudin, D. Lányi, C. Bekas and T. Laino, 'Found in Translation': Predicting Outcomes of Complex Organic Chemistry Reactions Using Neural Sequence-to-Sequence Models, *Chem. Sci.*, 2018, **9**(28), 6091–6098.
14. T. B. Kimber, S. Engelke, I. V. Tetko, E. Bruno and G. Godin, Synergy Effect between Convolutional Neural Networks and the Multiplicity of SMILES for Improvement of Molecular Prediction, *arXiv*, 2018, DOI: arXiv:1812.04439v1.
15. E. J. Bjerrum and R. Threlfall, Molecular Generation with Recurrent Neural Networks (RNNs), *arXiv:1705.04612*, 2017.
16. R. Gómez-Bombarelli, J. N. Wei, D. Duvenaud, J. M. Hernández-Lobato, B. Sánchez-Lengeling, D. Sheberla, J. Aguilera-Iparraguirre, T. D. Hirzel, R. P. Adams and A. Aspuru-Guzik, Automatic Chemical Design Using a Data-Driven Continuous Representation of Molecules, *ACS Cent. Sci.*, 2018, **4**(2), 268–276.
17. M. H. S. Segler, T. Kogej, C. Tyrchan and M. P. Waller, Generating Focused Molecule Libraries for Drug Discovery with Recurrent Neural Networks, *ACS Cent. Sci.*, 2018, **4**(1), 120–131.
18. M. Olivecrona, T. Blaschke, O. Engkvist and H. Chen, Molecular De-Novo Design through Deep Reinforcement Learning, *J. Cheminf.*, 2017, **9**(1), 48.
19. S. Hochreiter and J. Schmidhuber, Long Short-Term Memory, *Neural Comput.*, 1997, **9**(8), 1735–1780.
20. K. Cho, B. van Merrienboer, C. Gulcehre, D. Bahdanau, F. Bougares, H. Schwenk and Y. Bengio, Learning Phrase Representations Using RNN

Encoder-Decoder for Statistical Machine Translation, *arXiv preprint arXiv:1406.1078*, 2014, DOI: https://doi.org/10.3115/v1/D14-1179.

21. A. Gupta, A. T. Müller, B. J. H. Huisman, J. A. Fuchs, P. Schneider and G. Schneider, Generative Recurrent Networks for De Novo Drug Design, *Mol. Inf.*, 2018, **37**(1), 1700111.
22. M. J. Kusner, B. Paige and J. M. Hernández-Lobato, Grammar Variational Autoencoder, *arXiv:1703.01925*, 2017.
23. M. Abadi, A. Agarwal, P. Barham, E. Brevdo, Z. Chen, C. Citro, G. S. Corrado, A. Davis, J. Dean and M. Devin, *et al.*, TensorFlow: Large-Scale Machine Learning on Heterogeneous Distributed Systems, *arXiv:1603.04467*, 2016.
24. A. Paszke, G. Chanan, Z. Lin, S. Gross, E. Yang, L. Antiga and Z. Devito, Automatic Differentiation in PyTorch, *Adv. Neural Inf. Process. Syst.*, 2017, **30**, 1–4. No. Nips.
25. S. Bai, J. Z. Kolter and V. Koltun, An Empirical Evaluation of Generic Convolutional and Recurrent Networks for Sequence Modeling, *arXiv*, 2018, **89**, 131–139.
26. N. O'Boyle and A. Dalke, DeepSMILES: An Adaptation of SMILES for Use in Machine-Learning of Chemical Structures, *ChemRxiv*, 2018, **10**, 26434.
27. P. Ertl and A. Schuffenhauer, Estimation of Synthetic Accessibility Score of Drug-like Molecules Based on Molecular Complexity and Fragment Contributions, *J. Cheminf.*, 2009, **1**(1), 8.
28. D. Merk, L. Friedrich, F. Grisoni and G. Schneider, De Novo Design of Bioactive Small Molecules by Artificial Intelligence, *Mol. Inf.*, 2018, **37**(1), DOI: 10.1002/minf.201700153.
29. D. Polykovskiy, A. Zhebrak, D. Vetrov, Y. Ivanenkov, V. Aladinskiy, P. Mamoshina, M. Bozdaganyan, A. Aliper, A. Zhavoronkov and A. Kadurin, Entangled Conditional Adversarial Autoencoder for de Novo Drug Discovery, *Mol. Pharm.*, 2018, **15**(10), 4398–4405.
30. A. Gaulton, A. Hersey, M. Nowotka, A. P. Bento, J. Chambers, D. Mendez, P. Mutowo, F. Atkinson, L. J. Bellis and E. Cibrián-Uhalte, *et al.*, The ChEMBL Database in 2017, *Nucleic Acids Res.*, 2017, **45**(D1), D945–D954.
31. R. Visini, M. Awale and J. L. Reymond, Fragment Database FDB-17, *J. Chem. Inf. Model.*, 2017, **57**(4), 700–709.
32. M. Awale, F. Sirockin, N. Stiefl and J. Reymond, Drug Analogs from Fragment Based Long Short-Term Memory Generative Neural Networks, *ChemRxiv*, 2018, **17**(1), 1–18.
33. M. Popova, O. Isayev and A. Tropsha, Deep Reinforcement Learning for de Novo Drug Design, *Sci. Adv.*, 2018, **4**(7), eaap7885.
34. G. L. Guimaraes, B. Sanchez-Lengeling, C. Outeiral, P. L. C. Farias and A. Aspuru-Guzik, *Objective-Reinforced Generative Adversarial Networks (ORGAN) for Sequence Generation Models*, 2017, DOI: arXiv:1705.10843v3.
35. B. Sanchez-Lengeling, C. Outeiral, G. L. Guimaraes and A. Aspuru-Guzik, Optimizing Distributions over Molecular Space. An Objective-Reinforced

Generative Adversarial Network for Inverse-Design Chemistry (ORGANIC), *ChemRxiv*, 2017, 1–18.
36. C. A. Lipinski, F. Lombardo, B. W. Dominy and P. J. Feeney, Experimental and Computational Approaches to Estimate Solubility and Permeability in Drug Discovery and Development Settings, *Adv. Drug Delivery Rev.*, 2012, **64**(Suppl.), 4–17.
37. J. Besnard, S. Muresan, A. L. Hopkins, G. V. Paolini and G. R. Bickerton, Quantifying the Chemical Beauty of Drugs, *Nat. Chem.*, 2012, **4**(2), 90–98.
38. E. Putin, A. Asadulaev, Y. Ivanenkov, V. Aladinskiy, B. Sanchez-Lengeling, A. Aspuru-Guzik and A. Zhavoronkov, Reinforced Adversarial Neural Computer for de Novo Molecular Design, *J. Chem. Inf. Model.*, 2018, **58**(6), 1194–1204.
39. E. Putin, A. Asadulaev, Q. Vanhaelen, Y. Ivanenkov, A. V. Aladinskaya, A. Aliper and A. Zhavoronkov, Adversarial Threshold Neural Computer for Molecular de Novo Design, *Mol. Pharm.*, 2018, **15**(10), 4386–4397.
40. H. King, Y. Zwols, P. Blunsom, K. M. Hermann, M. Reynolds, S. G. Colmenarejo, J. Agapiou, K. Kavukcuoglu, A. Grabska-Barwińska and E. Grefenstette, *et al.*, Hybrid Computing Using a Neural Network with Dynamic External Memory, *Nature*, 2016, **538**(7626), 471–476.
41. T. Blaschke, M. Olivecrona, O. Engkvist, J. Bajorath and H. Chen, Application of Generative Autoencoder in De Novo Molecular Design, *Mol. Inf.*, 2018, **37**(1), DOI: 10.1002/minf.201700123.
42. H. Dai, Y. Tian, B. Dai, S. Skiena and L. Song, Syntax-Directed Variational Autoencoder for Structured Data, *arXiv:1802.08786*, 2018.
43. Z. Xu, S. Wang, F. Zhu and J. Huang, Seq2seq Fingerprint: An Unsupervised Deep Molecular Embedding for Drug Discovery, *Proc. 8th ACM Int. Conf. Bioinformatics, Comput. Biol. Heal. Informatics – ACM-BCB*, 2017, **17**, DOI: 10.1145/3107411.3107424.
44. J. Lim, S. Ryu, J. W. Kim and W. Y. Kim, Molecular Generative Model Based on Conditional Variational Autoencoder for de Novo Molecular Design, *J. Cheminform.*, 2018, **10**(1), DOI: 10.1186/s13321-018-0286-7.
45. P. Pogány, N. Arad, S. Genway and S. D. Pickett, De Novo Molecule Design by Translating from Reduced Graphs to SMILES, *J. Chem. Inf. Model.*, 2018, acs.jcim.8b00626.
46. B. Sattarov, I. I. Baskin, D. Horvath, G. G. Marcou, E. J. Bjerrum and A. Varnek, De Novo Molecular Design by Combining Deep Autoencoder Recurrent Neural Networks with Generative Topographic Mapping, *J. Chem. Inf. Model.*, 2019, DOI: 10.1021/acs.jcim.8b00751.
47. D. D. Johnson, Learning Graphical State Transitions, *Int. Conf. Learn. Represent.*, 2017, 1–19.
48. Y. Li, O. Vinyals, C. Dyer, R. Pascanu and P. Battaglia, Learning Deep Generative Models of Graphs, *arXiv preprint arXiv:1803.03324*, 2018, 1–16.
49. Y. Li, L. Zhang and Z. Liu, Multi-Objective de Novo Drug Design with Conditional Graph Generative Model, *J. Cheminf.*, 2018, **10**(1), 1–24.

50. T. N. Kipf and M. Welling, Semi-Supervised Classification with Graph Convolutional Networks, *arXiv:1609.02907*, 2016.
51. J. You, B. Liu, R. Ying, V. Pande and J. Leskovec, Graph Convolutional Policy Network for Goal-Directed Molecular Graph Generation, *Advances in Neural Information Processing Systems*, 2018, 1–11.
52. W. Jin, R. Barzilay and T. Jaakkola, Junction Tree Variational Autoencoder for Molecular Graph Generation, *arXiv:1802.04364*, 2018.
53. M. Rarey and J. S. Dixon, Feature Trees: A New Molecular Similarity Measure Based on Tree Matching, *J. Comput.-Aided Mol. Des.*, 1998, **12**(5), 471–490.
54. K. Yakoub, S. Jung, C. Sattler, H. Damerow, J. Weber, A. Kretzschmann, A. S. Cankaya, M. Piel, F. Rösch and A. S. Haugaard, *et al.*, Structure-Function Evaluation of Imidazopyridine Derivatives Selective for δ-Subunit-Containing γ-Aminobutyric Acid Type A (GABAA) Receptors, *J. Med. Chem.*, 2018, **61**(5), 1951–1968.
55. I. J. Goodfellow, J. Pouget-Abadie, M. Mirza, B. Xu, D. Warde-Farley, S. Ozair, A. Courville and Y. Bengio, Generative Adversarial Networks, *Advances in Neural Information Processing Systems*, 2014, 1–9.
56. N. De Cao and T. Kipf, MolGAN: An Implicit Generative Model for Small Molecular Graphs, *arXiv*, 2018, DOI: arXiv:1805.11973v1.
57. J. Arús-Pous, T. Blaschke, S. Ulander, J.-L. Reymond, H. Chen and O. Engkvist, Exploring the GDB-13 Chemical Space Using Deep Generative Models, *J. Cheminf.*, 2019, **11**(1), 20.
58. Q. Liu, M. Allamanis, M. Brockschmidt and A. L. Gaunt, Constrained Graph Variational Autoencoders for Molecule Design, *Advances in Neural Information Processing Systems*, 2018, 1–12.
59. M. Simonovsky and N. Komodakis, GraphVAE: Towards Generation of Small Graphs Using Variational Autoencoders, *Lect. Notes Comput. Sci. (including Subser. Lect. Notes Artif. Intell. Lect. Notes Bioinformatics)*, 2018, 412–422, 11139 LNCS.
60. B. Samanta, A. De, G. Jana, P. K. Chattaraj, N. Ganguly and M. Gomez-Rodriguez, NeVAE: A Deep Generative Model for Molecular Graphs, *arXiv:1802.05283*, 2018.
61. A. Kadurin, S. Nikolenko, K. Khrabrov, A. Aliper and A. Zhavoronkov, DruGAN: An Advanced Generative Adversarial Autoencoder Model for de Novo Generation of New Molecules with Desired Molecular Properties in Silico, *Mol. Pharm.*, 2017, **14**(9), 3098–3104.
62. X. Yang, J. Zhang, K. Yoshizoe, K. Terayama and K. Tsuda, ChemTS: An Efficient Python Library for de Novo Molecular Generation, *Sci. Technol. Adv. Mater.*, 2017, **18**(1), 972–976.
63. K. Preuer, P. Renz, T. Unterthiner, S. Hochreiter and G. Klambauer, Fréchet ChemNet Distance: A Metric for Generative Models for Molecules in Drug Discovery, *J. Chem. Inf. Model.*, 2018, **58**(9), 1736–1741.
64. T. Unterthiner, A. Mayr, G. Klambauer, S. Hochreiter, M. Steijaert, J. K. Wegner, H. Ceulemans and D.-A. Clevert, Large-Scale Comparison

of Machine Learning Methods for Drug Target Prediction on ChEMBL, *Chem. Sci.*, 2018, **9**(24), 5441–5451.
65. N. Brown, M. Fiscato, M. H. S. Segler and A. C. Vaucher, GuacaMol: Benchmarking Models for De Novo Molecular Design, *J. Chem. Inf. Model.*, 2019, **59**(3), 1096–1108.
66. D. Polykovskiy, A. Zhebrak, B. Sanchez-Lengeling, S. Golovanov, O. Tatanov; S. Belyaev, R. Kurbanov, A. Artamonov, V. Aladinskiy and M. Veselov, *et al.*, Molecular Sets (MOSES): A Benchmarking Platform for Molecular Generation Models, 2018, DOI: https://doi.org/arXiv:1811.12823v1.
67. E. Kussul and T. Baidyk, Improved Method of Handwritten Digit Recognition Tested on MNIST Database, *Image Vis. Comput.*, 2004, **22**(12 SPEC. ISS), 971–981.
68. T. Sterling and J. J. Irwin, ZINC 15 – Ligand Discovery for Everyone, *J. Chem. Inf. Model.*, 2015, **55**(11), 2324–2337.
69. G. Landrum, RDKit: Open-source Cheminformatics, 2014, http://www.rdkit.org/.
70. G. Schneider, Automating Drug Discovery, *Nat. Rev. Drug Discovery*, 2018, **17**(2), 97–113.

CHAPTER 14

Active Learning for Drug Discovery and Automated Data Curation

D. REKER

Department of Biomedical Engineering, Duke University, Durham, NC, USA
Email: daniel.reker@duke.edu

14.1 Introduction

Active learning is a sub-discipline in machine learning where the algorithm is put in charge of selecting unlabeled examples for labeling and inclusion in the training data (Figure 14.1).[1,2] Originally, this concept was pushed forward in domains such as text and image classification,[3–5] where large corpora of unlabeled data are easily available but generation of training data for supervised machine learning applications requires time- and resource-consuming annotations by human experts ("oracles"). In such situations, a careful selection of examples to generate a meaningful training set with minimum effort can reduce required resources to generate high-quality models at lower costs.[2] To this end, active machine learning workflows build upon the implementation of a selection function that quantifies the desirability of a possible new experiment.[1] This can be based on different

Drug Discovery Series No. 75
Artificial Intelligence in Drug Discovery
Edited by Nathan Brown

Published by the Royal Society of Chemistry, www.rsc.org

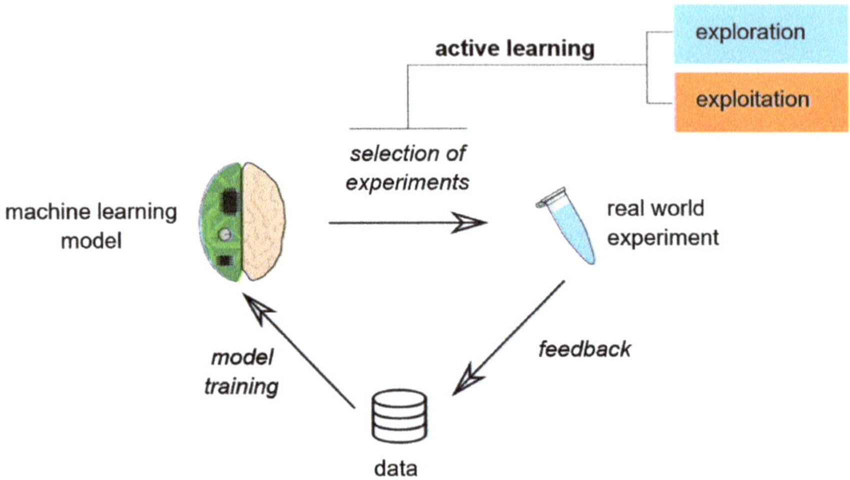

Figure 14.1 Schematic of the active learning concept. Active learning research develops selection functions that enable adaptive, rational experimental design with improved learning efficiency (exploration) or increased hit rates (exploitation).
Adapted from ref. 1 with permission from Elsevier, Copyright 2015.

objectives such as, for example, potential impact of this data point on the machine learning architecture, novelty compared to the previous training data, or promise of the newly selected example to belong to a desired or underrepresented class. The choice of selection functions will dramatically impact the future performance characteristics of the machine learning model, the trajectory of selected experiments, and the speed of hit identification and model improvement (Figure 14.2).[1] A large body of theoretical research has been devoted to tailor selection functions to different types of established machine learning models to achieve the desired behavior.[2] This selection function is then iteratively applied to grow the training dataset through performing experiments or annotations for the requested data points. Thereby, performing costly labeling tasks is confined to the most informative regions of the unlabeled dataset.[1]

About fifteen years ago, first applications of this concept to drug discovery emerged.[6] These studies highlighted that pharmaceutical research and development faced challenges well-suited to be tackled with active learning: millions of chemical structures are commercially-available for screening and are poised to provide novel drug hits and inform researchers about the underlying structure–activity relationship, but their activity (label) is *a priori* unknown and often expensive and time-consuming to acquire.[1,7] Therefore, in parallel to developments such as the generation of informer compound sets,[8] focused library design,[9–11] and rational *de novo* design,[12,13] active learning has emerged as a key computational technology to guide biochemical screening efforts by adaptively designing compound libraries with larger hit rates and improved molecular data for the discrimination between

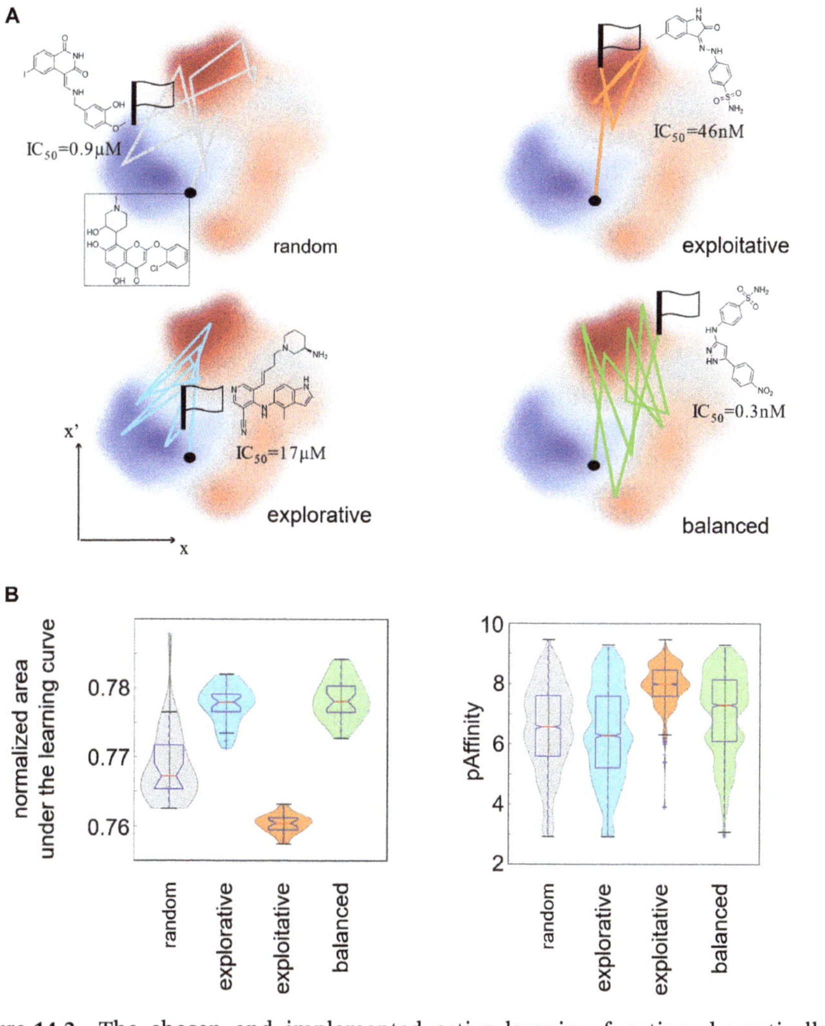

Figure 14.2 The chosen and implemented active learning function dramatically impacts the learning trajectory and performance outcomes[1,29] **A** Chemical space visualizations of CDK2 inhibitor space (ChEMBL20) traversed by random sampling, exploitative learning (maximum predicted pAffinity), explorative sampling (maximum uncertainty) and balanced sampling (maximum prediction + maximum uncertainty) using random forest prediction models (scikit-learn). Exploitative selection immediately focuses on the most promising regions of chemical space (red area), while explorative sampling alternates active (red) and inactive (blue) areas to improve the structure–activity relationship model. Balanced learning also samples chemical space broadly but focuses exclusively on regions of chemical space with active compounds (red and orange). **B** After 100 iterations of this active learning process, the different selection functions show different outcomes in their learning efficiency (area under the learning curve) and the activity of retrieved compounds (pAffinity distribution).
Adapted from ref. 1 with permission from Elsevier, Copyright 2015.

active and inactive compounds. A slowly but constantly growing number of researchers have applied the active learning concept to identify small molecular inhibitors against a wide range of different protein targets as well as studied active learning for other applications in chemistry and material science. In this chapter, we will summarize these studies and identify key observations and trends that will provide an overview of the opportunities and challenges for this technology.

14.2 Active Learning for Drug Discovery, Chemistry, and Material Science

Given the long-standing appreciation of rapid synthesis–test cycles in the medicinal chemistry community, active learning suggests itself as an easily deployable technology within established discovery platforms to assist scientists in their molecular reasoning and experimental design.[1] Furthermore, with the advent of automated systems such as robot-driven laboratories or lab-on-a-chip systems,[14–18] the autonomous generation and refinement of molecular hypotheses can hasten the understanding of structure–activity relationships as well as the discovery of molecular matter with desired properties by avoiding the bottleneck of human reasoning.[1,7,14,19–21]

Indeed, multiple promising studies have been published that highlight the power of active learning for improved model building and increased hit rates. A majority of the previously published applications are currently centered around retrospective screening campaigns for small molecular modulators of established protein targets[6,22–27] with a handful of prospective studies.[20,28–30] These applications highlight the ability of active learning to identify novel bioactive scaffolds that would have not been explored following other, classical screening strategies.[1,23,29,31] Another key result from these studies is the efficiency of such active learning workflows, showing significantly increased hit rates and more rapid convergence towards accurate structure–activity relationship models – as measured by various retrospective benchmarks.[23,32–34] In the context of polypharmacology,[35,36] a few studies have highlighted the ability of active learning to navigate complex pharmacological networks[32,34,37] and to identify hits with desired selectivity and property profile.[20,28] Further studies will be necessary to delineate whether active learning can also be productively applied to other stages of the drug discovery and development process, but some pioneering applications in domains such as formulation design[38] and combination therapy development[39] have showcased first promising results for the application of this technology to other challenges in pharmaceutical research and development.[1]

Most recently, a handful of active learning studies have shown how active learning pipelines can positively impact both material discovery[40,41] as well as the optimization of chemical reaction conditions (Table 14.1).[18,42,43] Such studies not only highlight the potential of active learning to impact other

Table 14.1 Summary of studies applying active machine learning for drug discovery and related fields. Adapted from ref. 1 with permission from Elsevier, Copyright 2015.

Target	Model	Type of study	Exploration	Exploitation	Descriptor	Batch	Ref.
Thrombin (& CDK2)	Jury of Perceptrons; SVM	Retrospective	Uncertainty sampling	Maximum confidence	Molecular shape features	Naïve	Warmuth *et al.*[6,22] (2003)
GPCRs	QBag	Retrospective (+ 1 prospective screen)	Uncertainty sampling	—	MDL Key + physico-chemical properties	Naïve	Fujiwara *et al.*[23] (2008)
Anticancer drug screen (NCI60)	Gaussian process	Retrospective	Uncertainty sampling	Variance corrected predictions and expected improvement	OpenBabel FP2	—	De Grave *et al.*[24,25] (2008)
12 human targets	Gaussian process	Retrospective	—	Expected improvement	MOE 2D descriptors	—	Ahmadi *et al.*[26] (2013)
Narcotic analgesics	KGCB	Retrospective	—	Expected improvement	Free-Wilson model	—	Negoescu *et al.*[27] (2011)
GPCR polypharma-cology and blood-brain-barrier penetration	Bayesian model	Prospective	Sampling by genetic algorithms	Maximum prediction	ECFP6	Naive	Besnard *et al.*[28] (2012)
Abl kinase	Random forest	Prospective	Undersampled building block	Maximum prediction	ECFP6 + physico-chemical properties	Naïve	Desai *et al.*[20] (2013)
Gyrase	QMOD	Retrospective	Binding mode analysis	Maximum prediction	—	Naïve	Varela *et al.*[31] (2012)
Chemogenomics	Random forest	Retrospective	Query by committee	Maximum confidence	ECFP & dipeptide	—	Reker *et al.*[32,57] (2017)
Various	SVM	Retrospective	Uncertainty sampling	—	ECFP	—	Lang *et al.*[34] (2016)
Phenotypic protein localization	Structure learning	Prospective	Minimize imputable experiments	—	—	Grouping of compounds & targets	Naik *et al.*[30,81] (2016)
CXCR4	Random Forest	Prospective	Top predictions with high uncertainty and low forest similarity	Variance corrected predictions and forest similarity	CATS2 + Morgan fingerprints + RDKit properties	Forest similarity regularizer	Reker *et al.*[29] (2016)

Table 14.1 (*Continued*)

Target	Model	Type of study	Exploration	Exploitation	Descriptor	Batch	Ref.
HIV-PR, BRD4-BD1, Microtubles	SVR	Retrospective	Predictive variance	—	COMBINE	Naive	Fusani *et al.*[33] (2018)
Reaction conditions	Random forest	Prospective	Uncertainty sampling	Maximum prediction	Conditions	—	Reker *et al.*[43] (2018)
Reaction conditions	BNN	Retrospective	Distance to training data	Maximum predictions	Conditions	Multiple selection functions	Haese *et al.*[42] (2018)
Materials	SVM	Prospective	Probabilistic uncertainty sampling	—	Ratios of reagents	Similarity regularizer	Duros *et al.*[40] (2017)
Materials	SVM	One prospective screen	—	Maximum confidence	Physicochemical descriptors + reaction conditions	Naïve	Raccuglia *et al.*[41] (2017)
Molecular properties	Group additivity	Retrospective	Various	—	Pathway fingerprints	—	Li *et al.*[62] (2019)
Not disclosed	SVM	Retrospective	—	Maximum confidence	Signature descriptors	—	Buendia *et al.*[82] (2019)
Various	Deep learning	Retrospective	Uncertainty sampling	—	Neural Graph Fingerprints	Naive	Zhang *et al.*[58] (2019)
Reaction conditions	LDA	Prospective	Random search	Maximum likelihood	One-hot encoding	—	Granda *et al.*[18] (2018)

aspects of the computational life sciences, but also provide important insights into possible algorithmic developments, performance estimations, as well as expected challenges when applying active learning for the optimization and discovery of innovative chemistry.[7,14] In spite of essential differences regarding the underlying data, applied algorithms, and goals of these studies, patterns have started to emerge that point out key opportunities and hurdles in the maturation and large-scale application of active learning for future drug discovery. In the following sections, we will summarize the key findings from these studies and highlight the most important considerations for researchers striving to deploy active learning in their research pipelines.

14.2.1 Exploitation *vs.* Exploration

Selection functions can be broadly classified into two groups: explorative selection functions strive to add novel knowledge to the model and improve the machine learning model while exploitative selection functions are designed to harness the current model to retrieve desired solutions.[1,42] A majority of theoretical and applied active learning research has focused on explorative approaches,[2] proposing various selection functions that strive to identify the most impactful experiments to generate useful training data.[32] Most studies follow a data-driven approach and quantify the desirability of a specific experiment. Such approaches quantify the novelty of an experiment or the uncertainty of the machine learning model regarding the possible outcome of an experiment. The latter are the most commonly implemented and best understood active learning strategies and are broadly referred to as "uncertainty sampling". The implementation of these selection functions is tailored towards the applied model to enable quantifying the model's predictive confidence. For example, the formalization of a decision boundary allows to sample experiments with high proximity to this hyperplane.[6,22,24,40] Similarly, ensemble approaches enable to quantify the disagreement of the individual models ("query by committee") and thereby strive to actively reduce the version space.[23,29,32] Conversely, exploitative approaches aim to perform the experiments with the highest confidence in a positive outcome. To this end, the selection function queries the predicted outcome of an experiment. In the simplest form, such approaches are referred to as "iterative screening", and they can show dramatically improved hit rates compared to a single large screen by adaptively re-focusing the search for desired compounds.[18,44–46] Measures of expected improvement or predictive confidence can contextualize such predictions further and yield elaborate exploitative selection functions striving to maximize chances of identifying novel and potent inhibitors against the desired targets.[24–26] Importantly, researchers have realized that the hit rate of such methods is closely linked to the performance of the machine learning model,[24–26,41] such that measures of the applicability domain might become key in steering the selection of most promising candidates.[47–49] Consequently, it has proven fruitful to couple exploitative and explorative workflows by running several rounds of explorative learning to improve the machine learning model and expand its applicability domain before exploitative selection will harness the acquired knowledge to retrieve active compounds.[20,29]

In most active learning applications in drug discovery, the exploitation and exploration functions are diametrical opposites, their mathematical formulations often corresponding to inverted scoring functions. Either selection strategy can rely on ranking novel molecular candidates according to the same formalization of predictive confidence and selecting experiments by choosing compounds from the bottom or top of that list. However, such extremes commonly show undesirable behavior: while explorative functions actively improve the machine learning model,[30,32,34] they can show reduced hit rates or devote resources to better understand the structure–activity relationship of less promising compounds.[1,20] Exploitative strategies, on the other hand, often show superb hit rates[24–26] but can struggle to find innovative molecular matter or to improve the machine learning model.[1,32] Therefore, most recent active learning application have moved towards balancing exploration, exploitation, and other objectives in their experimental design functions to select the most useful experiments according to multiple objectives (Figure 14.2).[1,20,29]

14.2.2 Balancing Different Objectives

Balancing exploration and exploitation in selecting new compounds for testing can ensure that resources are devoted to retrieving hits while also new knowledge is added through sampling less understood compounds.[31] This can enable synergies of both selection criteria in dual-strategy active learning.[42] In naïve implementations of this concept, explorative and exploitative selection can be performed in parallel by including examples for both objectives. Varela *et al.* added equal numbers of most promising compounds and most novel compounds to their test batch, thereby enriching their training data and identifying structurally more diverse hits after multiple learning iterations.[31] Similarly, Desai *et al.* have shown that alternating explorative and exploitative selection strategies can be a viable strategy to focus the search on "hot spot" regions of chemical space.[20] Advanced implementations of such concepts enable to study multiple objectives at the same time,[50] for example by identifying pareto-optimal solutions.[51] Aspuru-Guzik and colleagues have implemented an additional parameter in their model that allows to fine-tune the behavior of the selection function and thereby gradually emphasize exploration or exploitation.[42] Through such balancing selections, active learning is alternating picking of active or information-rich compounds for robust machine learning model improvement while enriching the training data with novel active small molecular modulators.[20,31,42]

By following a balanced selection approach that strives to include compounds that are predicted actives with high uncertainty, the training data can be actively enriched to better understand the structure–activity relationship focusing on the most promising parts of chemical space (Figure 14.2). Thereby, such approaches seek to combine the best of both worlds of exploitation and exploration: adding informative data to improve the model while focusing the search of candidates on the regions of chemical space that are rich in active compounds. To this end, the active learning pipeline developed by Reker *et al.* searches the most potently predicted antagonists of CXCR4 but focuses the selection towards compounds that had both a high predictive uncertainty as well as a low similarity to the training data.[29] Thereby, the model was forced to extrapolate rather than retrieve active compounds with high confidence predictions due to trivial chemical modifications. This resulted in the identification of multiple novel antagonists (Figure 14.3) while also dramatically improving the predictive confidence for the screening campaign (Figure 14.4). The utility of such active learning

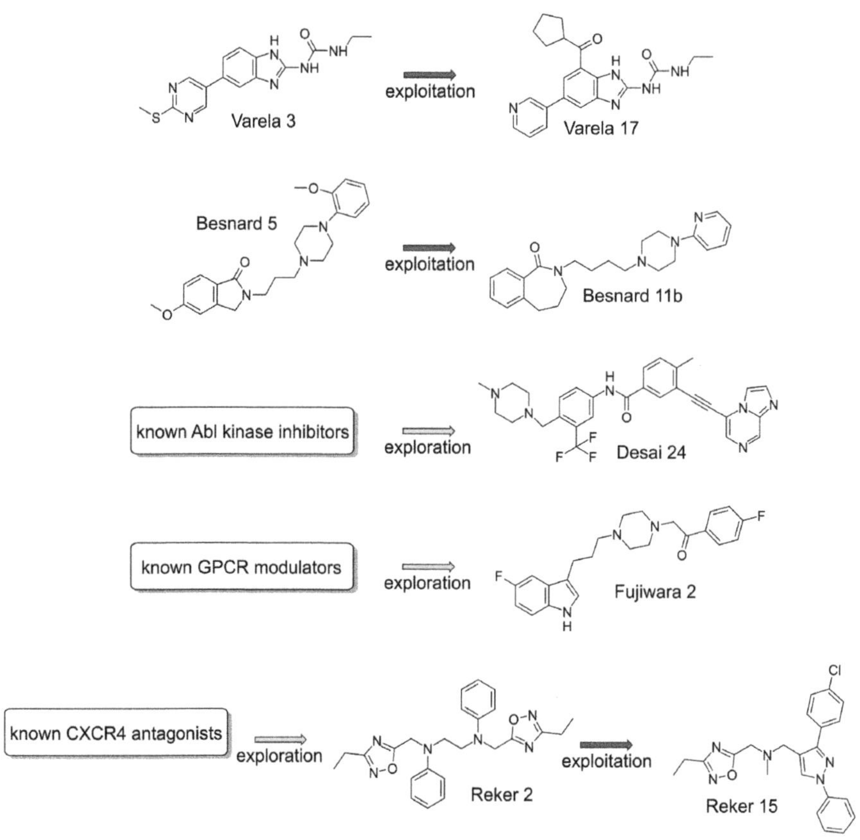

Figure 14.3 Novel chemical entities with desired biological activities identified by active learning. Compounds were chosen from respective publications according to reported activity and relative importance for the study. Data extracted from Varela *et al.*,[31] Besnard *et al.*,[28] Desai *et al.*,[20] Fuijwara *et al.*[23] and Reker *et al.*[29]
Adapted from ref. 1 with permission from Elsevier, Copyright 2015.

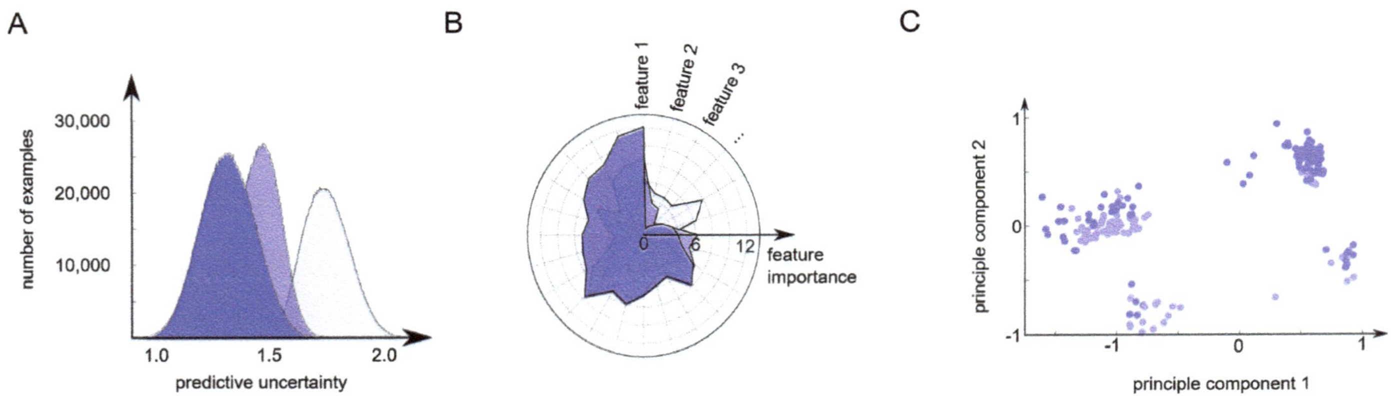

Figure 14.4 Internal model characteristics that aid in tracking model development. Initial model is plotted in light blue, the model resulting after one round of active learning is shown in blue, the model resulting after two iterations of active learning is shown in dark blue. **A** Predictive uncertainty of model on screening set after multiple iterations of active learning. As can be seen, the model's certainty monotonously increases when adding informative training data. Only 30 data points were added per iteration in a balanced fashion. **B** Change in random forest feature importance through active learning. While some features loose importance (*e.g.* feature 2–5), others remain equivalent (*e.g.* feature 9) or increases in importance (*e.g.* feature 1) during learning. **C** Position of most promising compounds in chemical space. Dimension reduction generated through PCA on most important features (*cf.* B). Two additional clusters were discovered during active learning, hinting at the ability of active learning to discover new bioactive chemotypes.
Adapted from ref. 29 with permission from the Royal Society of Chemistry.

strategies was further highlighted through the low initial predictive ranks of the final hits, indicating that such structures would not have been selected for testing without the additional data acquired through active learning. At the same time, the model retrieved inactives that closely resembled some of the known antagonists of CXCR4, highlighting the ability of the predictive uncertainty to refine our understanding of seemingly promising candidates. Thereby, the model is able to home in on the structure–activity relationship and enrich the understanding of the activity of the most promising chemical scaffolds.

The appropriate weighting of the different strategies can be fine-tuned using retrospective evaluations, tailoring the active learning campaign to achieve the desired performance statistics.[29] In some intriguing extensions of this concept, the weighting between different objectives can be automatically learned and adjusted dynamically on the dataset by rewarding objectives that improved the model.[52] Strategies that perform well can then be followed in the next iteration and will be constantly re-evaluated to ensure the sustainability of following the previously established multi-objective selection. In a gamification of this concept, Cronin and colleagues have developed independent chemical robots competing to discover novel chemistry.[16] While such advanced strategies have increased requirements in terms of necessary data[52] and hardware,[16] their implementation enables the active learning workflows to self-adapt and avoid following an unsuccessful strategy.[1] A key finding from previous active learning campaigns is that the quality of learning and retrieved compounds is often quite robust to small changes in the implemented strategy,[20,23,24,29,31] such that the direct deployment of active learning in current pharmaceutical, biotechnological, and academic workflows might not yet require such cutting-edge algorithms and setups.[1,43]

Additionally to the importance of balancing exploration and exploitation, a challenge for active learning in the life sciences will be the multi-objective nature of its desired solutions.[28,50,53] For example, researchers commonly strive to better understand the polypharmacology of compounds[28,54] or aim to simultaneously optimize on-target activity and physicochemical properties important for drug discovery and delivery.[14,50] In such cases, multiple machine learning models will be deployed in concert to anticipate the property profile of novel candidates.[35,36,50,53] The challenge then becomes to exploit multiple models at the same time for the rapid identification of tailored solutions with the desired properties. A naïve solution could implement filtering of proposed experiments by a single orthogonal machine learning algorithm, for example to avoid rule-of-five violations[28] and solubility issues.[29] Instead of such binary filtering, Besnard *et al.* have suggested to consider the distance to the "optimal achievement point" to accomplish the desired on-target activity profile while maintaining appropriate ADME properties.[28] Ample research in multi-objective optimization for drug discovery has provided other promising solutions – including genetic algorithms, simulated annealing, and iterative optimization strategies.[50]

Such technologies might be rapidly coupled to active learning workflows for improved design campaigns and improved machine learning performance.[1] A currently open question is whether the exploratory learning phase would equally benefit from considering all necessary machine learning models simultaneously. Similar to findings from combining exploratory and exploitative objectives in dual active learning strategies,[20,29,42] such stratifications might enable learning to be focussed only on areas that have the potential to provide improved understanding of the regions of chemical space that fit all desired objectives. However, we may find that the machine learning models might be more effectively trained independently to achieve good predictive models irrespective of the current application. The appropriate strategy will most likely depend on the investigated property, chemotype, and the general importance of this property for other projects with potentially different objectives. Further research into such multi-objectives active learning workflows will enable us to outline different learning strategies and when to stop learning and start exploiting the actively-trained model.[1]

14.2.3 When to Stop – Say When!

An important challenge for active learning is the decision for which type of selection function to apply and when a model might be good enough to stop exploring and start exploiting its predictive architecture. These questions will be tightly bound to the current quality and performance of the predictive model as well as the expected benefit of adding additional data. As commonly discussed in the context of retrospective evaluations, the challenge will be that the true performance might be inaccessible since activities of a relevant external test set might not be known or expensive to acquire. This might be particularly true for the cases where researchers decide to apply active learning to reduce screening costs.[1] Fortunately, Lang *et al.* have shown that estimation of the model performance on a small test set can be indicative of overall performance of an actively trained model if representative samples are drawn.[34] Further extending this concept, Buendia *et al.* have shown that a Venn–ABERS predictor can be calibrated on an additional large training set to provide accurate estimates of the expected hit rate in an active learning iteration.[55] Such approaches enable researchers to estimate the performance of an actively trained machine learning model and the expected hit rate in the next iteration to guide decisions on when to stop the active learning process[1] but require potentially large numbers of additional tests on external test sets to enable representative estimates.

If acquisition of additional data for an external test set is not an option, Murphy and colleagues have shown that an orthogonal machine learning model can be trained on simulated data to estimate the current quality of the active machine learning model from parameters that are easily derived from the available dataset.[56] These conservative estimates were shown to be lower bounds of actual model performance, which therefore allows researchers to ensure a lowest acceptable model quality before stopping the learning process. Alternatively, instead of fitting a machine learning model on simulated data, Brown and colleagues have shown that the decrease in model error through active learning on large chemogenomic datasets consistently resembles an exponential decay process with normally-distributed variations, which enables the analytic estimation of future model performance and quantifying the benefits of including additional data and resources for active learning.[32,37,57] Such assumptions can dramatically simplify the estimation of active learning performance, but might not hold true for all applications. For example, in the limit of small datasets, the initialization of the active learning strategy can heavily impact the learning performance.[33] Investigating the extreme limit, Lee and colleagues have shown that active learning performance can suffer when the initial training set is limited to a single scaffold.[58] Such results are fully in line with other studies that have tied active learning performance to the quality of the employed machine learning model,[26,37] indicating that learning rates might heavily depend on models and datasets investigated and potentially need to be delineated for a specific project of interest.

In cases where such assumptions on the predictability of the performance of the model cannot be made, it might be worthwhile to investigate indicators that can be directly derived from the (actively trained) machine learning model itself. For example, models such as Gaussian processes and ensemble models such as random forests inherently capture predictive uncertainties,[59,60] and monitoring these uncertainties on the screening set can be indicative of the model quality and assist in deciding when to stop learning.[29,61] Changes in model architecture have also been considered as benchmarks for model development, commonly by querying the changes in random forest feature importance.[29,43] Alternatively, results from the last round of learning can be indicative of multiple important quality assessments of the model such as the ability to retrieve novel scaffolds,[23] the predictive performance,[62] and the expected hit rate.[20] De Grave *et al.* have suggested an approach to estimate the probability of finding a compound that is more potent than the current best solution.[25] Such intrinsic measures are particularly accessible when hybrid-selection functions are used, which can enable tracking hit-rates of the exploitative aspects and the learning rate of the explorative aspects of the selection function.[20,29,31] While many prospective learning studies manually switch their selection strategies to showcase the power of the newly trained models,[18,20,28,29,43] a few studies of hybrid-selection functions have argued that their broad inclusion of various compounds according to different objectives will make switching of strategies unnecessary[42] and researchers have implemented technologies to automatically switch strategies according to intrinsic performance measures without the need to manually halt learning.[52,63] Irrespective of such advances, the accurate tracking of model development and performance will remain an important field of research with significant implications for the tractability of active learning and its implementation into research practice.

14.2.4 Batch Selection

Many of the currently available assay technologies have been optimized to be performed in high-throughput mode, rapidly testing dozens to thousands of different compounds in parallel. Given these setups, performing a single experiment cherry-picked by an active learning selection function is often not economical.[1] Even if performing a single experiment is feasible, re-fitting a complex model after adding a single data point might be intractable for computationally expensive models such as (deep) neural networks.[58,64] Similarly, batch-selected computational experiments might benefit from parallel experimental or computing architectures.[42] However, in the absence of re-training a model after adding a single example, multiple examples need to be selected from the pool of possible experiments utilizing the same machine learning model and desirability function.[1]

Commonly, researchers naively select the top candidates from the combination of machine learning model and desirability function.[6,22,23,58] However, it has been shown that this will introduce redundancy in the selected examples,[29,64] potentially defying the purpose of selecting these examples in the first place and reducing learning performance:[42] evidence has been provided that increasing the size of naively-selected batches adversely impacts the active learning efficacy.[23] This attests to the importance of avoiding redundancy for the active learning campaign and the importance to improve batch selection techniques.[1] In some select cases, redundancy reduction has been achieved through the application of orthogonal selection methods either ensuring diversity before presenting the compounds to the machine learning model or *post-hoc* on the prioritized examples. For example, Besnard *et al.* used random subsamples of chemical space through genetic algorithms while ensuring novelty and chemical accessibility through filtering.[28] Desai *et al.* used a combinatorial chemical library and forced their selection function to focus on previously under-investigated building blocks.[20] Raccuglia *et al.* forced their model to select compounds with varying similarity to the training data.[41] Varela *et al.* utilized a computational method to anticipate binding modes of the investigated compounds and prioritize structures that explore novel protein interactions.[31] Such approaches have proven to increase the diversity of retrieved hits and improve the machine learning model more rapidly compared to approaches that do not include any type of redundancy reduction.[1,31] However, strictly speaking, none of these methods are true batch selections, since they are not true prioritization of sets of compounds but rather force a selection from a subset of the data by the machine learning model to reduce redundancy *via* decreased data density.

More advanced batch selection methods will actively strive to enrich the dataset by ensuring that the selection function will prioritize a set of knowledge-rich experiments that will provide insights from the complete pool of possible experiments.[2] For example, Naik *et al.* suggested to cluster compounds according to their activity profile and ensured that the selection will select a minimal number of compounds from the same group.[30] Such approaches will ensure that the model is broadly exploring chemical space.[32] Molecular similarity rather than result similarity might also be productively employed to increase the scaffold-diversity of the batch and thereby ensure broader applicability of the actively trained machine learning model in terms of novel chemotypes.[65] In an application to material science, Cronin and colleagues have added such a similarity-based approach to their probabilistic uncertainty sampling to re-score experiments during batch selection to decrease the likelihood of performing similar experiments.[40] Such data-driven approaches can reduce the redundancy in the batch of selected compounds to ensure diverse data will be added in the learning iteration.[1,30,40]

Ideally, batch selection methods actively and adaptively consult the model to ensure that the added compounds are knowledge-rich for the specific model at the specific learning iteration.[66] Reker *et al.* have implemented an approach that queries the random forest similarity of two compounds as a measure for whether the model interprets these two examples differently.[29] Briefly, for every decision tree of the random forest model, the algorithm evaluates whether two prospective examples are predicted utilizing the same leaf of the tree, thereby rendering them equivalent according to all feature categorizations that allow the tree to distinguish active from inactive data.[67] By counting the number of trees that consider two examples to be equivalent, a percentage can be derived for every pair of compounds. This percentage corresponds to the fraction of the ensemble model that deems two compounds equivalent. These similarities can be rapidly re-calculated for the screening set after adding an additional example to the next batch without model re-training and thereby enable to iteratively re-score the screening pool for batch selection. The random forest similarity acts as a regularizer, ensuring that the collection of promising or information-rich compounds selected for testing are not informative to the model for the same reasons – thereby dramatically reducing the redundancy in the batch of new experiments (Figure 14.5).[29]

Alternatively, batch selection can be avoided by following multiple desirability functions and allowing every function to include one example in the new batch of experiments. The Phoenics optimizer elegantly implements this through a tuning parameter, which balances the explorative *vs.* the exploitative behavior of their model and thereby enables to sample a gradient of new experiments that differ in their novelty and promise.[42] The high performance of this method resembles results from other studies that included compounds for two distinct objectives in their batches to enable enriched sampling,[31] but effectively creates an infinite amount of such objective functions through its tuning parameter. Therefore, this method can theoretically scale the multi-objective selection procedure to any batch-size necessary for the desired experiments or calculations.[42]

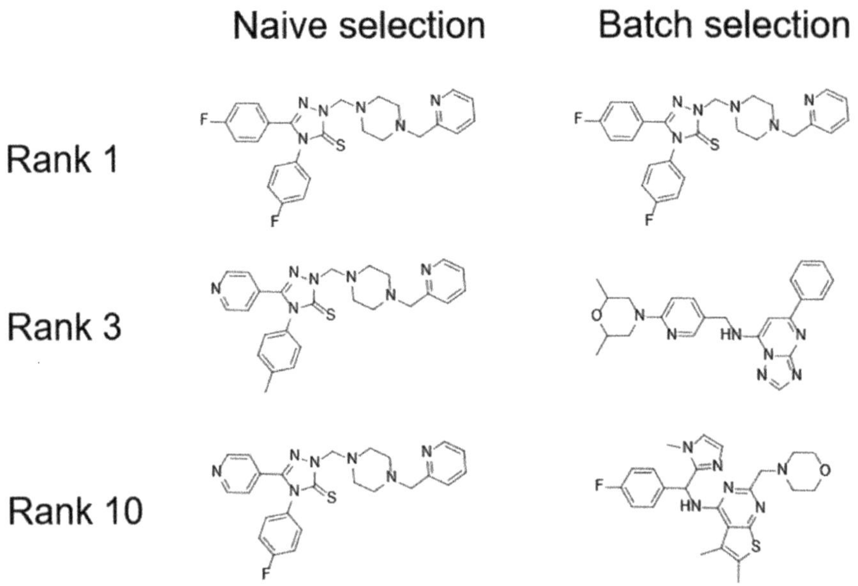

Figure 14.5 Batch selection is important to enable the efficient selection of multiple compounds for testing. In case of naïve selection strategies, a high redundancy between the top candidates can be observed (left). Through smart batch regularization, redundancy can be reduced and a more informative batch of compounds can be selected for further testing (right).
Adapted from ref. 29 with permission from the Royal Society of Chemistry.

14.2.5 Benchmarking the Learning

Evaluation of the performance of different active machine learning pipelines is a key step to enable selecting the correct experimental design strategies for a particular project and adjusting expectations for prospective drug discovery campaigns. Commonly, standard performance metrics are evaluated on different models during active learning campaigns to tract learning rates and extrapolate utility of learning.[23,29,32,34] Specific metrics have been developed to quantify and statistically compare different selection strategies. For example, the area under the learning curve (AULC) enables to quantify model improvement to automatically and quantitatively rank multiple active learning strategies.[68] These performance quantifications can enable statistical comparisons of different selection functions to identify strategies that perform significantly different.[25,26,32]

The most intriguing performance evaluations compare the ability of such algorithms to retrieve desired molecules and materials with the performance of human experts.[40,43,69] Such evaluations can hint at the ability of automated experimental design to surrogate or replace human-guided optimizations. The only benchmark reported for a drug discovery project so far is a retrospective evaluation: in this comparison, Reker *et al.* estimated human performance by sorting reported CXCR4 antagonists from the literature according to their publication date.[29] Compared to this baseline, multi-objective active learning was approximately three-fold faster at identifying the most ligand efficient antagonist.[70] Although this comparison might be considered unfair given that chemists would have considered a larger range of chemical modifications than the ones that have been eventually tested and reported, this result still attests to the efficiency of active learning to navigate focused structure–activity relationships without the need for full series enumeration.[29] Furthermore, the performance of the algorithm was not strongly impacted when distracting the algorithm by adding 50 000 random, assumed inactive compounds to the screening set[29] – hinting at the ability of active learning to rapidly disregard irrelevant subspaces of chemical matter.

Full prospective comparisons have been conducted in the context of material science and chemistry. For example, Norquist and colleagues have compared the effectiveness of their SVM-based exploitation model for predicting crystallization of template vanadium selenites with the effectiveness of human intuition.[41] Their model showed a success rate of 89% and outperformed human intuition at 78%. Human intuition was implemented by following rule-based experimental design established within the hydrothermal synthesis community.[41] While such rule-based approaches formalize intuition and enable reproducibility, their strict application potentially restricts the optimization.[69,71] Instead of restricting human optimization to rule-based approaches, two research projects led by Rodrigues or Cronin have compared the behavior of human experts to active learning algorithms.[40,43] Cronin and colleagues had two human experts compete against an actively trained random forest model in finding experimental conditions leading to the self-assembly and crystallization of polyoxometalates.[40] Both human experts only narrowly explored condition space and generated data that was not generalizable with limited predictive power. Their uncertainty-based active learning strategy explored the condition space broadly and was able to generate data that enabled accurate model building. However, the number of retrieved favorable conditions by the machine learning algorithm and the human expert varied strongly, with one of the experts outcompeting the machine learning algorithm through sampling of similar conditions. This fact does not diminish the success of the study since the employed active learning selection function was aiming at improving the model and not retrieving hits. To evaluate hit rates, human performance needs to be compared to exploitative active learning functions.[18] To that end, Rodrigues and colleagues implemented a random forest-based exploitation algorithm to identify favorable conditions with high yields for the synthesis of a small molecule following a Ugi 3 component reaction scheme.[43] They tasked multiple human experts to optimize conditions towards large conversion amounts but found that the machine learning algorithm was not only faster at identifying favorable conditions but also discovered conditions with a better performance than any of the queried human experts.

Although clearly impressive showcases of automated optimization, such head-to-head comparisons still suffer from some shortcomings that might impact the outcome of these evaluations. For example, the often limited number of human respondents might not yet reflect representative samples of the research community.[40,43] However, certain trends have already emerged that warrant further attention. One important finding is the high variation among the performance of human researchers.[72] Even when experts are queried with similar experience, training, and access to the same resources and data, participants have shown significant differences in their hit finding ability, their optimization trajectories, as well as their reasoning.[40,43] Cronin and colleagues have provided detailed descriptions of the optimization path and reasoning of their experts and found large differences.[40] Similarly, Rodrigues and colleagues found that their human experts followed different optimization strategies and valued the importance of various reaction parameters differently.[43] Such trends pinpoint towards large disparities in human behavior and the non-deterministic nature of human optimization campaigns.[73,74] Another intriguing observations of these benchmarks is that human strategies commonly resemble greedy strategies in terms of a high similarity of acquired results to previously known hits and a reduced exploration of novel solution.[7,40,43] This advocates for the utilization of active learning strategies when innovative hits are required and derivatization of currently available molecular matter is insufficient.[1,7] Furthermore, coupled with the observation that such greedy strategies adversely impact learning efficiency and model performance (Figure 14.2),[32,57] this finding potentially suggests that human-generated data might be suboptimal to train efficient machine learning models.[29,40,43] This hypothesis supports the implementation of active learning strategies early in novel projects and suggests that retrospective application of active learning strategies might enable automated curation of datasets for improved machine learning models.[32–34,75]

14.3 Active Learning for Data Curation

The most prominent current conceptualizations of active learning were designed with prospective applications in mind.[20,23,28–30] For example, the application of selection strategies to novel screening collections for experimental testing is poised to enable the identification of novel hits and augment machine learning models through additional data.[20,28,29] Conversely, researchers have started to explore retrospective applications of active learning on datasets with known labels such as known bioactivities of compounds. Labels are virtually hidden from the model and revealed after the model chooses the compound for (virtual) testing. The main motivation of such approaches is the rapid testing of different model architectures and picking strategies to optimize active learning workflows before applying them in costly prospective studies.[23,24,26] However, researchers soon realized that such retrospective pipelines might not only be helpful to better understand the computational workflow but also improve the understanding of the dataset to which the active learning campaign is applied to.[32,34,76]

Through this, active learning has now presented first promising results as an automated data curation technique.[57,76] Researchers have shown that active learning can reduce the amount of necessary data to generate accurate prediction models:[32,37,62,76] a model trained with only a fraction of the original training data can achieve the same predictability on an external test set compared to a model using the whole dataset when subsampling with an active learning strategy (Table 14.2). This behavior seems to be independent of the applied machine learning function and was shown to be effective using, for example, support-vector machine, deep neural networks, or random forest models with different model architectures.[32,34,37,58] Similarly, active learning-based data curation was also successful in multiple domains, including biological and chemical properties.[32,34,62,76] Finally, this behavior also seems to be robust with regards to the exact applied active learning selection functions.[34,57,62]

14.3.1 Reduced Redundancy and Balanced Data

It has been proposed that the ability to learn from reduced data stems from active learning avoiding redundant information and thereby avoiding the inclusion of compounds that do not provide novel insights into the structure–activity relationship.[34,57] This reasoning was derived from two observations: firstly, even a random subsample of the training data is often able to yield accurate machine learning models, hinting at not all data points being necessary to derive accurate prediction models.[23,32,34] Secondly, commonly-implemented active learning selection functions choose compounds according to maximal predictive uncertainty (Table 14.1).[1] Pivoting this observation suggests that compounds that are not chosen by the active learning function are data points that are already understood by the machine learning model based on previous training data and thereby would not add any novel knowledge.

Table 14.2 Reported reduction in dataset sizes thanks to active learning-based data curation strategies.

Dataset	Percentage of data	Criteria	Ref
Various	10–60%	Similar test set performance compared to models trained on full training dataset	Lang *et al.*[34] (2016)
Phenotypic protein localization	29%	Budget constraints	Naik *et al.*[30,81] (2016)
ChEMBL Kinase SARfari, GPCR SARfari, GLASS	10–20%	MCC > 0.8	Reker *et al.*[32,57] (2017)
ANI1	10%	Similar performance to 5-fold cross validation	Smith *et al.*[76] (2018)
QM7, NIST, catalysis	15%, 25%, 45%	Similar performance to 5-fold cross validation	Li & Rangarajan[62] (2019)

Fusani and Cabrera have provided first evidence that active learning is more effective at this data compression than simple diversity-selection strategies,[33] providing evidence that model-based approaches are key to efficiently navigate the data space and cannot simply be replaced by purely data-driven strategies. Furthermore, Roitberg and colleagues have shown that active subsampling enables to enrich the training data from a wider and more comprehensive training dataset, ultimately outperforming full model evaluation on much larger training sets based on a narrower chemical subspace.[76]

To further investigate why active learning can generate highly-effective models from subsets of the originally-used data, Brown and colleagues investigated the fraction of active and inactive compounds chosen by a random forest-based uncertainty sampling in chemogenomic data.[32] Interestingly, even if the original dataset was highly imbalanced, the subsample of training data chosen by the machine learning workflow contained the same number of data points labeled as active and inactive. This is remarkable given that the model did not have *a priori* knowledge regarding the labeling of the chosen datapoint nor regarding the fraction of positive and negative examples in the original dataset.[57] Nevertheless, the model was able to pick (virtual) experiments that resulted in a dataset that re-balances the objective classes. Interestingly, the balance of chosen examples was often not ensured in the very first active learning iterations but rapidly converged towards an even split of labeled training instances.[32] This provided first evidence that the model requires some predictive power to be able to choose balanced training data. In a follow-up study, it was shown that the ability of models to pick balanced training data was tightly-bound to the complexity of the employed machine learning model and it's predictive capability (Figure 14.6),[37] providing further evidence that the model uses its understanding of the structure–activity relationship to enable choosing balanced

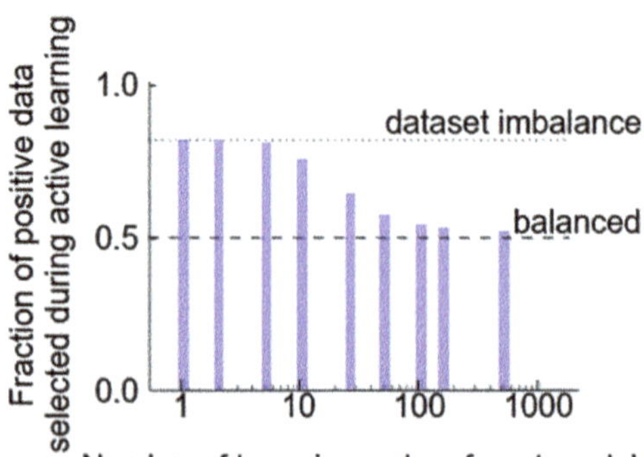

Figure 14.6 Balanced training data selection by random forest-based uncertainty selection. With increasing model complexity, the random forest model can increasingly select an equal number of positive and negative examples from the learning data.
Data from Rakers *et al.*[37]

data. However, given the superior performance of active learning compared to random sampling even in case of inherently balanced datasets,[32] it can be concluded that balancing of the dataset is a major but not the only characteristic of active learning that enables dataset curation.[37] Some active learning work points towards the ability of active learning to not only balance class labels but also other dataset characteristics such as scaffold content[1,23] or target proteins in chemogenomic datasets.[32,57] Follow up work will be necessary to further delineate the characteristics of the chosen datapoints to improve our understanding of the power of active learning, with important implications also for prospective applications.

Taken together, this work indicates that active learning can automatically curate datasets and remove internal biases such as imbalanced classes, scaffold content, and redundant information.[32,57] While advanced machine learning models might be able to circumvent such biases, simpler models might suffer from such data and could benefit from automatically curated datasets.[58] Furthermore, curated datasets could be amenable to simpler machine learning models that can be computed more rapidly or might be more interpretable.[37] Even for more complex models, multiple studies have provided evidence that data selected by an algorithm is more effective at generating an accurate machine learning model compared to historic or human-selected data,[29,40,43] attesting to the importance of providing clean data for accurate machine learning model development.

14.3.2 Reactive Learning

Pushing the concept of active machine learning for data curation further, the term reactive learning has been established for workflows in which examples are selected from the training data for re-testing.[77] Such approaches recognize the fact that experimental data contains errors, artifacts, and noise and puts machine learning in charge of identifying potentially erroneous data for re-testing.[1,21,78] Multiple machine learning studies have associated predictive confidence of machine learning methods with predictive accuracies[58] and first evidence suggests that some predictive inaccuracies might stem from errors in the data rather than an inaccurate machine learning model.[78] In such cases, active learning might be able to point to data that warrants retesting or investigating in secondary assays. This could be done by using predictive uncertainty measures to select data points from the original training data, by statistically-quantifying the difference in predicted behavior from the historically-measured values, or by coupling active learning workflows to computational models anticipating experimental errors stemming from various sources such as colloidal aggregation, membrane interaction, and reactivity.[79,80] Formally, such approaches could be rapidly implemented by applying the selection function to the original training data. However, many of the currently studied selection functions implicitly or explicitly capture the similarity of selected experiments to the already available training data, such that novel approaches might be necessary to enable reactive learning. Given that the current classification or label for these datapoints is available, this could potentially enable access to a novel generation of active learning approaches that consider these labels in their selection functions. Steinmetz *et al.* have contributed an active learning-like selection function for experiments whose predicted outcomes differ from previously measured conditions.[78] Their function selects experiments for re-testing when the apparently wrong predictions had a high predictive confidence and the considered compound had a high similarity to the remaining training data,[29] possibly as a surrogate for the application domain of the model. This strategy enabled correcting 27% of the re-tested training data.[78] Such results, coupled to the increasing awareness for errors in current screening datasets,[79] suggests value in harnessing active learning as a data curation technique.[1] Additional information such as source of data, cost of re-acquisition, as well as importance of this training data for predicting novel hits might even further improve the application of such methods and, even more importantly, allow the rapid and automated curation of historic data to harness its full potential in future drug discovery and development.

14.4 Conclusions

Active learning is a thoroughly studied theoretical concept[2] but applications in the life sciences are still sparse.[1,75] However, an increasing number of studies have shown that active learning can assist in multiple challenges of early drug discovery, including hit identification[20,22,23,25,29] and chemistry optimization.[42,43] Related promising studies have shown successful applications in material design.[40,41] Although these studies clearly attest to the promise of active machine learning for automated optimization, further increasing amounts of available data and innovations in automated laboratory technology are likely to ultimately reveal the true power of this computational technology.[14,15,19] Although expert intuition and human creativity will remain key to successful discovery platforms, the ability of active machine learning to drive reproducible, deterministic, and big data-guided optimization is poised to hasten various aspects in drug and material discovery.[1] Some of the most intriguing benchmarks have compared the performance of active learning algorithms to human experimenters and found that active learning is commonly more systematic, effective, and explorative in comparison to human experts,[29,40,41,43] which has been associated with algorithms generating data that enables more accurate predictions compared to algorithms trained on human-picked or randomly selected experiments.[29,32–34,40] As a logical consequence of this finding, active machine learning has now shown first results as an automated data curation technique. Through retrospective applications of the active machine learning technology on big datasets, the algorithm can compress the training data by up to a factor of ten, automatically reducing redundancy and eliminating bias in the data.[32,34,37,62,76] Both retrospective and prospective investigations will further improve our understanding and the tractability of active machine learning workflows, which will aid its maturation and the broad range application in research and development pipelines for machine-guided pharmaceutical drug discovery.

Acknowledgements

Daniel Reker is a Swiss National Science Foundation Fellow (Grant P2EZP3_168827 and P300P2_177833). His work is partly supported by the MIT-IBM Watson AI Lab and the MIT SenseTime Alliance. Daniel Reker is grateful to his PhD advisor Prof. Gisbert Schneider and his postdoctoral advisors Prof. Robert S. Langer and Prof. Giovanni Traverso for invaluable guidance and mentorship.

References

1. D. Reker and G. Schneider, *Drug Discovery Today*, 2015, **20**(4), 458–465.
2. B. Settles *Active Learning*, Morgan & Claypool Publishers, **Vol. 6**, 2012.
3. A. Ram; and L. Hunter, in *Machine Learning Proceedings 1991*, 2014, pp. 265–269.
4. K. Lang, in *Machine Learning Proceedings 1995*, 2014, pp. 331–339.
5. K. J. Lang; and E. B. Baum, *Ijcnn 1992*, 1992, pp. 335–340.
6. M. K. Warmuth, J. Liao, G. Rätsch, M. Mathieson, S. Putta and C. Lemmen, *J. Chem. Inf. Comput. Sci.*, 2003, **43**(2), 667–673.
7. P. S. Gromski, A. B. Henson, J. M. Granda and L. Cronin, *Nat. Rev. Chem.*, 2019, 119–128.
8. S. Paricharak, A. P. IJzerman, J. L. Jenkins, A. Bender and F. J. Nigsch, *Chem. Inf. Mod.*, 2016, **56**(9), 1622–1630.
9. B. R. Beno and J. S. Mason, *Drug Discovery Today*, 2001, **6**(5), 251–258.
10. M. E. Welsch, S. A. Snyder and B. R. Stockwell, *Curr. Opin. Chem. Biol.*, 2010, **14**(3), 347–361.
11. J. M. Jansen, G. De Pascale, S. Fong, M. Lindvall, H. E. Moser, K. Pfister, B. Warne and C. J. Wartchow, *Chem. Inf. Mod.*, 2019, **59**(5), 1709–1714.
12. P. Schneider and G. Schneider, *J. Med. Chem.*, 2016, **59**(9), 4077–4086.
13. G. Schneider, *De Novo Molecular Design*, Wiley-VCH, Weinheim, 2013.
14. G. Schneider, *Nat. Rev. Drug Discov.*, 2018, **17**(2), 97–113.
15. T. Rodrigues, P. Schneider and G. Schneider, *Angew. Chem., Int. Ed.*, 2014, **53**(23), 5750–5758.
16. D. Caramelli, D. Salley, A. Henson, G. Aragon Camarasa, S. Sharabi, G. Keenan and L. Cronin, *Nat. Comm*, 2018, **9**, 3406.
17. A. C. Bédard, A. Adamo, K. C. Aroh, M. G. Russell, A. A. Bedermann, J. Torosian, B. Yue, K. F. Jensen and T. F. Jamison, *Sience*, 2018, **361**(6408), 1220–1225.
18. J. M. Granda, L. Donina, V. Dragone, D. L. Long and L. Cronin, *Nature*, 2018, **559**(7714), 377–381.

19. D. M. Parry, *ACS Med. Chem. Lett.*, 2019, **10**(6), 848–856.
20. B. Desai, K. Dixon, E. Farrant, Q. Feng, K. R. Gibson, W. P. van Hoorn, J. Mills, T. Morgan, D. M. Parry and M. K. Ramjee, *J. Med. Chem.*, 2013, **56**(7), 3033–3047.
21. S. D. Pickett, D. V. S. Green, D. L. Hunt, D. A. Pardoe and I. Hughes, *ACS Med. Chem. Lett.*, 2011, **2**(1), 28–33.
22. M. K. Warmuth, G. Rätsch, M. Mathieson, J. Liao and C. Lemmen, *In NIPS*, 2001, 1449–1456.
23. Y. Fujiwara, Y. Yamashita, T. Osoda, M. Asogawa, C. Fukushima, M. Asao, H. Shimadzu, K. Nakao and R. J. Shimizu, *Chem. Inf. Mod.*, 2008, **48**(4), 930–940.
24. K. De Grave, J. Ramon and L. De Raedt, in *Benelearn 08, the Annual Belgian-Dutch Machine Learning Conference*, **vol. 2008**, 2008, pp. 55–56.
25. K. De Grave, J. Ramon and L. De Raedt, in *Discovery Science Conference*, Springer, 2008, pp. 185–196.
26. M. Ahmadi, M. Vogt, P. Iyer, J. Bajorath and H. J. Fröhlich, *Chem. Inf. Mod.*, 2013, **53**(3), 553–559.
27. D. M. Negoescu, P. I. Frazier and W. B. Powell, *INFORMS J. Comput.*, 2011, **23**(3), 346–363.
28. J. Besnard, G. F. Ruda, V. Setola, K. Abecassis, R. M. Rodriguiz, X.-P. Huang, S. Norval, M. F. Sassano, A. I. Shin, L. A. Webster, F. R. C. Simeons, L. Stojanovski, A. Prat, N. G. Seidah, D. B. Constam, G. R. Bickerton, K. D. Read, W. C. Wetsel, I. H. Gilbert, B. L. Roth and A. L. Hopkins, *Nature*, 2012, **492**(7428), 215–220.
29. D. Reker, P. Schneider and G. Schneider, *Chem. Sci.*, 2016, **7**, 3919–3927.
30. A. W. Naik, J. D. Kangas, D. P. Sullivan and R. F. Murphy, *eLife*, 2016, **5**, e10047.
31. R. Varela, W. P. Walters, B. B. Goldman and A. N. Jain, *J. Med. Chem.*, 2012, **55**(20), 8926–8942.
32. D. Reker, P. Schneider, G. Schneider and J. Brown, *Future Med. Chem.*, 2017, **9**(4), 381–402.
33. L. Fusani and A. C. Cabrera, *J. Comput.-Aided Mol. Des.*, 2019, **33**(2), 287–294.
34. T. Lang, F. Flachsenberg, U. Von Luxburg and M. J. Rarey, *Chem. Inf. Mod.*, 2016, **56**(1), 12–20.
35. C. Lipinski and A. Hopkins, *Nature*, 2004, **432**(7019), 855–861.
36. A. L. Hopkins, J. S. Mason and J. P. Overington, *Curr. Opin. Struct. Biol.*, 2006, **16**(1), 127–136.
37. C. Rakers, D. Reker and J. B. Brown, *J. Comput.-Aided Chem.*, 2017, **8**, 124–142.
38. F. Caschera, G. Gazzola, M. A. Bedau, C. Bosch Moreno, A. Buchanan, J. Cawse, N. Packard and M. M. Hanczyc, *PLoS One*, 2010, **5**(1), e8546.
39. B. G. Small, B. W. McColl, R. Allmendinger, J. Pahle, G. López-Castejón, N. J. Rothwell, J. Knowles, P. Mendes, D. Brough and D. B. Kell, *Nat. Chem. Biol.*, 2011, **7**(12), 902–908.

40. V. Duros, J. Grizou, W. Xuan, Z. Hosni, D.-L. Long, H. N. Miras and L. Cronin, *Angew. Chem., Int. Ed.*, 2017, **56**(36), 10815–10820.
41. P. Raccuglia, K. C. Elbert, P. D. F. Adler, C. Falk, M. B. Wenny, A. Mollo, M. Zeller, S. A. Friedler, J. Schrier and A. J. Norquist, *Nature*, 2016, **533**(7601), 73–76.
42. F. Häse, L. M. Roch, C. Kreisbeck and A. Aspuru-Guzik, *ACS Cent. Sci.*, 2018, **4**(9), 1134–1145.
43. D. Reker, G. Bernardes and T. Rodrigues, *Chemrxiv* 2018.
44. J. Bajorath, *Nat. Rev. Drug Discov.*, 2002, **1**(11), 882–894.
45. G. Schneider and A. Schüller, *Methods Mol. Biol.*, 2009, **572**, 135–147.
46. T. Miyao and K. J. Funatsu, *Chem. Inf. Mod.*, 2019, **59**(6), 2626–2641.
47. S. Weaver and M. P. Gleeson, *J. Mol. Graphics Model.*, 2008, **26**(8), 1315–1326.
48. H. Dragos, M. Gilles and V. J. Alexandre, *Chem. Inf. Mod.*, 2009, **49**(7), 1762–1776.
49. R. P. J. Sheridan, *Chem. Inf. Mod.*, 2013, **53**(11), 2837–2850.
50. C. A. Nicolaou and N. Brown, *Drug Discov. Today Technol.*, 2013, **10**(3), e427–e435.
51. M. Zuluaga, G. Sergent, A. Krause and M. Püschel, in *30th International Conference on Machine Learning*, 2013, pp. 462–470.
52. Y. Baram, R. El-Yaniv and K. Luz, *JMLR*, 2004, **5**, 255–291.
53. M. Reutlinger, T. Rodrigues, P. Schneider and G. Schneider, *Angew. Chem., Int. Ed.*, 2014, **53**(16), 4244–4248.
54. A. L. Hopkins, *Nat. Biotechnol.*, 2007, **25**(10), 1110–1111.
55. R. Buendia, T. Kogej, O. Engkvist, L. Carlsson, H. Linusson, U. Johansson, P. Toccaceli and E. J. Ahlberg, *Chem. Inf. Mod.*, 2019, **59**(3), 1230–1237.
56. M. Temerinac-Ott, A. W. Naik and R. F. Murphy, *BMC Bioinf.*, 2015, **16**(1), 213.
57. D. Reker and J. B. Brown, in *Methods in Molecular Biology*, 2018.
58. Y. Zhang and A. Le, *Chem. Sci.*, 2019, **10**, 8154–8163.
59. L. Breiman, *Mach. Learn.*, 2001, **45**(1), 5–32.
60. M. N. Radford in *Bayesian Statistics*, ed. J. Bernardo, J. Berger, A. Dawid and A. Smith, Oxford University Press, **vol. 6**, 1998, pp. 475–501.
61. B. Li and S. Rangarajan, *arXiv Prepr. arXiv1906.10273*, 2019.
62. B. Li and S. Rangarajan, *arXiv Prepr. arXiv1906.10273*, 2019.
63. P. Donmez, J. G. Carbonell and P. N. Bennett, in *Machine Learning: ECML 2007*, Springer, 2007, pp. 116–127.
64. J. S. Smith, B. Nebgen, N. Lubbers, O. Isayev and A. E. Roitberg, *J. Chem. Phys.*, 2018, **148**, 24.
65. G. Schneider, W. Neidhart, T. Giller and G. Schmid, *Angew. Chem., Int. Ed.*, 1999, **38**(19), 2894–2896.
66. B. Settles *Active Learning Literature Survey*, University of Wisconsin–Madison, Vol. Computer S, 2010.
67. V. Svetnik, A. Liaw, C. Tong, J. C. Culberson, R. P. Sheridan and B. P. Feuston, *J. Chem. Inf. Comput. Sci.*, 2003, **43**(6), 1947–1958.

68. I. Guyon, G. Cawley, G. Dror and V. Lemaire, JMLR: Workshop and Conference Proceedings, 2011, **16**, 19–45.
69. M. H. S. Segler, M. Preuss and M. P. Waller, *Nature*, 2018, **555**(7698), 604–610.
70. A. L. Hopkins, G. M. Keserü, P. D. Leeson, D. C. Rees and C. H. Reynolds, *Nat. Rev. Drug Discovery*, 2014, **13**(2), 105–121.
71. R. Gómez-Bombarelli, J. N. Wei, D. Duvenaud, J. M. Hernández-Lobato, B. Sánchez-Lengeling, D. Sheberla, J. Aguilera-Iparraguirre, T. D. Hirzel, R. P. Adams and A. Aspuru-Guzik, *ACS Cent. Sci.*, 2018, **4**(2), 268–276.
72. T. J. Smith, R. Henning, M. G. Wade and T. Fisher *Variability in Human Performance*, 2014.
73. P. S. Kutchukian, N. Y. Vasilyeva, J. Xu, M. K. Lindvall, M. P. Dillon, M. Glick, J. D. Coley and N. Brooijmans, *PLoS One*, 2012, **7**(11), e48476.
74. V. Schnecke and J. Boström, *Drug Discovery Today*, 2006, **11**(1–2), 43–50.
75. R. F. Murphy, *Nat. Chem. Biol.*, 2011, **7**(6), 327–330.
76. J. S. Smith, B. Nebgen, N. Lubbers, O. Isayev and A. E. J. Roitberg, *J. Chem. Phys.*, 2018, **148**(24), 241733.
77. C. H. Lin, Mausam and D. S. Weld, JMLR, 2015, **37**, 1–5.
78. F. P. Steinmetz, C. Petersson, U. Zanelli and P. Czodrowski, *Appl. Vitr. Toxicol*, 2019, **5**(2), 86–91.
79. D. Reker, G. J. L. Bernardes and T. Rdorigues, *Nat. Chem.*, 2019, **11**, 402–418.
80. J. B. Baell and G. A. Holloway, *J. Med. Chem.*, 2010, **53**(7), 2719–2740.
81. A. W. Naik, J. D. Kangas, C. J. Langmead and R. F. Murphy, *PLoS One*, 2013, **8**, e83996.
82. R. Buendia, T. Kogej, O. Engkvist, L. Carlsson, H. Linusson, U. Johansson, P. Toccaceli and E. J. Ahlberg, *Chem. Inf. Mod.*, 2019, **59**(3), 1230–1237.

Section 6: Synthesis Planning

CHAPTER 15

Data-driven Prediction of Organic Reaction Outcomes

CONNOR W. COLEY

Department of Chemical Engineering, Massachusetts Institute of Technology, Cambridge, MA 02139, USA
Email: ccoley@mit.edu

15.1 Introduction

15.1.1 The Role of Reaction Prediction

When running a chemical reaction in the lab, there are at most two goals in mind: to produce a molecule of interest, or to obtain information about the reaction itself. If we had a complete understanding of chemical reactivity and computational tools that provide the same information as experiments, that second goal would be obsolete. And while there will always be a need to synthesize compounds for physical use or evaluation, the first goal would be streamlined by eliminating the need for trial-and-error screening during synthetic route development.

Efforts to predict the outcomes of organic reactions, both quantitatively—as rates and yields—and qualitatively—as the identities of the major products—have been ongoing for decades. The importance of this prediction task became especially apparent as computer-aided synthesis planning (CASP) came into vogue.[1–4] CASP tools accelerate the synthesis of new molecules by proposing synthetic routes that map a target compound back to commercially-available starting materials through a series of plausible

Drug Discovery Series No. 75
Artificial Intelligence in Drug Discovery
Edited by Nathan Brown

Published by the Royal Society of Chemistry, www.rsc.org

synthetic steps. Due to the combinatorial nature of the synthesis planning problem, however, the number of proposals can be astronomical.

It was recognized early on, at least by Gasteiger and Ihlenfeldt in their workbench for the organisation of data for chemical applications (WODCA), that the *verification* of computer-recommended reactions is essential to prune the search space and to ensure chemist users are shown reasonable suggestions (Figure 15.1).[1,5] The purpose of this verification is to ask whether there exists some set of reaction conditions where the suggested reactants can form the desired product. That is, can the chemist make this work in the laboratory? The ability to answer this question accurately would provide a tremendous boost to the credibility and utility of CASP programs by eliminating false positives.

The prediction of reaction outcomes, even if done imperfectly, has many applications beyond CASP. In process development, it can accelerate the

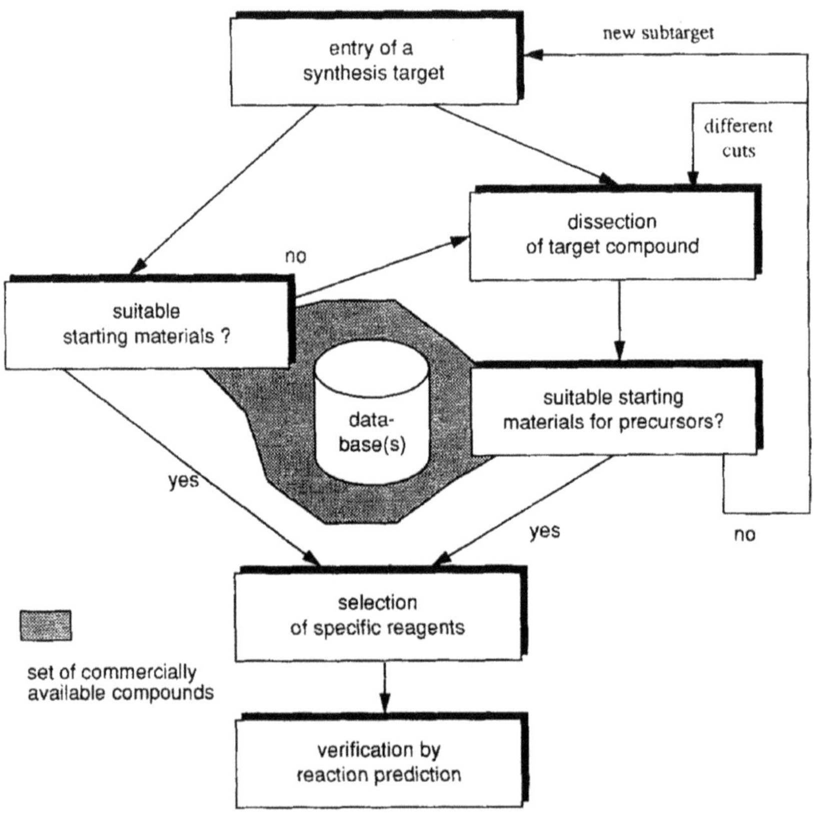

Figure 15.1 Workflow for synthesis planning using WODCA, showing an explicit step of "verification by reaction prediction" (bottom) to ensure chemically reasonable reaction suggestions.
Reproduced from ref. 1 with permission from John Wiley and Sons, © 1995 by VCH Verlagsgesellschaft mbH, Germany.

identification of impurities; this can subsequently inform the development of purification strategies or flag the presence of undesirable (*e.g.*, genotoxic) species that might prompt the use of a different synthetic strategy. In reaction screening, it can simplify the characterization of a complex product mixture where only retention times and m/z ratios are known from UPLC/MS analysis. In virtual screening for drug discovery, it can be used to generate massive libraries of synthetically-accessible compounds by virtually reacting commercially-available starting materials together.[6,7] Ideally, it could also be used to expand our synthetic toolbox and lead to the discovery of entirely new chemical reactions.

15.1.2 Non-data Driven Heuristic Systems

Early approaches to reaction prediction were *not* data-driven. Most sought either to treat the generation of reaction outcomes as a formal enumeration process or to codify expert chemist knowledge in the form of reaction heuristics. The formal (mathematical) treatment of chemistry was championed by Ugi in his concept of bond electron (BE) and reaction (R) matrices.[8] R-matrices describe the redistribution of valence electrons in a chemical system an can be added to reactants' BE-matrices to yield transformed products. To rein in the combinatorial number of possible products, concepts in physical organic chemistry were applied to evaluate reaction enthalpies, bond disassociation energies, *etc.* as part of the EROS program.[9]

The combinatorially-large space of possible outcomes can also be addressed by restricting analysis and enumeration to only the most reactive functional groups. CAMEO, for example, estimates pK_a values of common functional groups and identifies electrophile/nucleophile pairs that might undergo a reaction.[10] Similarly, Sello's Beppe program perceives nucleophiles and electrophiles, hydrides and protons, and acids and bases to identify the most reactive pairs through expert-encoded heuristics.[11] ROBIA takes a slightly different mechanistic approach to reaction prediction;[12] eleven common transformations are defined by a set of expert rules that recognize specific substructural motifs. When these transformations apply, a set of intermediate structures (according to the known/assumed reaction mechanism) are generated; the likelihood of that reaction pathway is evaluated based on quantum mechanical calculations of the intermediates' energies.

15.2 Data-driven Approaches

Data-driven methods, in principle, offer several key advantages over expert heuristic methods, summarized in the following list.

a) **Objectivity.** Chemists don't always agree on what reactions look plausible, so asking different expert chemists to generalize how molecules react will lead to different heuristics. Taking the intersection of multiple chemists' opinions would produce an overly conservative program that can only make predictions within a narrow set of common, well-trodden reactions. Algorithmic processing of experimental reaction data is able to provide a more objective analysis, though there can be bias in what trends the algorithm looks for and how it is implemented.

b) **Transparency.** As a related benefit to objectivity, data-driven approaches have the potential to be more transparent about *how* generalization occurs. Our intuition for reactivity is based on our experiences with formal courses, reading journal articles and textbooks, and performing experiments. Except for some specific cases grounded in physical organic chemistry (*e.g.*, Hammett coefficients[13]), any heuristics we construct will be challenging to trace back to quantitative, experimental data. In practice, however, the interpretability of statistical models, particularly deep neural networks, can present separate obstacles to transparency.

c) **Scalability.** The ideal reaction prediction program would cover the "full" range of our knowledge of synthetic chemistry. Reactions are being published at an exponentially-increasing rate, with modern databases already comprising tens of millions of reactions. Any approach to reaction informatics that relies on human analysis cannot possibly keep up with the literature. Moreover, data-driven methods have the potential to pick up on subtler trends in the underlying data that may be less obvious to human readers.

In the remainder of this section, we describe data-driven approaches to the prediction of organic reaction outcomes. This task has been approached from multiple perspectives—many recent ones incorporating machine learning techniques—and evaluated using various datasets—some covering a narrow range of reactivites, and some covering a broad range.

15.2.1 Focused Analyses of Specific Reaction Classes

Restricting the prediction task to specific reaction classes can assist in the problem formulation. One reason for this is that the output of statistical models (*e.g.*, neural networks) is typically numerical—it is more natural predict a scalar value (for a regression problem) or probabilities of class membership (for a classification) than to predict a *molecule*.

Previous studies have therefore trained models on experimental reaction data to address well-defined questions. Which aliphatic bond in a molecule is most likely to undergo a polar heterolysis reaction?[14] Which enantiomer is the preferred product of a catalytic reaction usingdifferent chiral amino alcohol catalysts?[15] Which of seven reaction types are these reactants likely to undergo under irradiation?[16] Is this substrate or functional group stable in the presence of butylammonia?[17] What will the yield of this C–N bond forming reaction be in the presence of potentially-interfering species?[18]

These questions are consistent with traditional problem formulations in QSAR/QSPR.[19,20]

15.2.2 At the Mechanistic Level

One class of approaches to reaction prediction includes those operating at the mechanistic level. Reactions mechanisms obtained through recursive prediction of individual mechanistic steps provide a physically-meaningful explanation for how certain products could have been formed. Because each mechanistic step should be governed by the same physical/chemical laws, one might expect that a model trained at this level might generalize more effectively to new types of global reactions.

The lack of databases containing mechanistic information complicates this approach. To overcome this, Baldi and coworkers developed a rule-based expert system, Synthesis Explorer, to define mechanistic pathways.[21,22] These rules are similar to those used by ROBIA, but cover a broader set of reaction categories (*ca.* 1500 transformation rules) and include additional consideration of reaction conditions (*ca.* 80 reagent models). While originally used in an educational setting, Synthesis Explorer later evolved into the ReactionPredictor tool.[23,24]

As shown in Figure 15.2, ReactionPredictor aggregates a number of individual modules to produce each recommendation. First, potential reactive sites are proposed, then coarsely filtered (using a pseudo-molecular orbital representation), and ranked. Reactions comprised of these sites (*e.g.*, a pair consisting of an electron source and an electron sink for polar reactions) are proposed and further ranked. This analysis is done for polar reactions, pericyclic reactions, and radical reactions separately before the three predictions are combined into an overall proposal for the most likely mechanistic step. The polar reaction model was further refined in ref. 25.

The ranking models are trained on synthetic data to differentiate between productive and unproductive mechanistic steps.[25] Reaction rules are used to enumerate pathways consisting of "productive" mechanistic steps, whereas "unproductive" mechanistic steps can be generated through combinatorial enumeration of potential interactions that deviate from the recorded pathway. Elementary reactions were hand curated and supplemented with pathways from advanced organic chemistry textbooks. A combination of physicochemical and topological atom-level descriptors are used to represent potential electron sources and sinks; these and additional reaction-level features are used for the final ranking of proposals.

The quantitative performance of the model is incredibly promising: on a benchmark set of 289 reactions, the polar reaction model achieved top-1, 3, and 5 accuracies of 31.5%, 75.8%, and 90.0% when predicting the electron source/sink pair. It is difficult to contextualize these metrics since most other methods operate on global reactions. Nevertheless, Fooshee *et al.* demonstrate successful predictions for a number of exemplary reactions,

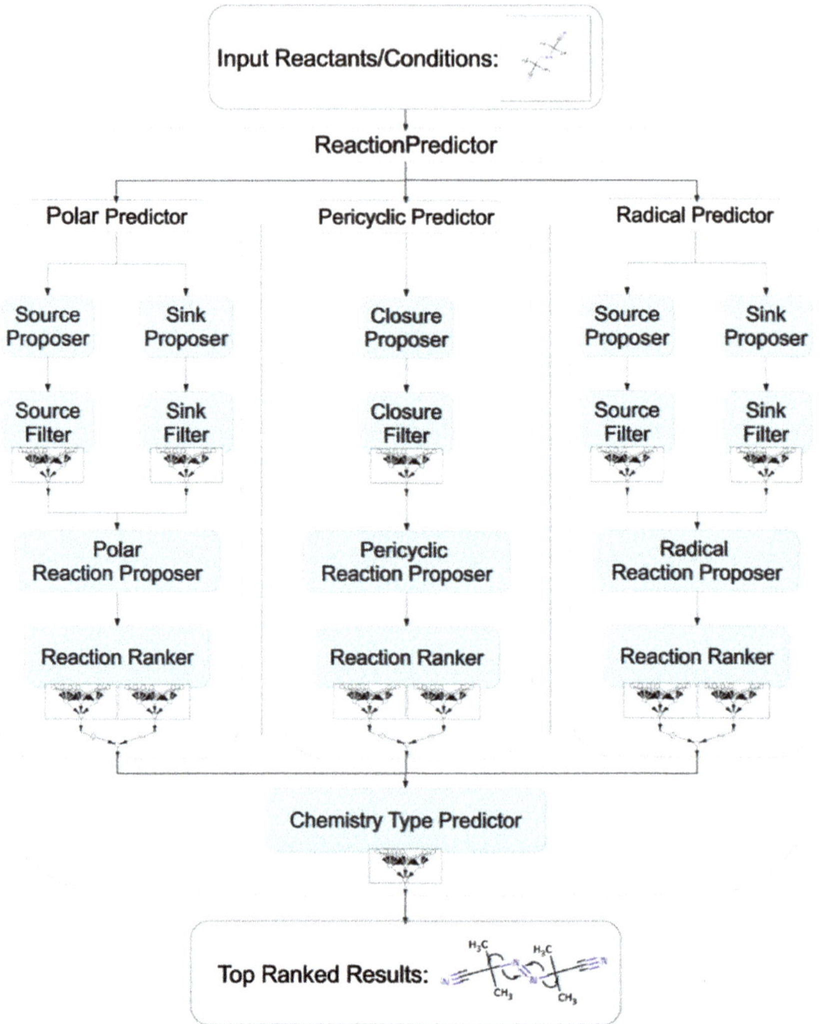

Figure 15.2 Workflow for reaction prediction using Kayala and Baldi's ReactionPredictor.
Reproduced from ref. 24 with permission from American Chemical Society, Copyright 2012.

including complex intramolecular rearrangements, and illustrate ReactionPredictor's application to impurity prediction.[25]

15.2.3 *Via* Reaction Templates

The backbone of many CASP programs are retrosynthetic templates: subgraph patterns that define how a structural motif in one species (*e.g.*, a product molecule) can be transformed into another (*e.g.*, reactant molecules). The same concept applies to synthetic templates that operate in the forward direction to convert reactant molecules into product molecules by matching specific substructures or functional groups in the reactants. They are commonly encoded as SMARTS strings.

Satoh and Funatsu's SOPHIA is arguably one of the first data-driven approaches for reaction prediction.[26,27] It makes use of a database of known chemical reactions by abstracting the transformation in terms of the structural characteristics of the reaction site (*i.e.*, a template). For a new set of reactants, SOPHIA will attempt to match every combination of bonds to its knowledge base with a user-adjustable degree of specificity; matching only the atom types of that bond, matching the structural characteristics, or matching farther out to the α, β, or γ positions.[†] The series of bond connections/disconnections is applied according to the matching precedent reactions to generate potential outcomes, which are optionally filtered by heuristic analysis of reaction conditions.

Explicitly extracted synthetic reaction templates can generate libraries of potential outcomes. This strategy is commonly used with expert-encoded reaction rules to generate virtual compound libraries.[29–32] One challenge is understanding when templates should or should not apply to a set of reactants. Even if each expert rule specifies the substrate scope of the transformation, several potential product molecules will be generated. Identifying which of the templates and products most accurately reflect what the "real" reaction would produce requires additional analysis.

The relevance of reaction templates to a particular set of reactants can be learned. Predicting which template out of a fixed library that should apply to a given molecule is simply a classification problem. Similar to Zhang and Aires-de-Sousa's work in 2005 looking at which of seven irradiation reactions is most likely,[16] Wei *et al.* describe a neural network model that is designed to predict one of 16 alkyl halide and alkene reactions on the basis of reactant and reagent molecular fingerprints.[33] To generate the dataset, they enumerated 1 277 329 synthetic reactions from 16 expert-encoded SMARTS and a library of small building block compounds, though a much smaller number was actually used for training/testing. A neural network model with a single hidden layer was trained on the simulated data and applied to a small external test set of textbook reactions.

This proof of concept was scaled to a much larger set of 3.5 million reactions from the Reaxys database by Segler and Waller soon thereafter.[34] The formulation of the problem is identical to Wei *et al.*: learn to predict the most probable reaction rule based on the molecular structure of the reactants (Figure 15.3).

The learned relevance approach was demonstrated using multiple template sets including 103 hand coded rules and 8720 automatically extracted rules. The 8720 templates represent a subset of what can be extracted from the Reaxys dataset of 4.9 million published reactions, requiring each

---

[†]This type of generalization is similar to what was later used for algorithmic extraction of reaction templates for retrosynthesis; see ref. 28.

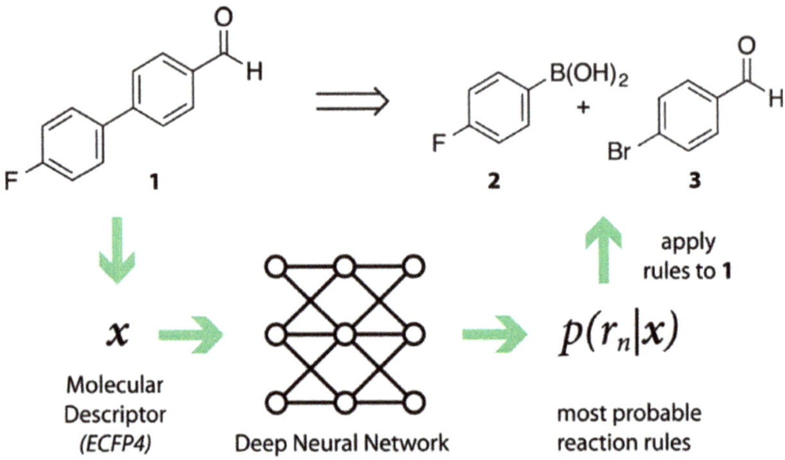

Figure 15.3 Approach to learning the relevance of reaction templates; shown in the retrosynthetic direction (product input, reactants generated by applying the template), but equally applicable in the forward direction (reactants input, product generated by applying the template).
Reproduced from ref. 34 with permission from John Wiley and Sons, © 2017 Wiley-VCH Verlag GmbH & Co. KGaA, Weinheim.

example to have at least 100 precedent reactions. For the larger template set, each reactant matched an average of 44.5 rule per query, highlighting the need to meaningfully prioritize the products they generate. A neural network model was trained for this 8720-way classification problem by treating the matching template as the "true" outcome for the corresponding reactant(s) structure, represented as a fingerprint. In a random 20% split for testing, the reaction prediction model selected the matching template 78% of the time with a top-10 accuracy of 98%.

One limitation of this evaluation is that predicting one reaction template does not correspond to predicting one reaction outcome. Applying the predicted template(s) to the reactant molecules will result in one or more product molecules. A single template can produce multiple product species; consider, for example, a template corresponding to an aromatic C–H functionalization defined by the SMARTS string [c:1][cH:2][c:3].[Br:4]≫[c:1][cH0:2]([Br:4])[c:3]. A reactant molecule is likely to have several possible sites for that template to apply, perhaps dozens if it has multiple aromatic rings.

To address this problem, Coley *et al.* describe a different template-based approach to reaction prediction that directly scores candidate outcomes rather than the templates used to produce them (Figure 15.4).[35] In this approach, synthetic reaction templates are used to define the scope of potential outcomes (candidate reactions), which are then scored and ranked by a neural network model. A dataset of 15 000 open source reactions from the United States Patent and Trademark Office (USPTO) were used to train/test the model. First, 1.1 million reactions were algorithmically processed to

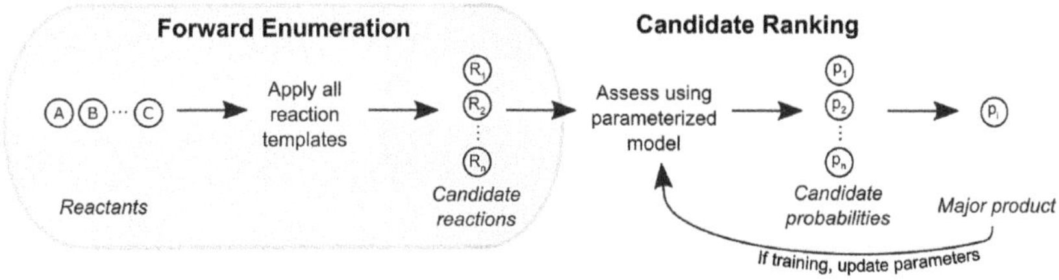

Figure 15.4 Approach to product prediction using reaction templates only to define the scope of possible outcomes.
Reproduced from ref. 35, https://doi.org/10.1021/acscentsci.7b00064, with permission from American Chemical Society, Copyright 2017.

produce reaction templates, 1689 of which were supported by at least 50 precedents (corresponding to a coverage of 76%). These templates were exhaustively applied to the 15 000 sets of reactants to generate a library of potential products for each. The number of products generated this way has a medium of 246 and mean of 353.

An "edit-based" representation of the reaction center was used in lieu of a more standard fingerprint representation to emphasize the bond changes required for the reaction to occur. Atom- and bond-level structural and functional features were used to represent the bonds lost and gained during the transformation. A standard feed forward neural network maps these features onto a scalar score for each candidate reaction. The model is trained using a categorical crossentropy loss function to ensure that the true (recorded) product is scored higher than any of the other candidate products. In a 5-fold cross validation, the true product was proposed 71.8% of the time and was in the top-5 predictions 90.8% of the time.

The approach of enumerate-then-rank suffers from the same limitation as predicting template relevance in that predictions cannot be made outside of the scope of the template library. If exhaustively applying all templates to a set of reactants does not generate the true product, then neither method will ever be able to predict that product. This effectively means that "new chemistry" (as defined by having a previously-unseen combination of reactant and product substructures at the reaction center) cannot be described.

A separate study by Segler and Waller begins to address this limitation by focusing specifically on binary reactions using exactly two reactant species (neglecting reagents, catalysts, and solvents).[36] From several million binary reactions from the USPTO, the authors extract two half-reactions that describe the changes in connectivity from each of the reactant molecules to the product. Known chemicals and reactions are stored in a knowledge graph. Reaction products are predicted from two reactant molecules by searching the knowledge graph for a path where the two are linked through mutual reaction partners. Provided the half-reactions are also analogous, a new reaction is proposed based on the half-reactions each reactant molecule undergoes in the precedent reactions. Including a similarity constraint allows the knowledge graph model to correctly predict the true product in its set of candidates 67.5% of the time, though the medium number of predicted products is 3 and the mean is 5.3.

Brute force enumeration of candidate products from the set of extracted half-reactions enables the prediction of 78% of the products in a held-out test set (compared to the 76% from Coley *et al.*[35] using full reaction templates); this leads to a median of 62 and mean of 311 products per example. The benefit to generalizing reactions with half-reaction templates rather than full reaction templates is that "new chemistry"—as defined above—can be described as long as the constituent half-reactions have been observed previously. Segler and Waller demonstrate this for a small number of reactions in a time-split validation.

15.2.4 Without Reaction Templates: Graphs

Overcoming the coverage limitation of reaction templates (full- or half-) requires a change in strategy. Jin *et al.* and Coley *et al.* describe a template-free approach to predicting reaction products that represents a chemical reaction as a set of edits to a graph;[37,38] atoms are nodes, bonds are edges. Most reactions relevant to synthesis of organic small molecules from smaller building blocks (*i.e.*, not including complexation reactions, acid/base neutralizations, *etc.*) can be described as a handful of bond breaking/forming edits to a reactant graph. Therefore, the prediction of reaction products can be formulated as the prediction of a set of graph edits, where application of those graph edits is sufficient to convert the reactant molecules into the major product molecule.

The authors use a Weisfeiller–Lehman Network (WLN) model architecture, a type of graph convolutional neural network, to learn this mapping. The prediction of the set of graph edits is decomposed into multiple stages (Figure 15.5): (1) representing reactants as a single graph; (2) predicting the most likely individual graph edits; (3) enumerating combinations of the highest-ranked graph edits; (4) scoring and ranking the products resulting from these combinations. A supervised learning approach to (2) is enabled by atom-mapping, so the exact set of bond changes for a recorded experimental reaction in the USPTO is known or at least assigned in a consistent manner. The enumeration and filtering stage is designed to enforce chemical valence rules. The final scoring/ranking is trained in a manner analogous to ref. 35, where model parameters are updated to maximize the score assigned to the "true" (recorded) product.

On a test of 40 000 reactions from the USPTO, this model achieves a top-1 accuracy of 85.6% and top-3 of 92.8%.[38] As a data-driven method, performance is higher for reactions that have a larger number of precedent reactions; its accuracy sharply drops off when test reactions have fewer than 10 similar precedents in the training set. The authors also emphasize that the model begins to address the question of model interpretability, as it can provide a crude heatmap of how each atom's reactivity is perceived to

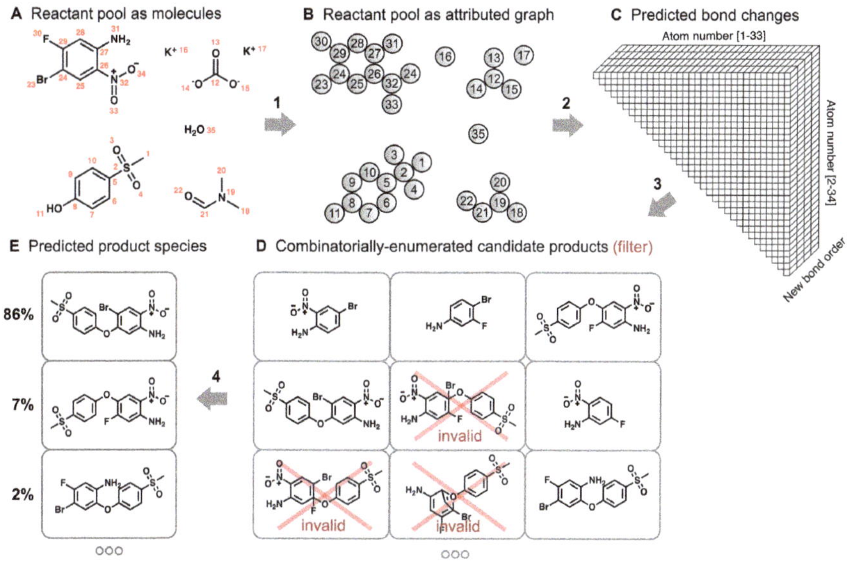

Figure 15.5 Approach to product prediction that treats a chemical reaction as a set of edits to a molecular graph of reactant molecules.
Reproduced from ref. 38 with permission from the Royal Society of Chemistry.

depend on other atoms through the use of a global attention mechanism. These explanations aren't nearly as informative as a full mechanistic explanation like that obtained by Baldi and coworkers, but the benefit is that it only relies on global reactivity data for training.

Bradshaw *et al.* try to find a middle ground between learning to predict a global reaction from experimental data and predicting a mechanistic pathway that is more informative to chemist users.[39] They examine a specific subset of reactions that can be described as "single electron paths"–pairs of electrons that move through a linear sequence. This is sufficient to describe many of the polar reactions that Kayala and Baldi address that are composed of electron source/sink interactions.

The authors make a further assumption that there is a canonical ordering to the electron pair movement. Heuristics estimate the direction of the electron path using each atom type's electronegativity to determine the beginning and end of the pseudo-mechanistic steps. A gated graph neural network model is used to predict a probability distribution over the action space of adding/removing bonds or stopping to reconstruct this sequence. Within their dataset, the model achieves a top-1 accuracy of 87.0% and a top-5 accuracy of 95.9%.

There are two assumptions in this approach that limit its utility. First, 27% of the USPTO dataset used by Jin *et al.*[37] cannot be described by these linear electron paths. This way of representing reactions provides more structure to the prediction task, but does so at the expense of coverage. The authors do not suggest any way to know *a priori* whether a set of reactants should be expected to be able to be described by a linear reaction path. The second is that the pseudo-mechanisms the model learns to provide are entirely based on expert heuristics and are not learned from the data. Figure 15.6 shows an example prediction for a Suzuki coupling, where the authors acknowledge that the pseudo-mechanism is unrelated to the true mechanism that relies on the presence of a Palladium catalyst.

A recent preprint examines whether the sequential prediction of graph edits can be formulated as a Markov Decision Process and trained using reinforcement learning.[40] While the results are slightly worse than previous graph-based formulations of chemical reaction prediction, predicting graph edits sequentially in this manner overcomes the assumptions made in Bradshaw *et al.* and the need for combinatorial enumeration in Coley *et al.*

15.2.5 Without Reaction Templates: Sequences

Another interesting and highly promising class of approaches to reaction prediction without using templates relies on SMILES representations of molecules. Treating reactants as SMILES strings and products as "translated" SMILES strings enables the application of language models that were originally developed for machine translation. An analogous usage for retrosynthetic reaction prediction performed by Liu *et al.* is shown in Figure 15.7.

The first proof of concept for this approach came from Nam and Kim in 2016.[42] They used TensorFlow's sequence-to-sequence model, consisting of multiple gated recurrent unit (GRU) cells, to define the model architecture. Roughly 1.1 million reactions from the USPTO served as a training set of experimental reactions; this was supplemented by synthetic training reactions following the process of Wei *et al.* The models were only tested on

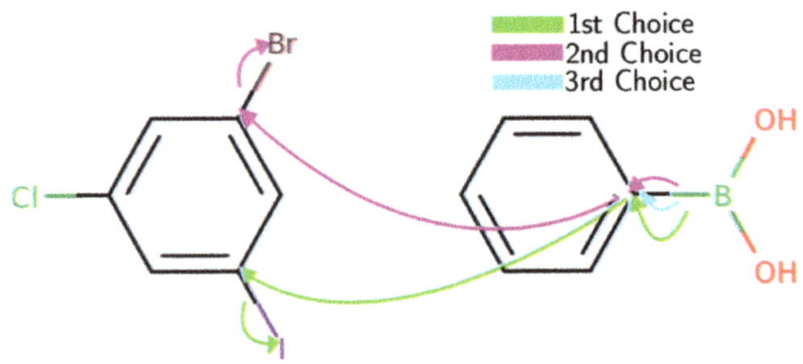

Figure 15.6 Example reaction prediction using a pseudo-mechanistic approach; the net movement of electron pairs can describe the global reaction, but does not reflect the true mechanism for this Pd-catalyzed reaction. Reproduced from ref. 39 with permission.

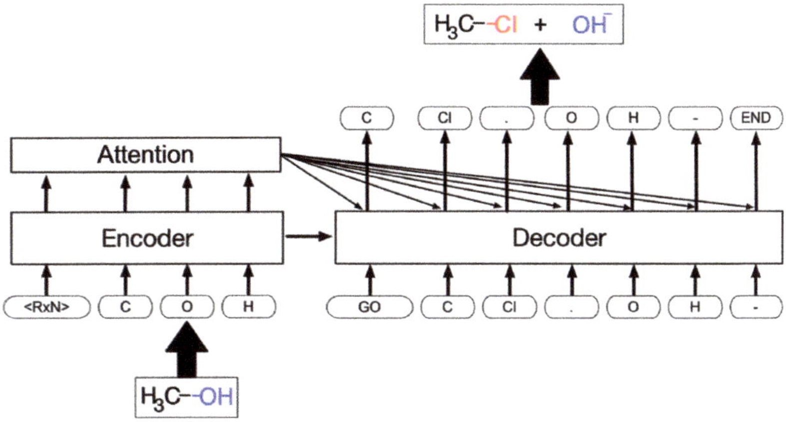

Figure 15.7 Treating chemical reactivity as a machine translation task using SMILES representations of reactants and products. Shown is the retrosynthetic direction, but the forward synthetic direction is formulated identically.
Reproduced from ref. 41 with permission from American Chemical Society, Copyright 2017.

the small number of textbook reactions (also used by Wei *et al.*) and on synthetic test data, but not on any subset of the USPTO dataset. Still, the results were quite promising and showed the potential for off-the-shelf language models to assist in reaction prediction.

The translation approach was further developed and refined by Schwaller *et al.* in two separate publications using different language models.[43,44] The first differed from Nam and Kim in its use of a particular attention mechanism, a more exhaustive hyperparameter optimization, a modified SMILES tokenizer, and evaluation on the USPTO dataset of real experimental reactions. The second took advantage of the Transformer architecture as a recent development in machine translation.[45] The Transformer model does not use recurrent units, but instead relies solely on positional encoding (augmented input features that contain information about which input tokens are adjacent) and a multi-headed attention mechanism.

The Transformer architecture proves remarkably successful for the prediction task, outperforming every previously-published approach to reaction prediction that has been tested on the USPTO benchmark dataset (*e.g.*, achieving a top-1 accuracy of 90.4% compared to 85.6% from Coley *et al.*[38]). One potential downside to this approach is that the model largely operates as a black box—it is not clear whether what is being learned is the underlying chemistry, or simply patterns in the SMILES strings. For many applications, however, that doesn't matter. The model is able to effectively generalize and can still prove tremendously useful in cheminformatics workflows, including assessing reaction feasibility for CASP.

15.3 Conclusion

15.3.1 Data Availability

Recent data-driven approaches have been enabled by an increasing availability of large reaction corpora, arguably moreso than advances in statistical model architectures and computing power. Purely data-driven methods should be expected to have a precipitous drop in performance when applied to reactions outside or close to their domain of applicability (Figure 15.8).

Two sources of reaction data dominate current studies: data from the USPTO (including an open source dataset originally extracted by Daniel Lowe[46] and a larger, well-curated dataset from NextMove Software's Pistachio[47]) and from Elsevier's Reaxys. All but the open source dataset present a barrier to full transparency and reproducibility. Unfortunately, there are no public databases of "ground truth" mechanistic reaction data, hence Baldi and coworkers use of expert rules to generate synthetic data and Bradshaw *et al.*'s use of heuristics to generate pseudo-mechanisms.

A feature of all of these data-driven models that is easy to overlook is that they will *always* try to predict a reaction outcome. In no situation is "no reaction" or 0% conversion a valid prediction, because that is not an

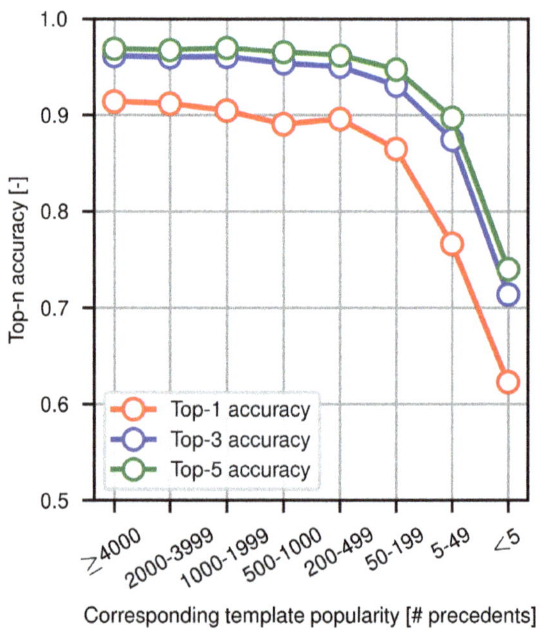

Figure 15.8 Model performance on unseen reactions as a function of how many training reactions undergo a similar structural transformation from Coley *et al.*
Reproduced from ref. 38 with permission from the Royal Society of Chemistry.

outcome that they have been trained to account for. Accelrys once offered a Failed Reaction database that included reactions with zero yields in addition to other examples where the observed product did not match what the chemist expected.[48] A modernized and open access version of such a database could prove valuable in providing low-conversion reaction examples and challenging test reactions.

15.3.2 Evaluation

Ostensibly, there is a single correct answer when predicting the outcome of a reaction.[‡] But this is only true when the prediction problem is well-posed and fully describes all of the input variables. No data driven method currently takes into consideration concentrations and equivalence ratios, temperature, or order of addition, partially because this information is not readily available in curated reaction databases. More complicated yet would be taking into account the rate of reagent addition (slow *versus* fast), the size/scale of the reaction, heat and mass transfer performance, and other subtler factors like exposure to ambient air and the humidity that day. These are all factors that we know can have a strong effect on the reaction outcome, yet there is no clear path toward capturing them in a data-driven model. For now, qualitative prediction of the major product (*i.e.*, an observed product with greater than 50% yield) has been the most detailed question one can ask while still making use of of hundreds of thousands of experimental reaction examples.

Even predicting the major product is not as straightforward as it might appear. One must be able to clearly define what we consider to be a unique molecular structure. Resonance structures are typically merged implicitly when sanitized in cheminformatics programs. Tautomers should be considered identical as well, but have distinct SMILES strings and are not automatically merged when using some programs, *e.g.*, RDKit. Acid/base chemistry adds another consideration—is a deprotonated carboxylic the same as the protonated acid? Or a free base amine *versus* its hydrochloride salt? To a chemist interpreting the results of the prediction manually, the answer is often yes; however, a strict comparison of the two would consider them not to match.

There is a trend to show results only within the scope of what can be predicted by a given method (this author is guilty of that as well). For example, ref. 34 reports template prediction accuracy only for reactions that can be described by their truncated template set, ref. 35 reports product ranking accuracy only when the true product is generated by its template library, and ref. 39 reports product prediction accuracy only for reactions that can be as a linear path of electron pairs. While there is no inherent issue

[‡]That is, to the extent that it is not a chaotic system and the outcome is deterministic.

with developing models with narrow domains of applicability, changing how accuracy is reported complicates the comparison of different methods.

15.3.3 Breadth *versus* Accuracy

There is a general question of whether reaction prediction models should try to maximize coverage (*e.g.*, describe any reaction found in the USPTO or Reaxys) or accuracy within specific reaction families. These are two distinct goals that are useful in different scenarios. The approaches described in this Chapter focus on breadth, which is useful in CASP workflows, predicting impurities, *etc.* as described in the Introduction. Studies with a narrower scope can embed domain expertise and prior knowledge more directly into the model, *e.g.*, considering the stability of a carbocation intermediate when predicting the regioselectivity of electrophilic aromatic substitution reactions.[49]

15.4 Model Types

The Transformer model is remarkably powerful despite not incorporating any sort of chemistry-motivated strategy except for SMILES augmentation and some restrictions on SMILES tokens during decoding to promote syntactic validity. However, it's not yet clear how well these systems generalize to new chemistries and identify the true underlying patterns of chemical reactivity, rather than patterns in the data. To be fair, it's not clear how *any* of these forward prediction methods generalize to new synthetic methodologies. The ability to discover new reactions was one of the motivating factors for Ugi's formal treatment of reactivity, which was successfully demonstrated.[50,51] Recent data-driven methods have not yet shown their ability to prospectively generate new synthetic methods; doing so would be compelling evidence that they are meaningfully generalizing reaction knowledge–much more compelling than achieving high accuracy in a time-split validation.

15.5 Conclusion

There has been a significant amount of progress in data-driven prediction of reaction outcomes even within the past decade. We can accurately predict the major products of reactions that are well-represented in our datasets as well as one can expect, given the data quality. What we *can't* do well is predict stereoselectivity, predict subtle cases of regioselectivity, or consider the effects of other experimental variables besides the identity of reactants and reaction conditions. As a consequence of this last point, we are, in general, unable to predict yields quantitatively unless working with datasets that hold all other variables constant. Other challenges are pushing the limits of low data learning, extracting meaningful explanations for model predictions (except for mechanistic predictors), and extrapolating to entirely new modes of reactivity. As access to data improves and more researchers enter the field, these challenges will surely be addressed.

Acknowledgements

The author would like to thank Wengong Jin, Regina Barzilay, Tommi Jaakkola, Timothy F. Jamison, William H. Green, Klavs F. Jensen, and all other members of the Machine Learning for Pharmaceutical Discovery and Synthesis Consortium who have contributed to discussions on this topic.

References

1. W.-D. Ihlenfeldt and J. Gasteiger, Computer-Assisted Planning of Organic Syntheses: The Second Generation of Programs, *Angew. Chem., Int. Ed. Engl.*, 1996, **34**, 2613–2633.
2. M. H. Todd, Computer-aided organic synthesis, *Chem. Soc. Rev.*, 2005, **34**, 247–266.
3. A. Cook, A. P. Johnson, J. Law, M. Mirzazadeh, O. Ravitz and A. Simon, Computer-aided synthesis design: 40 years on, *Wiley Interdiscip. Rev.: Comput. Mol. Sci.*, 2012, **2**, 79–107.
4. W. A. Warr, A Short Review of Chemical Reaction Database Systems, Computer-Aided Synthesis Design, Reaction Prediction and Synthetic Feasibility, *Mol. Inf.*, 2014, **33**, 469–476.
5. J. Gasteiger; and W. Ihlenfeldt, in *Software Development in Chemistry 4*, Springer, 1990, pp. 57–65.
6. W. P. Walters, Virtual Chemical Libraries, *J. Med. Chem.*, 2018, DOI: 10.1021/acs.jmedchem.8b01048.
7. J. Lyu, S. Wang, T. E. Balius, I. Singh, A. Levit, Y. S. Moroz, M. J. O'Meara, T. Che, E. Algaa, K. Tolmachova, A. A. Tolmachev, B. K. Shoichet, B. L. Roth and J. J. Irwin, Ultra-large library docking for discovering new chemotypes, *Nature*, 2019, 1.
8. I. Ugi, J. Bauer, J. Brandt, J. Friedrich, J. Gasteiger, C. Jochum and W. Schubert, New Applications of Computers in Chemistry, *Angew. Chem., Int. Ed. Engl.*, 1979, **18**, 111–123.
9. J. Gasteiger and C. Jochum, in *Organic Compounds*, Springer-Verlag, Berlin/Heidelberg, **vol. 74**, 1978, pp. 93–126.
10. T. D. Salatin and W. L. Jorgensen, Computer-assisted mechanistic evaluation of organic reactions. 1. Overview, *J. Inorg. Chem.*, 1980, **45**, 2043–2051.
11. G. Sello, Reaction prediction: the suggestions of the Beppe program, *J. Chem. Inf. Model.*, 1992, **32**, 713–717.
12. I. M. Socorro, K. Taylor and J. M. Goodman, ROBIA: a Reaction Prediction Program, *Org. Lett.*, 2005, **7**, 3541–3544.
13. L. P. Hammett, The Effect of Structure upon the Reactions of Organic Compounds. Benzene Derivatives, *J. Am. Chem. Soc.*, 1937, **59**, 96–103.

14. V. Simon, J. Gasteiger and J. Zupan, A combined application of two different neural network types for the prediction of chemical reactivity, *J. Am. Chem. Soc.*, 1993, **115**, 9148–9159.
15. J. Aires-de-Sousa and J. Gasteiger, New Description of Molecular Chirality and Its Application to the Prediction of the Preferred Enantiomer in Stereoselective Reactions, *J. Chem. Inf. Comput. Sci.*, 2001, **41**, 369–375.
16. Q.-Y. Zhang and J. Aires-de-Sousa, Structure-Based Classification of Chemical Reactions without Assignment of Reaction Centers, *J. Chem. Inf. Model.*, 2005, **45**, 1775–1783.
17. G. V. S. M. Carrera, S. Gupta and J. Aires-de-Sousa, Machine learning of chemical reactivity from databases of organic reactions, *J. Comput.-Aided Mol. Des.*, 2009, **23**, 419–429.
18. D. T. Ahneman, J. G. Estrada, S. Lin, S. D. Dreher and A. G. Doyle, Predicting reaction performance in C-N cross-coupling using machine learning, *Science*, 2018, **360**, 186–190.
19. A. Cherkasov, E. N. Muratov, D. Fourches, A. Varnek, I. I. Baskin, M. Cronin, J. Dearden, P. Gramatica, Y. C. Martin, R. Todeschini, V. Consonni, V. E. Kuz'min, R. Cramer, R. Benigni, C. Yang, J. Rath- man, L. Terfloth, J. Gasteiger, A. Richard and A. Tropsha, QSAR Modeling: Where Have You Been? Where Are You Going To?, *J. Med. Chem.*, 2014, **57**, 4977–5010.
20. J. B. O. Mitchell, Machine learning methods in chemoinformatics, *Wiley Interdiscip. Rev.: Comput. Mol. Sci.*, 2014, **4**, 468481.
21. J. H. Chen and P. Baldi, Synthesis Explorer: A Chemical Reaction Tutorial System for Organic Synthesis Design and Mechanism Prediction, *J. Chem. Educ.*, 2008, **85**, 1699.
22. J. H. Chen and P. Baldi, No Electron Left Behind: A Rule-Based Expert System To Predict Chemical Reactions and Reaction Mechanisms, *J. Chem. Inf. Model.*, 2009, **49**, 2034–2043.
23. M. A. Kayala, C.-A. Azencott, J. H. Chen and P. Baldi, Learning to Predict Chemical Reactions, *J. Chem. Inf. Model.*, 2011, **51**, 2209–2222.
24. M. A. Kayala and P. Baldi, ReactionPredictor: Prediction of Complex Chemical Reactions at the Mechanistic Level Using Machine Learning, *J. Chem. Inf. Model.*, 2012, **52**, 2526–2540.
25. D. Fooshee, A. Mood, E. Gutman, M. Tavakoli, G. Urban, F. Liu, N. Huynh, D. V. Vranken and P. Baldi, Deep learning for chemical reaction prediction, *Mol. Syst. Des. Eng.*, 2018, **3**, 442–452.
26. H. Satoh and K. Funatsu, SOPHIA, a knowledge base-guided reaction prediction system-utilization of a knowledge base derived from a reaction database, *J. Chem. Inf. Comput. Sci.*, 1995, **35**, 34–44.
27. H. Satoh and K. Funatsu, Further Development of a Reaction Generator in the SOPHIA System for Organic Reaction Prediction. Knowledge-Guided Addition of Suitable Atoms and/or Atomic Groups to Product Skeleton, *J. Chem. Inf. Comput. Sci.*, 1996, **36**, 173–184.
28. J. Law, Z. Zsoldos, A. Simon, D. Reid, Y. Liu, S. Y. Khew, A. P. Johnson, S. Major, R. A. Wade and H. Y. Ando, Route Designer: A Retrosynthetic

Analysis Tool Utilizing Automated Retrosynthetic Rule Generation, *J. Chem. Inf. Model.*, 2009, **49**, 593–602.
29. Synthetically Accessible Virtual Inventory (SAVI) Database, Download Page https://cactus.nci.nih.gov/download/savi_download/ (accessed 02/12/2019).
30. Q. Hu, Z. Peng, J. Kostrowicki and A. Kuki, LEAP into the Pfizer Global Virtual Library (PGVL) space: creation of readily synthesizable design ideas automatically, *Methods Mol. Biol.*, 2011, **685**, 253–276.
31. F. Chevillard and P. Kolb, SCUBIDOO: A Large yet Screenable and Easily Searchable Database of Computationally Created Chemical Compounds Optimized toward High Likelihood of Synthetic Tractability, *J. Chem. Inf. Model.*, 2015, **55**, 1824–1835.
32. C. A. Nicolaou, I. A. Watson, H. Hu and J. Wang, The Proximal Lilly Collection: Mapping, Exploring and Exploiting Feasible Chemical Space, *J. Chem. Inf. Model.*, 2016, **56**, 1253–1266.
33. J. N. Wei, D. Duvenaud and A. Aspuru-Guzik, Neural networks for the prediction of organic chemistry reactions, *ACS Cent. Sci.*, 2016, **2**, 725–732.
34. M. H. S. Segler and M. P. Waller, Neural-Symbolic Machine Learning for Retrosynthesis and Reaction Prediction, *Chem. - Eur. J.*, 2017, **23**, 5966–5971.
35. C. W. Coley, R. Barzilay, T. S. Jaakkola, W. H. Green and K. F. Jensen, Prediction of Organic Reaction Outcomes Using Machine Learning, *ACS Cent. Sci.*, 2017, **3**, 434–443.
36. M. H. S. Segler and M. P. Waller, Modelling Chemical Reasoning to Predict and Invent Reactions, *Chem. - Eur. J.*, 2017, **23**, 6118–6128.
37. W. Jin, C. W. Coley, R. Barzilay and T. Jaakkola, Predicting Organic Reaction Outcomes with Weisfeiler-Lehman Network, in *Advances in Neural Information Processing Systems*, 2017, pp. 2607–2616.
38. C. W. Coley, W. Jin, L. Rogers, T. F. Jamison, T. S. Jaakkola, W. H. Green, R. Barzilay and K. F. Jensen, A graph-convolutional neural network model for the prediction of chemical reactivity, *Chem. Sci.*, 2019, **10**, 370–377.
39. J. Bradshaw, M. J. Kusner, B. Paige, M. H. S. Segler, J. M. Hernández-Lobato Predicting Electron Paths, *arXiv:1805.10970 [physics, stat]*, 2018.
40. K. Do, T. Tran and S. Venkatesh, Graph Transformation Policy Network For Chemical Reaction Prediction, in *Proceedings of the 25th ACM SIGKDD International Conference on Knowledge Discovery & Data Mining*, 2019, pp. 750–760.
41. B. Liu, B. Ramsundar, P. Kawthekar, J. Shi, J. Gomes, Q. Luu Nguyen, S. Ho, J. Sloane, P. Wender and V. Pande, Retrosynthetic Reaction Prediction Using Neural Sequence-to-Sequence Models, *ACS Cent. Sci.*, 2017, **3**, 1103–1113.
42. J. Nam, J. Kim, Linking the Neural Machine Translation and the Prediction of Organic Chemistry Reactions, *arXiv:1612.09529 [cs]*, 2016.

43. P. Schwaller, T. Laino, T. Gaudin, P. Bolgar, C. Bekas, A. A. Lee, Molecular Transformer for Chemical Reaction Prediction and Uncertainty Estimation, 2018, DOI: 10.26434/chemrxiv.7297379.v1.
44. P. Schwaller, T. Gaudin, D. Lanyi, C. Bekas and T. Laino, "Found in Translation": predicting outcomes of complex organic chemistry reactions using neural sequence-to-sequence models, *Chem. Sci.*, 2018, **9**, 6091–6098.
45. A. Vaswani, N. Shazeer, N. Parmar, J. Uszkoreit, L. Jones, A. N. Gomez, L. Kaiser and I. Polosukhin, Attention Is All You, *Need.arXiv:1706.03762 [cs]*, 2017.
46. D. M. Lowe, Patent reaction extraction: downloads, 2014, https://bitbucket.org/dan2097/patent-reaction-extraction/downloads.
47. NextMove Software – Pistachio https://www.nextmovesoftware.com/pistachio.html (accessed 04/04/2019).
48. Hindsight is an exact science http://pubs.acs.org/subscribe/archive/ci/31/i09/html/09potter.html (accessed 05/02/2019).
49. J. C. Kromann, J. H. Jensen, M. Kruszyk, M. Jessing and M. Jørgensen, Fast and accurate prediction of the regioselectivity of electrophilic aromatic substitution reactions, *Chem. Sci.*, 2018, **9**, 660–665.
50. D. Forstmeyer, J. Bauer, E. Fontain, R. Herges, R. Herrmann and I. Ugi, Reaction of Tropone with a Homopyrrole. The Result of a Computer-Assisted Search for Unique Chemical Reactions, *Angew. Chem., Int. Ed. Engl.*, 1988, **27**, 1558–1559.
51. R. Herges and C. Hoock, Reaction Planning: Computer-Aided Discovery of a Novel Elimination Reaction, *Science*, 1992, **255**, 711–713.

Section 7: Future Outlook

CHAPTER 16

ChemOS: An Orchestration Software to Democratize Autonomous Discovery

LOÏC M. ROCH,*[a,b,c,d,e] FLORIAN HÄSE*[a,b,c,d,e] AND ALÁN ASPURU-GUZIK*[b,c,d,e,f]

[a] ChemOS Sàrl, Lausanne, VD 1006, Switzerland; [b] Department of Chemistry, University of Toronto, 80 St George St, Toronto, ON M5S 3H6, Canada; [c] Department of Computer Science, University of Toronto, 214 College St, Toronto, ON M5T 3A1, Canada; [d] Vector Institute for Artificial Intelligence, 661 University Ave Suite 710, Toronto, ON M5G 1M1, Canada; [e] Department of Chemistry and Chemical Biology, Harvard University, Cambridge, MA 02138, USA; [f] Lebovic Fellow, Canadian Institute for Advanced Research (CIFAR), Toronto, ON M5G 1M1, Canada
*Emails: loic@chemos.io; hase.florian@gmail.com; alan@aspuru.com

16.1 Introduction

The key economic and societal challenges of the 21st century, including clean energy, global health and sustainability,[1] require the scientific discovery process to be streamlined and substantially accelerated. The current state-of-the-art Edisonian[2] approaches for the discovery and deployment of novel functional materials are inherently slow, capital intensive, and have arguably reached a plateau.[3] A transformative advance demands to rethink the current human-centered approaches to, thus, enable a *Moore's law* for scientific discovery.[3–6]

Drug Discovery Series No. 75
Artificial Intelligence in Drug Discovery
Edited by Nathan Brown

Published by the Royal Society of Chemistry, www.rsc.org

Established drug discovery processes, starting with the discovery of promising materials candidates or active pharmaceutical ingredients and culminating in a fully commercialized product, typically require 10 to 20 years of process optimization and development.[7–10] One of the reasons for this slow turnaround is that discoveries commonly rely on experimental trial-and-error methods, often referred to as *Edisonian* approaches, which are costly, human-intensive, and slow. Recently, the concept of empowering automated infrastructures for synthesis, characterization and fabrication with artificial intelligence (AI) has emerged, leading to autonomy in experimentation. While automated platforms constitute artificial machines designed to execute specific tasks, autonomous platforms also transfer the decision-making process for experiment design from the human to the machine, thus enabling unsupervised experimentation without human intervention. Self-driving laboratories can realize autonomous experimentation, and thus have potential to expedite scientific discovery by orders of magnitude.[5,11–17] The self-driving laboratories leverage advances in robotics, automation, AI, data science, and high-throughput characterization techniques to realize a closed-loop approach to experimentation, where experiments are iteratively designed on-the-fly by a machine-driven decision-maker based on previously conducted experiments.

Recent progress in computational materials screening augmented with machine learning (ML) methods that are trained on simulated or experimental data offers the potential to substantially accelerate discovery and process intensification.[18–26] Advances in reaction prediction and retrosynthesis will furthermore help to overcome the widely known gap from theoretical prediction to experimental implementation.[27–31] Further progress in the automation of synthesis, device preparation and characterization enables the development of the so-called self-driving laboratories. Such laboratories augment automated experimentation systems with ML strategies such as Bayesian learning and reinforcement learning for efficient experiment planning leading to an accelerated discovery process.

The drug discovery process typically requires a substantial number of experiments to discover and optimize promising lead candidates, identify delivery parameters, and run preclinical tests and clinical trials. These steps, from initial discovery to market adoption, accumulate considerable expenses estimated between US$0.5 billion and US$2.6 billion, with typical research timelines spanning 10–20 years.[32–34] As such, substantial acceleration of these steps would significantly reduce the overall costs of the drug discovery process. Currently, the discovery process relies on high-throughput strategies (HTS) to identify promising materials or potential drug candidates.[35] Moving away from these human-centered approaches to data-driven approaches *via* the use of a closed-loop orchestration software like ChemOS has the potential to simplify the implementation and the deployment of such complex workflows using self-driving laboratories.

This chapter provides an overview of automated approaches commonly used in materials science, chemistry and the life sciences. We cover well-established algorithmic strategies for experiment planning, highlighting their strengths and limitations through key examples from academia and industry. Then we describe and define autonomous strategies to scientific discovery. In this section, we cover single- and multi-objective optimization strategies, overview the early realization of autonomous laboratories and lay out a roadmap for deploying and orchestrating self-driving laboratories. Section 4 details the architecture and suite of applications composing the ChemOS software package. Section 5 highlights successful application of ChemOS. The specific use of ChemOS to accelerate drug discovery is detailed in Section 6. Finally we draw our conclusions and outline our vision in Section 7.

16.2 Automated Approaches to Scientific Discovery

Automation describes the process of designing and developing artificial machines to execute tasks that have formerly been performed by a human operator. It was introduced to improve efficiency, increase reliability and product quality, and ensure reproducibility of manufacturing processes. In 1673, Robert Boyle already pointed out that the reproducibility of the experiments "may disappoint your expectation",[36] calling for approaches to improve reliability, and quality.

Automation liberates scientists from repetitive, monotonous tasks and thus enables them to focus on more creative and complex activities.[37–39] Early mentions of automation in the physical sciences date back as far as 1875,[40] by Thaddeus M. Stevens. He reported an artificial device designed as a filter washer with a controlled aperture modulating the rate at which water drips over a filtrate. Automated laboratories enable automated process optimization: systematically varying the values of manipulatable process parameters within pre-defined bounds, collectively referred to as the *parameter space*, to identify the optimum response of the studied process that satisfies desired targets (*objectives*). Automation enables an extensive, systematic screening of the parameter space, which is oftentimes labour intensive and time consuming. Leveraging the aforementioned advantages of automation, automated laboratories therefore have the potential to increase the experimentation throughput while moderately increasing the cost of operation, thus providing opportunities to scale scientific discovery to more challenging problems with larger parameter spaces.

16.2.1 Algorithmic Strategies to Screen the Parameter Space

With automated experimentation platforms, the strategy to screen the parameter space is driven by humans. The choice of the strategy to drive the screening phase is commonly based on *prior* expectations and scientific intuition. A set of experiments, collectively referred to as the *experimental campaign*, is designed prior to the experimentation process accounting for the manipulatable experimentation parameters and the target. Once the experimental campaign is fully defined, the set of parameters are submitted to the automated solutions to evaluate the properties of interest. Three main strategies are commonly employed to extensively screen the parameter space.

One of the first strategies to designing experiments was introduced to laboratory environments in 1947 and was later referred to as one-factor-at-time (OFAT, Figure 16.1(A)) approach.[41,42] The OFAT procedure holds all parameters (or *factors*) but one fixed, to identify the optimal response when the one free input is varied. Once the optimal value is determined, this parameter is fixed, and another parameter is varied. This process is repeated until all parameters have been adjusted to yield the most promising response. As such, OFAT constitutes an optimization approach similar to line searches.[43] OFAT is applicable if the response surface of the process at hand is *flat* in all dimensions, *i.e.*, the optimum can be located regardless of the location of the starting point and the order in which the parameters are optimized. In such a surface, commonly referred to as a *main effect model*, all parameters are independent, *i.e.* there is no correlation between any two parameters. OFAT is not guaranteed to locate the optimum if correlations between factors exist, which explicitly poses challenges for this method for non-convex response surfaces.

High-throughput experimentation (HTE, Figure 16.1(B)) is a strategy that is widely used in situations where experiments are inexpensive such that data describing the experimental responses for individual parameters is cheap and abundant. HTE campaigns start with the design of a (possibly large) set of experiments for which parameters are sampled from the pre-defined parameter space, possibly reflecting the researchers'

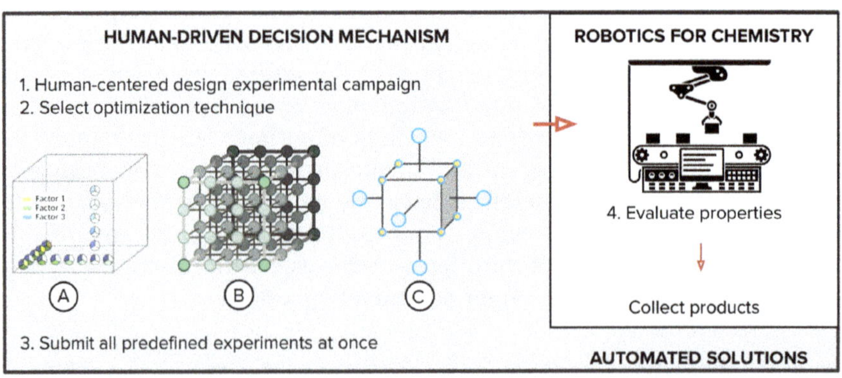

Figure 16.1 Schematic representation of experimentation with automated solutions highlighting two strategies to designing experiment: one-factor-at-a-time (OFAT, A), high-throughput experimentation (HTE, B), and design of experiment (DoE, C). Right panel © MicroOne/Shutterstock.

expectations about the location of the optimum. While HTE enables the massive parallelization of individual experiments, this approach in its pure form does not refine the chosen set of experiments on-the-fly based on recent measurements and thus does not implement a closed-loop. Continuous parameters in HTE experiments can, for example, be chosen from an extensive grid with fixed size increment steps, thus spanning a full-factorial campaign. However, parameters evaluated from a grid are correlated which can result in the omission of possibly occurring correlations between factors. In addition, grid-based search strategies require a substantial number of evaluations suffering from an exponential explosion of the search space with the number of probed parameters to capture relevant phenomena leading, eventually, to discovery.

In the spirit of HTE, another strategy to design the experimental campaign leverages quasi-random low-discrepancy sequences such as the Sobol sequence.[44] Originally introduced in 1967 by Sobol in the context of constructing approximations to integrals of real-valued functions on high-dimensional hypercubes, Sobol sequences aim to generate sequences of parameter points which enable the construction of a rapidly converging approximation to the integral.[45] To this end, Sobol sequences generate a set of parameter points which uniformly cover the parameter space (as opposed to purely random strategies) with little correlation between consecutive parameter points.

One step further, the integration of design of experiments (DoE, Figure 16.1(C)) to automation emerged as a more elaborate strategic approach to experimentation. The method was introduced by Sir Ronald Fisher in 1935.[46–48] Since then, several designs and variants of DoE have been suggested, each targeting different applications and balancing the required number of experiments with the resolution of the search space. Like OFAT, the choice of the design relies heavily on *prior* knowledge about the parameter space, and on the expected interaction between any two of the probed factors. Full factorial designs, for example, discretize each factor into a desired number of levels, thus creating a hypercube of experiments (similar to grid search).[49] However, the required number of experiments scales exponentially with the number of probed factors. Other designs, such as fractional factorial designs,[47] lower the required number of experiments but display the same scaling. The Plackett–Burman design reduces the requirements on the number of experiments but assumes negligible interactions between any two factors.[50] As a consequence, these designs might miss important aspects of the response surface if the prior assumptions are violated. Once a design has been chosen, experiments are typically conducted following the design layout without further refinement. As such, all experiments associated with the design layout can be parallelized. After completing all experiments, a new design can be proposed in the subsequent refinement step potentially including human decisions, accounting for the measurements of the previously conducted experiments.

16.2.2 Examples of Automation in Key Industrial Sectors and Academia

Several industries and scientific fields have already demonstrated the benefits of automated systems on a wide variety of applications by leveraging the advantages of the aforementioned strategies to designing experiments.

The automation and robotics revolution was intensively initiated and streamlined by the industrial sector when searching for intensifying chemical and pharmaceutical processes to increase productivity and improve quality.[37,51–57] The patenting of the first industrial robots in 1961 by Unimation, Inc.[†] accelerated the deployment of robotics and automation across industries, most notably in clinical laboratories. A couple years later, Zymark Corporation patented a user-programmable robotic arm. It was equipped with interchangeable hands allowing this new generation of robot to be adapted to a vast variety of applications notably to numerous assays and sample-handling approaches, pre-analytical sample preparation and to potency and stability testing in the pharmaceutical industry.

Early automated systems were introduced into research environments by Merrifield *et al.* in 1966 who reports an *Instrument for automated synthesis of peptides*,[58] a platform where all of the steps involved in the synthesis were fully automated. The sequential operations and timings were controlled by a 32 roller-type microswitch. A few years later, computers progressively started to digitize mechanical controls and operations to become a common instrument in laboratories.[59] During the last decades, fully automated approaches were applied to topics as diverse as the synthesis of small organic molecules,[37,60–63] to the screening of polymers,[64,65] to continuous-flow synthesis of peptides,[66,67] to real-time monitoring of chemical reactions unraveling kinetics of homogeneous and heterogeneous reactions,[68,69] to the identification of novel wide bandgap perovskites with higher stability,[70] and to the identification of colloidal nanocrystals.[71]

Automation in clinical laboratories started simultaneously with Sasaki in the early 1980s at the Kochi Medical School in Japan,[72] and with Felder at the University of Virginia in the US.[73] Starting from the mid 1990s, these two approaches, known as *total laboratory automation* (TLA) and as *point-of-car* (POC), were at the heart of robotic automation in the clinical laboratory.[74,75] Important advances in laboratory automation were achieved in the areas of gene sequencing in the Human Genome Project and in drug discovery studies in the pharmaceutical industry.[38,76,77] Most notably, with the boost experienced by the commercialization of robotized systems in the early 2000s,[39] large pharmaceutical companies started to explore applications of automation to the lead optimization phase of drug discovery to reduce cycle times by increasing efficiency in both flow of candidate drugs and target compound preparation.[78]

[†]G. C. Devol Jr., *U. S. Patent* 2988237 (1961).

As a matter of fact, advances in experimentation hardware, notably the automation of synthesis and characterization processes, have further accelerated the throughput of automated infrastructures, resulting in robotic systems for pharmaceutical applications which nowadays can screen more than 100 000 drug candidates per day in an ultra-high throughput approach.[79,80] The drastic increases of the achievable experimental throughput rendered HTE approaches that are massively parallelizable the most promising of the three experimentation strategies in the context of automated experimentation.

This shift to logging of experimental data and digitizing of drug development marked the beginning of the automation era in the pharmaceutical industry.[81] In 2005, Aventis and Accelab reported SynCar as an approach to automated synthesis. It consisted of seven independent workstations connected by a shuttle transfer system to process up to four reactions in parallel, with the ability to perform synthesis, filtration, liquid–liquid extraction, evaporation, weighing, solid-phase extraction, and HPLC/MS analysis.[82] Later on, Godfrey *et al.* at Eli Lilly, reported a remote-controlled automated chemical synthesis laboratory, with the goal to minimize the burden of repetitive, routine, rule-based operations that characterize more established chemistry workflows.[39,83]

16.2.3 Limitations of Automated Approaches

Despite remarkable successes of automated approaches to screen large parameter spaces inaccessible to human researchers, their applicability is limited by various factors.

First, a trial-and-error fabrication of materials and drug discovery yields a combinatorial explosion of the candidate space, which renders an exhaustive search of this space impractical. Second, single-factor methods and factorial designs require substantial prior knowledge of the response surface to be applied successfully. The inadequacy of these methods for analytical chemical methods to applications where this prior knowledge is not available has been reported as early as 1974.[84] Third, the quest for suitable software packages to control automated equipment from a central platform proved to be much less straightforward than initially anticipated.[39] These limitations have long been realized by the chemistry community, which has been exploring alternative search strategies for chemical processes for decades.[37,51,85–87]

Viable alternatives to HTS need to alleviate the curse of dimensionality, and typically formulate the discovery task as an optimization problem. Notably, these alternatives use guided searches, leveraging AI and ML, to selectively navigate vast candidate spaces. The optimization problem targets the identification of compounds which yield a set of desired properties (for example Lipinski's rule of five). Compounds which do not satisfy the property requirements can be ranked quantitatively based on the deviation of their properties from the target properties. This quantitative ranking allows the estimation of the merit of a compound and thus to make informed decisions for suggesting new compounds for future tests.[3] Such a guided approach significantly reduces the number of experiments yielding discovery, accelerating, thus, the discovery rate.[13]

Yet another limitation to the applicability of automation lies in the lack of standards for communications between automated equipment and characterization tools. Indeed, without such standards, laboratories cannot easily integrate equipment from different vendors, which is crucial to realize elaborate workflows, reduce deployment times, optimize processes, and control costs. As such, to further streamline the discovery process, the community, notably chemistry, biochemistry, material science, pharma and automation, is ready to move towards the next-generation of autonomous laboratories, which augment automation with AI algorithms; the self-driving laboratories.

16.3 Autonomous Approaches to Scientific Discovery

Self-driving laboratories leverage automated platforms along with the ability of AI/ML algorithms to efficiently navigate the candidate space in search for target properties, in a closed-loop approach. Such an approach is illustrated in Figure 16.2.

Self-driving laboratories employ algorithmic strategies to design experimental campaigns and leverage AI algorithms for efficient experiment planning. AI algorithms facilitate the rapid identification of optimal process parameters by inferring most informative design strategies on-the-fly from conducted experiments. To this end, a surrogate to a multi-dimensional and multi-objective response surface of the considered experiment is formulated, defined *via* the set of modulated experimental parameters and the objectives of the campaign. Surrogates can be constructed implicitly or explicitly based on previously conducted experiments or prior knowledge and quantify the

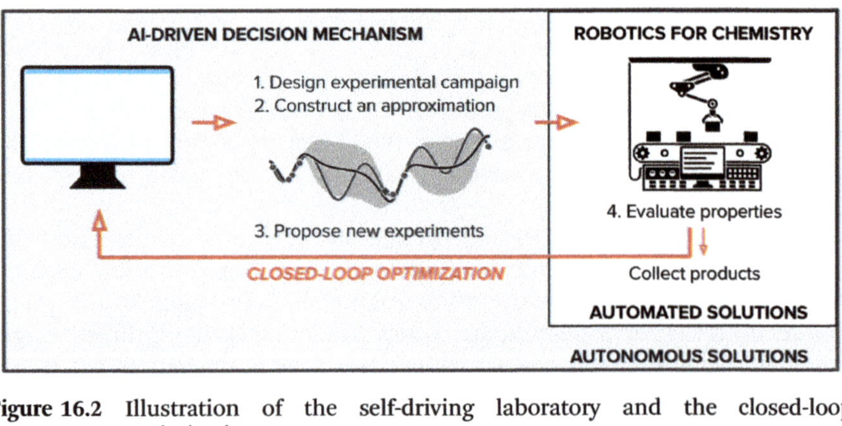

Figure 16.2 Illustration of the self-driving laboratory and the closed-loop optimization.
Adapted from ref. 13 with permission from Elsevier, Copyright 2019.

expected merit for all parameters in the parameter space. Based on these surrogates, the most informative experimental parameters are selected for experimental evaluation and submitted to the automated solutions to assess the performance of the experiment according to the defined objectives. On the one hand the products[‡] are collected and stored for later analysis and characterization. On the other hand, the measurements are reported back to the AI algorithms to refine the experimental campaign on-the-fly, thus enabling closed-loop experimentation without human intervention.

16.3.1 Algorithmic Strategies to Experiment Planning

We start this section by reviewing strategies developed and deployed in the context of single-objective optimization, and then overview approaches to Pareto (or multi-objective) optimization.

16.3.1.1 Single-objective Optimization

Assessing the merits of a proposed set of experimental parameters typically requires the execution of an experiment or an extensive computation. In the context of the closed-loop approach to experimentation, efficient experiment strategies should be targeted towards the following assumptions about the experiments:

(i) Conducting experiments is costly, measured with respect to any budgeted resources including time, money, reagents, *etc.* Thus, only a limited number of evaluations can be afforded.
(ii) Experiments are noisy due to uncontrollable environmental influences and imprecisions of the measurement device, such that the measured response for the same set of parameters can differ. The noise is expected to be stationary and unimodal.
(iii) In addition to noisy responses, response surfaces can be non-convex, which requires global search strategies.
(iv) Experiments do not yield singularities and are bounded in both the parameter space and the measurements.
(v) Different sets of parameters for the same experimental procedure can be evaluated sequentially or in parallel.

Gradient-based optimization strategies are widely used across different fields, bridging machine learning, mechanical engineering, quantum chemistry and biomolecular physics. While various flavors of gradient-based optimization strategies have been developed to date,[88,89] the core idea always constitutes a stepwise procedure that conditions the next step on the current location in the parameter space as well as derivatives of the surface at this

[‡]Note that in this context, *products* refers to the result of the process and can be a chemical, a material, a device, a design setup, *etc.*

location. Most gradient-based approaches, notably gradient descent,[90,91] determine both the stepping direction and the step size from the local gradient at the current point. As such, these approaches can reliably identify global optima on convex surfaces. Individual steps of the optimization procedure require several function evaluations to determine numerical approximations to the gradients. The number of function evaluations required in one step generally increases with the dimensionality of the parameter space. Higher-order methods condition the next step not only on local gradients but also employ the local Hessian such as Newton's method[92,93] or higher-order derivatives. While higher-order methods have the potential to converge in fewer steps, they require substantially more function evaluations per step to numerically approximate higher-order derivatives. Intermediate methods such as conjugate gradient[94] or L-BFGS[95,96] attempt to approximate the Hessian with first-order derivatives. The numerical approximation of gradients generally poses a challenge in the context of experimentation where the response surface is subject to noise. Due to noisy feedback for individual function evaluations, approximations to gradients can be highly inaccurate and thus substantially increase the number of required function evaluations to determine the optimum. Gradient-based optimization strategies are therefore only applied in isolated cases in the context of experimentation.[86]

Early algorithmic strategies used to plan experimental campaigns on a larger scale relied on direct search strategies, notably the simplex algorithm (Nelder–Mead method).[97] Several variants of the simplex method have been developed for the physical sciences to date,[98,99] all of which share the common idea of conditioning the experiment planning strategy on a heuristic stepping process but using slightly different heuristics and policies to span the initial simplex. Nelder–Mead suggests new parameters in the n-dimensional search space based on the previous $n+1$ experiments (spanning a simplex in the search space) following a set of human-defined rules. Simplex methods have the ability to overcome local optima but are not guaranteed to converge to the global optimum in the limit of infinitely many evaluations and thus do not constitute a reliable global search strategy. Nevertheless, the non-local policies to determine the next set of parameters can alleviate the challenges posed by noise, and simplex strategies have been successfully applied notably in the context of chemistry.[100–104]

Nature-inspired optimization algorithms that rely on non-local ensemble searches have the potential to converge to the global optimum. These methods are typically based on a set of simultaneously evaluated parameter points (*population*), and create new generations of this population based on heuristic rules conditioned on the responses observed for each parameter point.[105,106] The covariance matrix adaptation evolutionary strategy (CMA-ES),[107,108] for example, generates populations as random samples drawn from a high-dimensional multivariate distribution and has been successfully applied to microdroplet reactions.[109] Location and covariance matrix of the multivariate distribution are iteratively updated based on the fitness of the population. Other examples of this class of algorithms include particle-swarms[110,111] and differential evolution.[112,113] Genetic algorithms, as a special class of evolutionary strategies, are highly flexible and can be quite powerful, but also contain a large number of hyperparameters (>10) concerning the types of mutations and mutation rates by which populations are modified, and can thus rarely be applied out-of-the-box.

As experiments are costly, robust global search strategies which require less fine tuning are more desirable. Stable noisy optimization by branch and fit (SNOBFIT),[114,115] for example, describes a global optimization strategy for bounded, constrained, noisy and expensive objective functions on continuous parameter domains that typically performs well with default hyperparameter choices. Based on previous function evaluations, SNOBFIT fits local linear models to construct local surrogate models. New experimental parameters are proposed based on a recursive partitioning of the search space with the goal to (i) improve the accuracy of the approximative model, or (ii) find more desirable function values. To this end, SNOBFIT defines a fine grid of distinct parameter points from which parameters are proposed for evaluation and explicitly defines heuristics targeting the aforementioned goals. Solutions found by SNOBFIT have been shown to converge to the true global optimum in the limit of infinitely many distinct function evaluations. SNOBFIT has notably been successfully applied in chemistry.[104,116–118]

More recently, Bayesian optimization has experienced increased attention as a global search strategy, notably in the context of hyperparameter optimization of machine learning models. Bayesian optimization extends the idea outlined in SNOBFIT to condition the search strategy on the construction of a surrogate based on past experiments. Contrary to SNOBFIT, surrogates in Bayesian optimization are typically constructed from more flexible probabilistic machine learning models such as tree Parzan estimators (TPE),[119,120] Gaussian processes (GP),[121,122] random forests (RF)[123–125] or Bayesian neural networks (BNN).[126,127] By regressing collected measurements, probabilistic machine learning models can yield global non-parametric surrogates to the response surface that are robust to noise and overfitting. New parameters are proposed by balancing the predictions of the surrogates with quantitative estimates of their uncertainties. Bayesian optimization has only recently been applied to selected applications in materials science and chemistry,[127–130] but has shown promising performances compared to other search strategies. While Bayesian optimization strategies are more computationally involved than the aforementioned strategies, their runtime still rarely exceeds the typical execution times of experiments.

Reinforcement learning (RL) has recently emerged as a strategy for data-driven policy searches in various fields bridging robotic control[131,132] and molecule generation.[133,134] While the aforementioned search strategies rely on human-developed policies to suggest new experimental parameters, RL has the potential to infer efficient search policies from previous experiments. To this end, experiment planning, notably reaction optimization, is considered a policy-based decision-making process, where the next experiment to evaluate is determined based on previous evaluations. RL aims to identify data-driven experiment planning strategies which identify optimal experimental conditions in only a few steps. In this context, software agents trained with reinforcement learning have been shown to successfully optimize diverse sets of chemical reactions,[109] notably Pomeranz–Fritsch synthesis, Friedlander synthesis and the synthesis of ribose phosphate. While this approach has the potential to identify the optimal policy for a given reaction optimization task, many examples might be required to train the software agent and the transferability of a trained software agent to other applications with different parameters and optimization goals has yet to be demonstrated.

In addition to the aforementioned strategies several hybrid approaches have been developed, which are mostly application specific and, thus, not context-free. These hybrid strategies consist of two tiers: in tier one, a broad grid-search based survey of conditions is performed in parallel. This allows identification of the most promising starting points for tier two. In tier two, this promising region is used as a starting point for in-depth searches *via* different flavors of the simplex algorithm.[135] Most notably, these strategies were applied to the optimization of chemical reactions.[136–139] Note that other hybrid strategies exist and have been demonstrated in various contexts. They differ in the choice of optimization techniques in tier two from a suite of modules including decision-tree,[140] factorial design/grid search,[141] multidirectional search,[142] composite modified Simplex,[143] parallel Simplex search,[144] and successive focused grid search.[144,145]

16.3.1.2 Pareto Optimization

While the experiment planning strategies outlined in Section 3.1.1 consider a single objective in the design process, real-world applications are often composed of multiple, competing objectives. For example, the yield of a chemical reaction might need to be maximized, while the amount of an expensive catalyst used in this reaction is to be minimized. Considering multiple possibly competing objectives simultaneously in the design process has the potential to immediately identify promising drug candidates.

In the non-trivial case, where two or more objectives are competing with each other, *i.e.* the most desirable values cannot be reached with the same set of parameters, acceptable trade-offs and balances need to be defined. The set of points for which any given objective cannot be improved without necessarily degrading at least one other objective is generally referred to as the *Pareto surface*. All parameters which yield objectives on the Pareto surface can be considered to be equally optimal. Yet, some solutions might be preferred over others based on the application at hand. Trade-offs and balances between the points on the Pareto surface can express this preference information and guide optimization algorithms to identify the most preferred optimal solutions. *A posteriori* methods express trade-offs based on a data-driven estimate of the Pareto surface. *A priori* methods express trade-offs implicitly without knowing the Pareto surface. While *a posteriori* methods might thus find a more desirable balance between the objectives, they also require more experiments as the entire Pareto surface needs to be resolved and are thus impractical for larger parameter spaces and objectives.

Commonly employed *a priori* approaches consist in scalarizing the objectives of interest *via* analytic reduction operations, *e.g.* weighted sums or weighted products. Due to the analytic formulation of the reduction operation, these approaches are not guaranteed to reach every point of the Pareto surface, and choosing the appropriate set of parameters (*i.e.* weights) is non-trivial and highly dependent on the unknown Pareto surface. Other approaches rely on the definition of tight constraints on each objective, which enables optimization algorithms to reach every point of the Pareto surface but require substantial prior knowledge about the achievable values for each objective. A more flexible approach consisting of a combination of dynamic thresholds and importance hierarchies has recently been reported for closed-loop experimentation with self-driving laboratories.[146]

16.3.2 Roadmap for Deploying and Orchestrating the Self-driving Laboratories

Self-driving laboratories leverage automated platforms along with the ability of AI/ML algorithms to efficiently navigate parameter spaces in the search for global optima in the presence of noise and without requiring prior expectations on the response surfaces. Nonetheless, versatile software packages providing the layers indispensable to the deployment and control of self-driving laboratories are yet to be engineered.

The closed-loop approach to autonomous experimentation decomposes the experimentation process into AI-driven experiment planning and automated experiment execution. To accelerate the rate of scientific discovery, software packages for the deployment and orchestration of the self-driving laboratories need to simultaneously leverage the AI component and maximize the use of automated platforms to their full capacity. Despite promising attempts to design and build autonomous systems in the past (see Section 3.2) the implementation of a robust but versatile connection between design algorithms and robotic platforms requires elaborate software solutions and poses a challenge to the transition from automation to autonomy. Only recently, developments towards parallelized workflow managers have been demonstrated in the literature.[3,147]

To-date, existing software packages for chemical and materials science have focused on closed-loop optimization with specific experiments, robotic hardware, and algorithms in mind. One of the first examples is ARES, which was designed to study single-walled carbon nanotubes and introduced in 2016.[10] Since then, a handful of software packages were published in academic literature, such as NREL-HTEM for physical vapor deposition of inorganic thin films,[148] AIR-Chem for targeting inorganic perovskite quantum dots,[149] ESCALATE[150] for studying metal halide perovskite crystallization and Ada for optimizing thin film of hole-transport materials deposited on glass substrates,[15] as well as a number of solutions from Adams and Schubert for polymer chemistry,[151,152] and data handling solutions such as Uncountable to accelerate R&D, and Citrine Informatics for data-driven design of experiments.

Generalizable automation and data capture in biology is, however, more mature and includes both commercial and academic variants such as Wet Lab Accelerator,[153] Autoprotocol, laboratory automation such as Par-Par[153,154] and Roboliq,[155] full experimental workflow and design with integrated LIMS such as Aquarium,[156] and services offered by companies such as Emerald Cloud Lab, Transcriptic.[157]

Early examples of ML applied to drug discovery include techniques such as neural networks and support vector machines for the identification of biological targets and feature extraction, although prediction and generalization capabilities were limited due to constrained computational resources and experimental data.[158] Recently, commercial solutions to accelerate the *in silico* discovery of new drugs have been reported by AI-Therapeutics for predicting novel treatments to outsmart cancer and rare diseases, *In Vivo* AI for structure–activity prediction, identification of synergistic drug–drug interactions, molecular design and retrosynthesis planning, InSilico Medicine for discovering signatures of diseases and identifying promising targets. Lately, DeepMind submitted AlphaFold for the *Critical Assessment of Structure Prediction* (CASP) and ended up making the most accurate predictions of all contestants on average. At the end of 2018, efforts to leverage biological testing, informatics and automated chemical synthesis to accelerate drug discovery and map biologically active chemical space have been laid out by both the ASPIRE[159] Program, and by Faxian Therapeutics in New York City.

The *ultimate* software package needs to connect several layers indispensable for autonomous discovery. The most fundamental connection enables communication between the experiment design process and the execution of experiments. For maximized throughput, the experiment design process needs to be separated from the execution of the experiment. A formal decoupling of these two processes allows closed-loop experimentation to be approached with an interleaved strategy where both actions are executed in parallel. To this end, experiment parameters suggested by the experiment planner are stored in databases for later evaluation. Experiment planning steps are triggered when this database is exhausted or when new experimental feedback is available, to update the existing database. This method of caching ensures that the automated platforms can almost instantaneously continue the experimentation process upon completion of a single experiment, as long as new experiments can be suggested faster than they are executed. Connections to a database management system allow experimental results to be recorded as soon as they are available, and can be easily synchronized with remote servers for online data analysis. In addition, the database system facilitates the storage of experimental results in a standardized format, which can be used to extract chemical knowledge for future experimental procedures. At the same time, databases can be used to store raw experimental results, which can be post-processed in different ways depending on the application to compute the merit of a conducted experiment.

Most algorithmic experiment planning strategies developed to date (see Section 3.1) rely on a mostly sequential execution of experiments for a targeted application. Parallelization capabilities are typically limited due to the fundamental problem that prior knowledge is required to infer the next most informative experiment(s). This limitation poses a challenge to the maximized usage of the robotic equipment at full capacity and is in contrast to massively parallelizable strategies such as HTE. Nevertheless, the experimental throughput of a self-driving laboratory can still be increased by rethinking the experimentation approach: instead of parallelizing over individual experiments for one application, a suitable orchestration software could facilitate the parallelization of a few experiments for several applications, assuming that the applications are governed by the same set of controlled parameters. This has already been demonstrated in the context of the optimization of polymer blends for photostable organic photovoltaic devices.[160] Note, that the applications can differ in their objectives, as the merits of individual parameters are computed by the control software as well as in the bounds of the parameter spaces.

Implementing an experimental procedure requires the decomposition of a higher-level operation (*e.g.* spin-coat) to a set of lower level operations (*e.g.* place sample, start centrifuge, ...). With recent advances in reinforcement learning, there might be a chance to train an agent to execute these higher-level operations without explicitly providing details on how to execute this action. In combination with a camera, the agent could be trained to adapt to different hardware platforms. Additionally, the agent could be trained to slightly change the procedure searching for improved protocols.

16.3.3 Early Realization of Self-driving Laboratories

To the best of our knowledge, Winicov *et al.* published one of the very first closed-loop optimization systems, a "Chemical process optimization by computer—a self-directed chemical synthesis system", in *Analytica Chimica Acta.* in 1978. In this reported self-driving laboratory, all steps of the experimental procedure are controlled by a computer and driven by a variant of the Nelder–Mead algorithm to suggest new experiments.[51] A few years later, in 1988, Matsuda *et al.* reported an algorithmic closed-loop optimization of chemical reactions based on the same experiment planning strategy. The system consists of robotic platforms, automated characterization equipment, and a computer.[136] In 1996, Porte *et al.* reported *Contalab*, a chemical reactor built for reaction optimization. The system is controlled by a computer, and the optimization is driven by a modified version of the simplex algorithm.[137]

With the rise of the *Internet of Things* (IoT) in recent years, Skilton *et al.* reported advances towards *cloud chemistry* in 2015, where experiments are controlled remotely by humans or algorithmic operators.[161] A year later, in 2016, Fitzpartick *et al.* published an autonomous self-optimization platform for chemical synthesis monitored *via* the internet.[100] Notably, this platform was controlled and operated by a customized software package referred to as the *Ley Lab*, which is deployed locally and can be accessed remotely.

In 2016, Maruyama and coworkers report the use of AI-controlled closed-loop optimization to identify optimal experimental conditions for nanomaterials applications.[10,162] In 2016 and 2017, several groups report the use of AI to navigate high-dimensional parameter spaces in different application domains, spanning physics, chemistry and materials science.[16,163–165]

In January 2018, Mission Innovation—a global initiative of 24 countries and the European Commission (on behalf of the European Union)—published the expert report "Materials Acceleration Platforms".[5] This extensive report analyzes opportunities to accelerate the optimization and discovery of advanced materials with AI-controlled closed-loop approaches that leverage the benefits of automation, robotics, high-performance computing, databases and AI.[5] In the same year, Tabor *et al.* provided a vision for an integrated AI approaches towards autonomous materials discovery by integrating recent technological innovations in automation, robotics and computer science into standard experimental procedures in chemistry, materials synthesis and characterization.[4]

16.4 ChemOS to Orchestrate Next-generation Experimentation

ChemOS is a modular and versatile AI-driven software package. It digitizes operations and empowers the self-driving laboratories. The suite of modules (Figure 16.3) as implemented in ChemOS allows for fast deployment and efficient orchestration of the self-driving laboratories. Each of the modules is independent from each other. Such modularity allows for a seamless integration into a variety of existing or custom robotics and automated solutions and facilitates piecewise upgrades and additions to the existing software package. We detail each module depicted in Figure 16.3 hereafter in their respective section.

16.4.1 AI-aided Experiment Planning

The learning module is the key component to reach autonomy as it designs new experiments for the scientific procedures requested by the researcher. The purpose of this module is to support methods for parameter space searches and decision making. Typically, the learning procedures can be formulated as an optimization problem of a possibly high-dimensional parameter space, where the allowed ranges along each dimension are

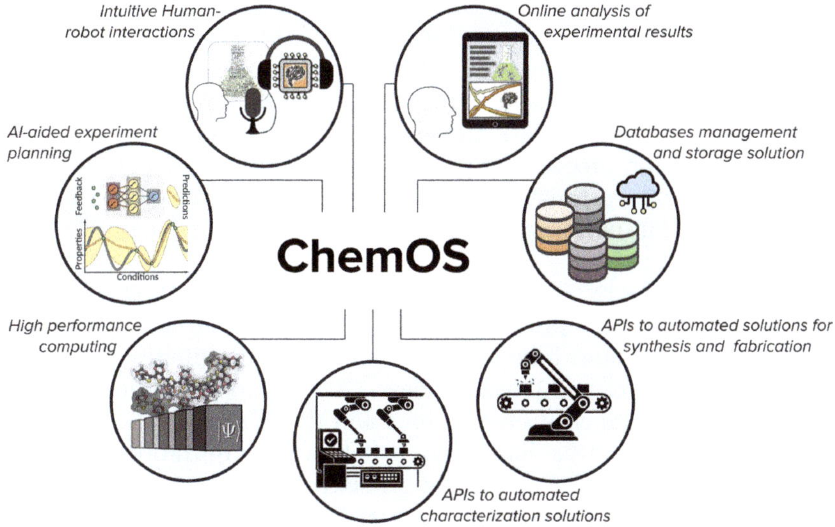

Figure 16.3 Representation of the modules as implemented in the ChemOS AI-driven software package: intuitive human–robot interactions, online analysis of experimental results, databases management and storage solution, APIs to auotmated solutions for synthesis and fabrication, APIs to automated characterization solutions, high-performance computing, and AI-aided experiment planning.
Adapted from ref. 13 with permission from Elsevier, Copyright 2019.

defined by the application space of the scientific procedure at hand. In agreement with the considerations about the experimentation process outlined in Section 3, ChemOS contains four different learning procedures for global optimization: random search, Phoenics,[127] SMAC,[123–125] and Spearmint.[121,166]

Random search refers to the exploration of a parameter space *via* uniform sampling of all parameters. Because this procedure only makes uninformed decisions and does not memorize observations, it can be used to explore the parameter space in an unbiased, and uncorrelated fashion.[120,167] Random search finds usage in benchmarking performance and detecting error and misbehavior of the robotic hardware.

Phoenics is a deep Bayesian optimizer developed by Häse *et al.* for global optimization. It is a probabilistic algorithm, which combines aspects from Bayesian optimization with ideas from Bayesian kernel density estimation. As a consequence, Phoenics shows a favorable linear scaling with the dimensionality of the parameter space as well as the number of observations. Phoenics was shown to be particularly well-suited for automated and parallel experimentation as it can propose parameter points with sampling behaviors explicitly biased towards exploration of the parameter space or exploitation of the previously collected experimental feedback. As a result, with a single (additional) observation, multiple new experimental conditions can be suggested in batches, thus maximizing parallelization.

SMAC is a Bayesian optimization package based on random forest (RF) models. RFs are computationally inexpensive regression models with a favorable linearithmic scaling with the number of observations and a linear scaling with the dimensionality of the parameter space. SMAC features a very efficient implementation of RF models, which keeps the computational cost of the optimization procedure to a minimum. However, model uncertainty for RFs needs to be estimated empirically which limits the performance of RF-based Bayesian optimization frameworks particularly for purely convex response surfaces.

Spearmint is a Bayesian optimization package based on Gaussian processes (GPs).[121,166] GPs provide a flexible way of finding analytic approximations to the objective function based on normal distributions associated with every parameter point. Nonetheless, GP based optimization scales cubically with the number of observations and the dimensionality of the parameter space. With this limitation, GP based optimization is typically applied to relatively low dimensional problems for which the optimum can be found in relatively few function evaluations.

The aforementioned machine-learning optimization algorithms can be combined with Chimera,[146] a general purpose achievement scalarizing function for multi-target optimization where evaluations are the limiting factor or when no prior information about the experimental response is available. It uses concepts of lexicographic methods to dynamically scalarize multiple objectives into a single-objective function based on a user-defined hierarchy in the objectives.

16.4.2 Intuitive Human–Robot Interactions

ChemOS provides an intuitive interface for researchers. This interface not only favors the interaction between researchers, the learning module, and the robotic hardware, but also facilitates processing new requests, and filtering/summarizing relevant results. To this end, we supplement ChemOS with a natural language processing (NLP) module in a chatbot framework. This module is based on neural network text classifiers, which ChemOS connects to application programming interfaces (APIs) of several common social media platforms and communication services *via* an abstract facade software interface.[168]

The chatbot framework is constructed from conversational intents defined in the Javascript Object Notation (JSON) format. The defined intents are processed using the natural language toolkit (NLTK)[169] in Python and transformed to bag-of-words arrays for processing in TensorFlow.[170] A multilayer perceptron is then trained on the defined conversational intents to map given inputs, *i.e.* plain text, to conversational classes, for which responses are pre-defined. Given a newly received message, the neural network determines the message category, and a response is sampled from the set of responses pre-defined for this category.

ChemOS supports the customization of responses based on information specific to the received message. This is achieved by using placeholders in the pre-defined responses sampled from the neural network classification. At a later stage, the response is replaced with the associated specific information by ChemOS before sending it to the researcher.

A number of adapters to common social media platforms and messaging services are implemented in ChemOS to address the personal preferences of researchers. This includes e-mail exchange *via* Gmail, private and public messaging *via* Twitter, and private messaging *via* Slack. In addition, ChemOS also supports communication *via* the command line interface, and can be triggered by file system events, *e.g. via* Dropbox. With these supported methods of communication, we intend to make ChemOS easily accessible to other computational procedures.

16.4.3 Online Analysis of Experimental Results

ChemOS features a simple analysis module for pre-processing results obtained from the experimental procedures. This enables researchers to quickly perceive the progress of the current experiment or summarize the results of a completed scientific procedure. ChemOS supports (i) the generation of time traces, which displays the progress of the ChemOS session, (ii) the distance distribution, which provides intuition for the sampled parameters in the high dimensional parameter space, and (iii) the distances to the best performing set of parameters observed so far.

16.4.4 Databases Management and Storage Solutions

Long-term storage of information in ChemOS is facilitated *via* database-management systems (DBMS). ChemOS features a facade connected to multiple adapters, which interact with the APIs of different DBMS. Currently we support SQLite, a widely used relational database management system embedded into the end program. Nevertheless, we provide interfaces for the integration of other types of DBMS in ChemOS.

ChemOS stores information in four distinct databases (DBs, see Figure 16.3), which serve specific purposes essential to an efficient workflow within ChemOS. All request entries are stored in the request DB, which contains unique identifiers for the scientific procedure associated with the experiment, and a unique key for the experiment itself. The parameter DB contains parameters for the experiment at hand with an indication flagging their previous usage, along with the identifier of the scientific procedure. The robot DB contains robot-specific information, such as the communication protocol, hardware configurations, and status. Finally, experimental results are stored in the feedback DB in addition to information on the scientific procedure and parameters used for a specific experiment.

All DBs operate on the first-in-first-out (FIFO) principle. Consequently, new requests are queued chronologically, and they are processed as soon as parameters and the robotic hardware are available. ChemOS automatically matches generated experimental parameters and qualified robotic hardware to a given request such that monitoring multiple distinct scientific procedures on different platforms with a single instance of ChemOS is possible.

16.4.5 Connection to Automated Solutions

The purpose of the robotics and characterization modules in ChemOS is to communicate high-level instructions defined in the scientific procedure to low-level elementary operations, executed on the robotic hardware. High-level instructions are researcher-oriented keywords commonly used in scientific procedures: *e.g.* mixing, heating, sample drawing, *etc.* The low-level instructions, however, define a set of elementary machine-oriented operations, executed by the robotic hardware at hand to perform a single high-level operation. Such a set of low-level operations could consist in encoding the movement of a robotic arm, operating pumps, or pre-processing experimental results.

It goes without saying that low-level instructions are specific to particular robotic and characterization hardware. Nevertheless, the robotics and characterization modules provide bridges between the ChemOS core and the actual robotic hardware. With these bridges, new robotic hardware can easily be integrated into ChemOS to directly benefit from the architecture in place.

There is nowadays a number of large-scale automated solutions designed for high-throughput experimentation in chemistry and materials science including notably the Material Acceleration Factory[171] (MIF) at the University of Liverpool, the Centre for Rapid Online Analysis of Reactions[172,173] (ROAR) at Imperial College London, the High-Throughput Molecular Screening Center[172] at Scripss Research, the High Throughput Screening Laboratory at the University of Kansas[174] (KU-HTSL) or the High-Throughput Experimentation platform at ETH Zurich (HTE@ETH).[175] Such large-scale solutions would directly benefit from an integration with an operating system like ChemOS to further streamline process optimization and accelerate materials discovery to find solutions to the Grand Challenges of the century at a faster pace.

16.5 Successful Applications of ChemOS to Scientific Challenges

ChemOS has already been applied to a variety of systems in materials science and chemistry to accelerate both the optimization procedure and the discovery process. Below we highlight three scientific challenges tackled by ChemOS.

16.5.1 Calibration of an Automated Setup for Real-time Reaction Monitoring

Real-time reaction progress monitoring enables unattended sample aliquoting and dilution, as well as immediate quantification and identification of reaction components. This type of information is of importance for the detailed understanding of the dynamics of multi-step reactions. For this purpose, the group of Prof. J. Hein at University of British Columbia, UBC, reported an automated analytical platform for real-time HPLC-MC reaction progress monitoring.[176]

The procedure starts with a robotic arm (N9) drawing a sample from a reaction vessel on the multiwell tray. The sample is then passed through a sample loop, inline mixer, and injection loop and is finally analyzed *via* high-pressure liquid chromatography (HPLC). The peak areas of the chromatogram determine the amount of sample delivered to the HPLC sensor. Overall, the automated platform is controlled by six independent parameters determining sample draws, wait times and push rates. Prior to monitoring reactions in real-time, these six parameters need to be optimized to calibrate the system and maximize the performance and efficiency of the platform.

The process was optimized with ChemOS suggesting values for each of the six parameters. Parameters were refined based on the integrated peak area automatically collected from the HPLC, in a closed-loop approach. The procedure could run overnight, allowing the researcher to start their day with a setup ready for experimentation.

In the procedure, ChemOS optimized the setup in full autonomy for 100 experiments per session, with five experiment planners: random search, design of experiment (DoE), Phoenics, SMAC, and Spearmint.[177] We found that DoE is outperformed by the random search strategy, which can be attributed to the high-dimensional nature of the search space. In addition, Spearmint displays a comparable performance to random search. Only SMAC and Phoenics demonstrate significantly better average performances, with an average acceleration of ~6.7 compared to DoE.

16.5.2 Discovery of Conductive Thin-film Materials

Organic hole transport materials (HTMs) are essential to perovskite solar cells (PSCs), as they facilitate the collection and transport of holes created in the light-harvesting material for an efficient charge separation. For this purpose, organic HTMs need to demonstrate high hole mobilities, which are influenced by structural disorder determined by the process conditions under which the HTM is deposited.

HTMs are typically incorporated in thin films, which are applied as layers to the light-harvesting layers of a solar cell. In this process, the HTM is mixed with a dopant, spin-coated on the substrate and then annealed to resolve structural disorders. Specifically, the experimental workflow involves (1) mixing an HTM-dopant-additive solution, (2) spin coating the solution onto a substrate, (3) thermally annealing for a variable amount of time, (4) imaging with visible-light camera, (5) acquiring UV–Vis–NIR spectra in reflection and transmission modes, and (6) measuring the I–V curve of the film with a 4-point probe. This application aims to optimize the hole mobility of the thin films in which the HTM (spiro-OMeTAD) is deposited. Specifically, the amount of dopant (cobalt(III)) as well as the annealing time are optimized to achieve high hole mobilities.

In the closed-loop approach to experimentation, ChemOS suggests the amount of dopant and the annealing time, based on which the thin film is first synthesized and then characterized.[15] Based on the measured absorption spectra and the probed conductivities, a surrogate to the hole mobility is computed, based on which the next most informative set of parameters is proposed. It has been reported that the closed-loop approach accelerates the experimentation process by approximately a factor of five over a grid-based HTE approach.

16.5.3 Formulation of Polymer Blends for Photo-stable Solar Cells

Organic photovoltaic (OPV) materials are emerging as a competitive approach to commercial solar cell devices, notably due to their lower energy payback time and the structural flexibility to tune device properties. However, desirable properties are often limited by energetic and structural disorder of the material blends, with one major obstacle being the rapid degradation of energy efficient materials under exposure to sunlight.

The photostability of polymer blends in organic solar cells is challenging to model computationally, and determining this property experimentally requires time-consuming and resource-demanding studies of how the material blends behave under sunlight. Recently, the benefits of the self-driving closed-loop approach to the design of photostable material compositions for OPV candidate have been reported and compared to conventional high-throughput experimentation (HTE).[178] This application aims to identify the most photostable blend of four different polymers to serve as the donor and acceptor materials.

While the massively parallelizable HTE approach enabled the simultaneous evaluation of 96 different polymer blends, ChemOS proposed four blends in one iteration. HTE evaluated a total of 859 different blend combinations, accounting for constraints in the ratios of donor materials to acceptor materials, while ChemOS identified material compositions of comparable photostability within after probing about 30 different blend compositions.

Consequently, the closed-loop approach requires about as many experimentation batches as HTE. However, the amount of materials consumed in the experimentation process is about a factor of 30 lower for the closed-loop approach. In addition, the study reports the possibility to probe multiple different blend systems simultaneously in one experimentation batch, all of which are simultaneously controlled by ChemOS. Consequently, ChemOS enables a substantial increase of the application throughput despite a moderate increase of the experimentation throughput due to the sequential nature of the experimentation process.

16.6 Application of ChemOS to Drug Discovery

The currently most adopted approaches to drug discovery require a substantial number of experiments to discover and optimize promising lead candidates, identify delivery parameters, and run preclinical tests and clinical trials. These steps accumulate considerable expenses in terms of both involved resources and development times. In fact, the typical cost to discover a viable drug candidate is estimated between US$0.5 billion and US$2.6 billion, with typical research timelines spanning 10–20 years.[32–34] Early stage drug discovery alone, notably lead identification and optimization, constitute about a quarter of these costs. A substantial acceleration of these early steps could significantly reduce the overall costs of the drug discovery process, but also enables the simultaneous screening of drug candidates for much broader applications.

The drug discovery process typically starts with the identification of a biological target, usually a protein or protein complex, which is speculated to modulate the targeted disease or disorder. This step typically requires a profound understanding of pathology at the molecular level. Modern advances in structural genomics, bioinformatics and high-throughput crystallography enable the efficient structural resolution of most protein complexes at the atomic level. Once the biological target, usually a protein, has been identified, drug candidates can be searched by evaluating the affinity of potential candidates related to the biological target, or more generally their medicinal activity against the disease. The drug discovery problem can therefore be formulated as an optimization problem, which aims to identify drug molecules with maximal affinity, in addition to other secondary desired properties. For example, most drug discovery efforts are focused on small organic molecules with a molecular weight $<$ 900 Daltons with typical sizes of 1 nm or less.[179] More specifically, Lipinski's rule of five has been established as an empirical rule to evaluate the drug-likeness of a given compound based on its absorption, distribution, metabolism, and excretion.[180,181]

Current approaches to drug discovery frequently employ high-throughput strategies (HTS) to identify promising materials or potential drug candidates.[35] Initially, the pool of potential drug candidates is generated as a relatively diverse library in order to identify first hit molecules, *i.e.* drug candidates with promising medicinal activity. With affinity being a necessary but not sufficient property for a viable drug candidate, additional factors (possibly based on Lipinski's rule of five) are included at a later stage of the search, which then focuses on less diverse, more focused molecular libraries consisting of molecules similar to the identified hits.[182] Early HTS approaches were accomplished with large-scale *in vitro* experiments, which are generally demanding in terms of resources and time. Fragment-based drug design (FBDD) aims to alleviate the combinatorial explosion of potential drug candidates by assembling drug candidates from functional, weakly-interacting fragments which can be tested independently.[183] HTS approaches for measuring the binding affinity of small organic compounds to proteins include differential scanning fluorimetry (DSF) and calorimetry (DSC), surface plasmon resonance (SPR), and nuclear magnetic resonance (NMR) which achieve medium to high throughput.[184]

A first acceleration of this capital intensive experimental HTS approach was achieved with the rise of sufficiently accurate computational methods and abundantly available computational resources, which enabled the *in silico* search of viable candidates *via* computational pre-screening. This process commonly referred to as virtual HTS or computer-aided drug design (CADD) is now part of most current drug discovery campaigns and is mostly based on phenomenological models or empirical data.[185] Instead of the *de novo* creation of chemical libraries, *in silico* drug discovery campaigns oftentimes start from a known drug and variations of it to generate the design space to be screened. This strategy is based on a combination of reasons, including highly non-linear relations between molecular structures and biological activity and the high costs of late-stage failures.

The emerging field of data-driven chemical design *via* deep generative models only recently experienced increased interest for drug discovery.[18] Contrary to HTS approaches, which require a pre-defined search space, generative models infer the probability distribution of molecular structures and targeted properties to directly generate candidate molecules with the desired properties. Although the opportunity to generate drug candidates without manually constructing a search space seems promising,[29,186–188] generated compounds are synthesized and experimentally verified only in isolated cases. Only recently, Zhavoronkov *et al.* introduced a deep generative tensorial reinforcement learning (GENTRL) approach for the *de novo* design of potent kinase inhibitors targeting discoidin domain receptor (DDR1).[189] Notably, this study tightly integrates their deep generative model into the drug design workflow and identified, synthesizes and experimentally verified six active compounds, each complying with Lipinski's rule, in less than two months. GENTRL's success can partially be attributed to the feedback loops on which the discovery workflow was based. Specifically, the search algorithm was rewarded whenever it generated novel molecules with high activities against kinases. Contrary to traditional HTS approaches, this feedback loop allows more promising candidates to be quickly generated, thus avoiding experiments on compounds with low expected merit.

A closed-loop orchestration software like ChemOS has the potential to simplify the implementation and the deployment of such complex workflows using self-driving laboratories. Given a discovery problem, feedback about the merit of a given compound or candidate can be collected from multiple approaches, including first principles calculations, heuristic or empirical computations, data-driven techniques and automated platforms. A flexible and versatile orchestration software has the potential to integrate all of these approaches into a multi-step workflow to leverage the advantages of every single approach.

16.7 Conclusion and Outlook

The convergence of key technologies, *i.e.*, artificial intelligence, machine learning, robotics, automation and high-performance computing, enabled the self-driving laboratories to prove successful. In particular, by leveraging the advantages of AI and automated experimentation platforms scientific discovery rate have substantially improved to even discover solutions that researchers would initially not find intuitive. Nonetheless, and although the self-driving laboratories seem to be on track to revolutionize process optimization and materials discovery, they are still isolated cases. Hereafter, we discuss what would be the next short-term advances required for the self-driving laboratories to fully embrace the vision of autonomous experimentation.

On the hardware front, equipment is becoming increasingly remotely operable, opening up opportunities with far-reaching consequences: large, and expensive equipment can be clustered *via* the IoT to form delocalized self-driving laboratories, which we see as the research centers of the future. Materials and device characterization techniques are crucial elements in the autonomous materials discovery approach outlined in Figure 16.2. They are crucial information (*e.g.*, materials composition, and structure, physical properties and device functionality) for the closed-loop approach as implemented in the self-driving approach. Oftentimes, experimental protocols require the use of multiple stations, each consisting of multiple independent steps running on different equipment, such that seemingly simple tasks can consist of a number of different equipment and actions which need to be executed sequentially. While one experimental procedure could require one specific protocol and set of elementary actions, another experimental procedure could require another protocol with different elementary actions. Hardcoding the actions and the equipment needed in such an experimental campaign comes with major disadvantages, leading to the execution of multi-step experimentation procedures by supplying them with a flexible and dynamic scheduling system.

Such a scheduling system yields three key advantages for self-driving laboratories:

(i) One piece of experimentation equipment can be used for multiple experimental procedures in parallel. Dynamic scheduling allows one piece of equipment to be assigned to one experimental procedure, and be re-assigned to another procedure at a later time. This reduces the hardware requirements of a self-driving laboratory.

(ii) The self-driving laboratory can dynamically allocate resources for different experimental procedures in parallel. This implies that the execution of one experimental procedure can be set on hold for a short period of time, while a task for another experimental procedure is completed on another piece of hardware.
(iii) An intelligent scheduling system can minimize the waiting times of individual pieces of equipment, and thus maximize the experimentation throughput.

To further increase the modularity of the self-driving laboratories, it would be desirable to have real-time recognition and interpretation of equipment to enable plug-and-play approaches to high-throughput hardware cloud, *via* the use of visual recognition. The increasing prevalence of digital camera to capture visible information had a profound impact on many aspects of our modern society, spanning facial recognition to self-driving cars.[190] Advances in convolutional neural network (CNN) algorithms have revolutionized pattern recognition, and are nowadays widely used for end-to-end learning in self-driving cars to identify patterns[191] such as roads, lines, traffic lights, pedestrians, vehicles, *etc.* Similarly, CNNs could be used in a laboratory environment to learn patterns such as glassware, location of starting materials, fabrication infrastructures, and characterization equipment.

On the AI/ML front, several opportunities of development to further catalyze the discovery process are identified. Materials are typically characterized with equipment varying in capabilities, levels of confidence and cost. For example, crude but fast evaluations of material properties could be estimated with methods as simple as taking an image of the material. Arguably, more involved methods can estimate material properties at a much higher degree of confidence, but oftentimes also require more time or resources. Crude estimates of materials properties, however, might still allow identifying poorly performing materials with sufficient confidence, and thus save more involved property measurements to accelerate the discovery process. Additionally, ML algorithms can be used to estimate the differences between fast and more accurate evaluators. This approach, introduced as delta-learning in the context of computing thermochemical properties of small molecules,[192] can be used to learn the correction to the fast evaluator during the experimentation process.

It would also be desirable to enable transfer learning abilities, where knowledge acquired from previously studied applications is transferred to new applications: tools to construct ML-based models for transfer learning within the self-driving laboratories. Efficient transfer learning strategies would require that only the relevant information is extracted from the recorded experimental results, to generate generalizable "data-driven chemical intuition" which still holds for new applications. Such an approach would allow us to construct informative priors for new applications based on findings of other applications. For example, prior experiments for a particular organic coupling reaction R on a substrate A can be used to hypothesize about the behavior of another substrate *B* under another reaction *S*, using information about the relation between the two substrates. Consequently, fewer experiments need to be carried out to study the behavior of substrate *B* under reaction *S*, which accelerates the discovery process. We believe that the power of transfer learning across a global set of laboratories will become a much-needed *panopticon* for Chemistry.

On the policy front, it is of importance to define common communications streams to enable sherability and transferability of the massive amount of (meta)data generated by the self-driving laboratories. Such a common stream would be required at all levels: institutional, national and international in order to streamline efficient transfer learning strategies.

Finally, on the societal front, the transition from automated approaches to scientific to autonomous experimentation has a direct impact on the role and responsibilities of researchers, making us redefine what it means to be a scientist. Most importantly, the deployment of the self-driving laboratories will free researchers from arduous, repetitive tasks allowing them to focus on creativity, innovation and intellectually stimulating tasks.[193] Hence, self-driving laboratories will empower researchers to approach real-life problems to tackle global challenges humanity is facing today. Although a number of processes will be automated within a self-driving laboratory, the researchers will be key to the deep scientific understanding of these challenges.[13]

References

1. 21st Century Challenges, *21st Century Challenges*, https://21stcenturychallenges.org/.
2. C. Pursell and T. P. Hughes, American Genesis: A Century of Invention and Technological Enthusiasm, 1870-1970, *Am. Hist. Rev.*, 1991, **96**, 247.
3. L. M. Roch, *et al.*, ChemOS: Orchestrating autonomous experimentation, *Sci. Robot.*, 2018, **3**, eaat5559.
4. D. P. Tabor, *et al.*, Accelerating the discovery of materials for clean energy in the era of smart automation, *Nat. Rev. Mater.*, 2018, **3**, 5–20.
5. A. Aspuru-Guzik and K. Persson, Materials Acceleration Platform: Accelerating Advanced Energy Materials Discovery by Integrating High-Throughput Methods and Artificial Intelligence, 2018, http://mission-innovation.net/wp-content/uploads/2018/01/Mission-Innovation-IC6-Report-Materials-Acceleration-Platform-Jan-2018.pdf.
6. J.-P. Correa-Baena, *et al.*, Accelerating Materials Development via Automation, Machine Learning, and High-Performance Computing, *Joule*, 2018, DOI: 10.1016/j.joule.2018.05.009.
7. J. A. DiMasi, R. W. Hansen and H. G. Grabowski, The price of innovation: new estimates of drug development costs, *J. Health Econ.*, 2003, **22**, 151–185.
8. S. K. Saikin, C. Kreisbeck, D. Sheberla, J. S. Becker and A. Aspuru-Guzik, Closed-loop discovery platform integration is needed for artificial

intelligence to make an impact in drug discovery, *Expert Opin. Drug Discovery*, 2018, 1–4.
9. Y. Cao, J. Romero and A. Aspuru-Guzik, Potential of quantum computing for drug discovery, *IBM J. Res. Dev.*, 2018, **62**, 6:1–6:20.
10. P. Nikolaev, *et al.*, Autonomy in materials research: a case study in carbon nanotube growth, *Npj Comput. Mater.*, 2016, **2**, 16031.
11. G.-Z. Yang, *et al.*, The grand challenges of Science Robotics, *Sci. Robot.*, 2018, **3**, eaar7650.
12. R. D. King, *et al.*, The automation of science, *Science*, 2009, **324**, 85–89.
13. F. Häse, L. M. Roch and A. Aspuru-Guzik, Next-Generation Experimentation with Self-Driving Laboratories, *Trends Chem.*, 2019, **1**, 282–291.
14. A.-C. Bédard, *et al.*, Reconfigurable system for automated optimization of diverse chemical reactions, *Science*, 2018, **361**, 1220–1225.
15. B. P. MacLeod *et al.* Self-driving laboratory for accelerated discovery of thin-film materials. *arXiv [physics.app-ph]* (2019).
16. P. B. Wigley, *et al.*, Fast machine-learning online optimization of ultra-cold-atom experiments, *Sci. Rep.*, 2016, **6**, 25890.
17. J. M. Granda, L. Donina, V. Dragone, D.-L. Long and L. Cronin, Controlling an organic synthesis robot with machine learning to search for new reactivity, *Nature*, 2018, **559**, 377–381.
18. B. Sanchez-Lengeling and A. Aspuru-Guzik, Inverse molecular design using machine learning: Generative models for matter engineering, *Science*, 2018, **361**, 360–365.
19. K. T. Butler, D. W. Davies, H. Cartwright, O. Isayev and A. Walsh, Machine learning for molecular and materials science, *Nature*, 2018, **559**, 547–555.
20. B. Liu, *et al.*, Retrosynthetic Reaction Prediction Using Neural Sequence-to-Sequence Models, *ACS Cent. Sci.*, 2017, **3**, 1103–1113.
21. A. Zunger, Inverse design in search of materials with target functionalities, *Nat. Rev. Chem.*, 2018, **2**, 0121.
22. A. P. Bartók, *et al.*, Machine learning unifies the modeling of materials and molecules, *Sci. Adv.*, 2017, **3**, e1701816.
23. P. Raccuglia, *et al.*, Machine-learning-assisted materials discovery using failed experiments, *Nature*, 2016, **533**, 73–76.
24. J. Westermayr *et al.*, Machine learning enables long time scale molecular photodynamics simulations, *arXiv* [physics.chem-ph], 2018.
25. S. Lu, *et al.*, Accelerated discovery of stable lead-free hybrid organic-inorganic perovskites via machine learning, *Nat. Commun.*, 2018, **9**, 3405.
26. H. S. Stein and J. M. Gregoire, Progress and prospects for accelerating materials science with automated and autonomous workflows, *Chem. Sci.*, 2019, **10**, 9640–9649.
27. C. W. Coley, *et al.*, A robotic platform for flow synthesis of organic compounds informed by AI planning, *Science*, 2019, **365**, eaax1566.

28. C. W. Coley, R. Barzilay, T. S. Jaakkola, W. H. Green and K. F. Jensen, Prediction of Organic Reaction Outcomes Using Machine Learning, *ACS Cent. Sci.*, 2017, **3**, 434–443.
29. M. H. S. Segler, M. Preuss and M. P. Waller, Planning chemical syntheses with deep neural networks and symbolic AI, *Nature*, 2018, **555**, 604–610.
30. M. H. S. Segler and M. P. Waller, Neural-Symbolic Machine Learning for Retrosynthesis and Reaction Prediction, *Chemistry*, 2017, **23**, 5966–5971.
31. P. Schwaller, T. Gaudin, D. Lanyi, C. Bekas and T. Laino, 'Found in Translation': predicting outcomes of complex organic chemistry reactions using neural sequence-to-sequence models, *Chem. Sci.*, 2018, 6091–6098.
32. J. Avorn, The $2.6 billion pill—methodologic and policy considerations, *N. Engl. J. Med.*, 2015, **372**, 1877–1879.
33. S. M. Paul, *et al.*, How to improve R&D productivity: the pharmaceutical industry's grand challenge, *Nat. Rev. Drug Discovery*, 2010, **9**, 203–214.
34. J. A. DiMasi, H. G. Grabowski and R. W. Hansen, Innovation in the pharmaceutical industry: New estimates of R&D costs, *J. Health Econ.*, 2016, **47**, 20–33.
35. R. Macarron, *et al.*, Impact of high-throughput screening in biomedical research, *Nat. Rev. Drug Discov.*, 2011, **10**, 188–195.
36. R. Boyle, Two Essays, Concerning the Unsuccessfulness of Experiments, in *The Works of Robert Boyle, The Sceptical Chymist and other publications of 1661*, **vol. 2**, DOI: 10.1093/oseo/instance.00246700, 1999.
37. J. S. Lindsey, A retrospective on the automation of laboratory synthetic chemistry, *Chemom. Intell. Lab. Syst.*, 1992, **17**, 15–45.
38. Boyd, Robotic laboratory automation, *Science*, 2002, **295**, 517–518.
39. C. Bernlind and C. Urbaniczky, An Efficient Laboratory Automation Concept for Process Chemistry, *Org. Process Res. Dev.*, 2009, **13**, 1059–1067.
40. T. M. Stevens, Rapid and Automatic Filtration, *Am. Chem.*, 1875, **6**, 102.
41. C. Daniel, One-at-a-Time Plans, *J. Am. Stat. Assoc.*, 1973, **68**, 353–360.
42. M. Friedman and L. J. Savage, Planning experiments seeking maxima, *Tech. Stat. Anal.*, 1947, 365–372.
43. R. Reinhardt, A. Hoffmann and T. Gerlach, *Nichtlineare Optimierung: Theorie, Numerik und Experimente*, Springer-Verlag, 2012.
44. I. M. Sobol', On the distribution of points in a cube and the approximate evaluation of integrals, *USSR Comput. Math. Mathe. Phys.*, 1967, **7**, 86–112.
45. L. Kocis and W. J. Whiten, Computational Investigations of Low-discrepancy Sequences, *ACM Trans. Math. Software*, 1997, **23**, 266–294.
46. R. A. Fisher, *The Design of Experiments Edinburgh and London*, Oliver and Boyd, 1937.

47. G. E. P. Box, J. S. Hunter and W. G. Hunter, *Statistics for Experimenters*, Wiley Series in Probability and Statistics, Wiley Hoboken, NJ, USA, 2005.
48. M. J. Anderson and P. J. Whitcomb, *DOE Simplified: Practical Tools for Effective Experimentation*, CRC Press, 3rd edn, 2016.
49. R. A. Fisher, The Arrangement of Field Experiments, *Breakthroughs in Statistics: Methodology and Distribution*, ed. S. Kotz and N. L. Johnson, Springer, New York, 1992, pp. 82–91.
50. R. L. Plackett and J. P. Burman, The Design of Optimum Multifactorial Experiments, *Biometrika*, 1946, **33**, 305–325.
51. H. Winicov, *et al.*, Chemical process optimization by computer — a self-directed chemical synthesis system, *Anal. Chim. Acta*, 1978, **103**, 469–476.
52. T. Sugawara and D. G. Cork, Past and present development of automated synthesis apparatus for pharmaceutical chemistry at Takeda Chemical Industries, *Lab. Rob. Autom.*, 1996, **8**, 221–230.
53. C. Simms and J. Singh, Rapid Process Development and Scale-Up Using A Multiple Reactor System, *Org. Process Res. Dev.*, 2000, **4**, 554–562.
54. Y. L. Dar, High-Throughput Experimentation: A Powerful Enabling Technology for the Chemicals and Materials Industry, *Macromol. Rapid Commun.*, 2004, **25**, 34–47.
55. T. Chapman, A structured approach, *Nature*, 2003, **421**, 661.
56. C. A. Nicolaou, I. A. Watson, H. Hu and J. Wang, The Proximal Lilly Collection: Mapping, Exploring and Exploiting Feasible Chemical Space, *J. Chem. Inf. Model.*, 2016, **56**, 1253–1266.
57. G. Schneider, Automating drug discovery, *Nat. Rev. Drug Discovery*, 2018, **17**, 97–113.
58. R. B. Merrifield, J. M. Stewart and N. Jernberg, Instrument for automated synthesis of peptides, *Anal. Chem.*, 1966, **38**, 1905–1914.
59. J. W. Frazer, Computer Automation in Chemistry, *Instrum. Sci. Technol.*, 1970, **2**, 271–295.
60. H. Okamoto and K. Deuchi, Design of a robotic workstation for automated organic synthesis, *Lab. Rob. Autom.*, 2000, **12**, 2–11.
61. E. M. Woerly, J. Roy and M. D. Burke, Synthesis of most polyene natural product motifs using just 12 building blocks and one coupling reaction, *Nat. Chem.*, 2014, **6**, 484–491.
62. R. F. Service and R. F. Service, The synthesis machine, *Science*, 2015, **347**, 1190–1193.
63. J. Li, *et al.*, Synthesis of many different types of organic small molecules using one automated process, *Science*, 2015, **347**, 1221–1226.
64. M. A. R. Meier, J.-F. Gohy, C.-A. Fustin and U. S. Schubert, Combinatorial synthesis of star-shaped block copolymers: host-guest chemistry of unimolecular reversed micelles, *J. Am. Chem. Soc.*, 2004, **126**, 11517–11521.
65. R. Hoogenboom, F. Wiesbrock, M. A. M. Leenen, M. A. R. Meier and U. S. Schubert, Accelerating the living polymerization of

2-nonyl-2-oxazoline by implementing a microwave synthesizer into a high-throughput experimentation workflow, *J. Comb. Chem.*, 2005, **7**, 10–13.
66. J. P. McMullen and K. F. Jensen, Integrated microreactors for reaction automation: new approaches to reaction development, *Annu. Rev. Anal. Chem.*, 2010, **3**, 19–42.
67. A. Adamo, *et al.*, On-demand continuous-flow production of pharmaceuticals in a compact, reconfigurable system, *Science*, 2016, **352**, 61–67.
68. A. J. Mijalis, *et al.*, A fully automated flow-based approach for accelerated peptide synthesis, *Nat. Chem. Biol.*, 2017, **13**, 464–466.
69. D. C. Patel, Y. F. Lyu, J. Gandarilla and S. Doherty, Unattended reaction monitoring using an automated microfluidic sampler and on-line liquid chromatography, *Anal. Chim. Acta*, 2018, **1004**, 32–39.
70. S. Chen, *et al.*, Exploring the Stability of Novel Wide Bandgap Perovskites by a Robot Based High Throughput Approach, *Adv. Energy Mater.*, 2018, **8**, 1701543.
71. E. M. Chan, *et al.*, Reproducible, high-throughput synthesis of colloidal nanocrystals for optimization in multidimensional parameter space, *Nano Lett.*, 2010, **10**, 1874–1885.
72. M. Sasaki, *et al.*, Total laboratory automation in Japan, *Clin. Chim. Acta*, 1998, **278**, 217–227.
73. R. A. Felder, J. Savory, K. S. Margrey, J. W. Holman and J. C. Boyd, Development of a robotic near patient testing laboratory, *Arch. Pathol. Lab. Med.*, 1995, **119**, 948–951.
74. R. A. Felder, Clinical laboratory robotics in the 1990s, *Chemom. Intell. Lab. Syst.*, 1992, **17**, 111–118.
75. R. A. Felder, *et al.*, Robotics in the Clinical Laboratory, *Clin. Lab. Med.*, 1988, **8**, 699–712.
76. M. Nettekoven and A. W. Thomas, Accelerating drug discovery by integrative implementation of laboratory automation in the work flow, *Curr. Med. Chem.*, 2002, **9**, 2179–2190.
77. International Human Genome Sequencing Consortium, Finishing the euchromatic sequence of the human genome, *Nature*, 2004, **431**, 931–945.
78. W. J. Coates, D. J. Hunter and W. S. MacLachlan, Successful implementation of automation in medicinal chemistry, *Drug Discovery Today*, 2000, **5**, 521–527.
79. M. M. Hann and T. I. Oprea, Pursuing the leadlikeness concept in pharmaceutical research, *Curr. Opin. Chem. Biol.*, 2004, **8**, 255–263.
80. I. Caraus, A. A. Alsuwailem, R. Nadon and V. Makarenkov, Detecting and overcoming systematic bias in high-throughput screening technologies: a comprehensive review of practical issues and methodological solutions, *Brief. Bioinform.*, 2015, **16**, 974–986.
81. H.-J. Federsel, Handing over the baton: connecting medicinal chemistry with process R&D, *Drug News Perspect.*, 2008, **21**, 193–199.
82. A. Weber, E. von Roedern and H. U. Stilz, SynCar: An Approach to Automated Synthesis, *J. Comb. Chem.*, 2005, 7(2), 178–184.

83. A. G. Godfrey, T. Masquelin and H. Hemmerle, A remote-controlled adaptive medchem lab: an innovative approach to enable drug discovery in the 21st Century, *Drug Discovery Today*, 2013, **18**, 795–802.
84. S. L. Morgan and S. N. Deming, Simplex optimization of analytical chemical methods, *Anal. Chem.*, 1974, **46**, 1170–1181.
85. R. W. Wagner, F. Li, H. Du and J. S. Lindsey, Investigation of Cocatalysis Conditions Using an Automated Microscale Multireactor Workstation: Synthesis ofmeso-Tetramesitylporphyrin, *Org. Process Res. Dev.*, 1999, **3**, 28–37.
86. A. Lucia and J. Xu, Chemical process optimization using Newton-like methods, *Comput. Chem. Eng.*, 1990, **14**, 119–138.
87. T. J. Berna, M. H. Locke and A. W. Westerberg, A new approach to optimization of chemical processes, *AIChE J.*, 1980, **26**, 37–43.
88. S. Boyd and L. Vandenberghe, *Onvex Optimization*, Cambridge University Press, 2004.
89. S. Bubeck, Convex Optimization: Algorithms and Complexity, *Found. Trends Mach. Learn.*, 2015, **8**, 231–357.
90. A. Cauchy, Méthode générale pour la résolution des systemes d'équations simultanées, *Comp. Rend. Sci. Paris*, 1847, **25**, 536–538.
91. H. Robbins and S. Monro, A Stochastic Approximation Method, *Ann. Math. Stat.*, 1951, **22**, 400–407.
92. I. Newton, *De analysi per aequationes numero terminorum infinitas*, 1669, W. Jones, London, 1711.
93. P. Deuflhard, *Newton Methods for Nonlinear Problems: Affine Invariance and Adaptive Algorithms*, Springer Science & Business Media, 2011.
94. M. R. Hestenes and E. Stiefel, *Methods of Conjugate Gradients for Solving Linear Systems*, NBS Washington, DC, **vol. 49**, 1952.
95. R. Byrd, P. Lu, J. Nocedal and C. Zhu, A Limited Memory Algorithm for Bound Constrained Optimization, *SIAM J. Sci. Comput.*, 1995, **16**, 1190–1208.
96. C. Zhu, R. H. Byrd, P. Lu and J. Nocedal, Algorithm 778: L-BFGS-B: Fortran Subroutines for Large-scale Bound-constrained Optimization, *ACM Trans. Math. Software*, 1997, **23**, 550–560.
97. J. A. Nelder and R. Mead, A Simplex Method for Function Minimization, *Comput. J.*, 1965, **7**, 308–313.
98. S. N. Deming, L. R. Parker and M. Bonner Denton, A Review of Simplex Optimization in Analytical Chemistry, *Crit. Rev. Anal. Chem.*, 1978, **7**, 187–202.
99. D. E. Long, Simplex optimization of the response from chemical systems, *Anal. Chim. Acta*, 1969, **46**, 193–206.
100. D. E. Fitzpatrick, C. Battilocchio and S. V. Ley, A Novel Internet-Based Reaction Monitoring, Control and Autonomous Self-Optimization Platform for Chemical Synthesis, *Org. Process Res. Dev.*, 2016, **20**, 386–394.
101. S. A. Pergantis, W. R. Cullen and A. P. Wade, Simplex optimization of conditions for the determination of arsenic in environmental samples

by using electrothermal atomic absorption spectrometry, *Talanta*, 1994, **41**, 205–209.
102. D. Friedrich, E. Mangano and S. Brandani, Automatic estimation of kinetic and isotherm parameters from ZLC experiments, *Chem. Eng. Sci.*, 2015, **126**, 616–624.
103. J. P. McMullen, M. T. Stone, S. L. Buchwald and K. F. Jensen, An integrated microreactor system for self-optimization of a heck reaction: From micro-to mesoscale flow systems, *Angew. Chem., Int. Ed.*, 2010, **49**, 7076–7080.
104. J. P. McMullen and K. F. Jensen, An Automated Microfluidic System for Online Optimization in Chemical Synthesis, *Org. Process Res. Dev.*, 2010, **14**, 1169–1176.
105. X.-S. Yang, *Nature-Inspired Optimization Algorithms*, Elsevier, 2014.
106. I. Fister Jr., X.-S. Yang, I. Fister, J. Brest and D. Fister, A Brief Review of Nature-Inspired Algorithms for Optimization. *arXiv [cs.NE]* (2013).
107. N. Hansen and A. Ostermeier, Completely derandomized self-adaptation in evolution strategies, *Evol. Comput.*, 2001, **9**, 159–195.
108. N. Hansen, S. D. Müller and P. Koumoutsakos, Reducing the time complexity of the derandomized evolution strategy with covariance matrix adaptation (CMA-ES), *Evol. Comput.*, 2003, **11**, 1–18.
109. Z. Zhou, X. Li and R. N. Zare, Optimizing Chemical Reactions with Deep Reinforcement Learning, *ACS Cent. Sci.*, 2017, **3**, 1337–1344.
110. R. Eberhart and J. Kennedy, A new optimizer using particle swarm theory, in *MHS'95. Proceedings of the Sixth International Symposium on Micro Machine and Human Science*, 1995, pp. 39–43, ieeexplore.ieee.org.
111. Y. Shi and R. Eberhart, A modified particle swarm optimizer, in *1998 IEEE International Conference on Evolutionary Computation Proceedings. IEEE World Congress on Computational Intelligence (Cat. No.98TH8360)*, 1998, pp. 69–73, ieeexplore.ieee.org.
112. R. Storn and K. Price, Differential evolution–a simple and efficient heuristic for global optimization over continuous spaces, *J. Global Optimiz*, 1997, **11**, 341–359.
113. K. Price, R. M. Storn and J. A. Lampinen, *Differential Evolution: A Practical Approach to Global Optimization*, Springer Science & Business Media, 2006.
114. W. Huyer and A. Neumaier, Benchmarking of SNOBFIT on the noisy function testbed, *Proceedings of the 11th Annual Conference*, 2009.
115. W. Huyer and A. Neumaier, Snobfit-stable noisy optimization by branch and fit, *ACM Trans. Math. Software*, 2006, **35**, 200.
116. N. Holmes, *et al.*, Online quantitative mass spectrometry for the rapid adaptive optimisation of automated flow reactors, *React. Chem. Eng.*, 2016, **1**, 96–100.
117. S. Krishnadasan, R. J. C. Brown, A. J. deMello and J. C. deMello, Intelligent routes to the controlled synthesis of nanoparticles, *Lab Chip*, 2007, **7**, 1434–1441.

118. R. A. Skilton, A. J. Parrott, M. W. George, M. Poliakoff and R. A. Bourne, Real-time feedback control using online attenuated total reflection Fourier transform infrared (ATR FT-IR) spectroscopy for continuous flow optimization and process knowledge, *Appl. Spectrosc.*, 2013, **67**, 1127–1131.
119. J. Bergstra, D. Yamins and D. Cox, Making a Science of Model Search: Hyperparameter Optimization in Hundreds of Dimensions for Vision Architectures, *Int. Conference Mach. Learn.*, 2013, 115–123.
120. J. S. Bergstra, R. Bardenet, Y. Bengio and B. Kégl, Algorithms for Hyper-Parameter Optimization, in *Advances in Neural Information Processing Systems 24*, ed. J. Shawe-Taylor, R. S. Zemel, P. L. Bartlett, F. Pereira and K. Q. Weinberger, Curran Associates, Inc., 2011, pp. 2546–2554.
121. J. Snoek, H. Larochelle and R. P. Adams, Practical Bayesian Optimization of Machine Learning Algorithms, in *Advances in Neural Information Processing Systems 25*, ed. F. Pereira, C. J. C. Burges, L. Bottou and K. Q. Weinberger, Curran Associates, Inc., 2012, pp. 2951–2959.
122. The GPyOpt authors, *GPyOpt: A Bayesian Optimization Framework in Python*, 2016.
123. F. Hutter, H. H. Hoos and K. Leyton-Brown, Sequential Model-Based Optimization for General Algorithm Configuration, *Learning and Intelligent Optimization*, Springer, Berlin Heidelberg, 2011, pp. 507–523.
124. F. Hutter, H. Hoos and K. Leyton-Brown, Bayesian Optimization With Censored Response Data, *arXiv [cs.AI]*, 2013.
125. F. Hutter, H. Hoos and K. Leyton-Brown, An Evaluation of Sequential Model-based Optimization for Expensive Blackbox Functions, *Proceedings of the 15th Annual Conference Companion on Genetic and Evolutionary Computation*, ACM, 2013, pp. 1209–1216.
126. J. Snoek, O. Rippel, K. Swersky, R. Kiros and N. Satish, Scalable bayesian optimization using deep neural networks, in *International Conference on Machine Learning*, 2015, pp. 2171–2180.
127. F. Häse, L. M. Roch, C. Kreisbeck and A. Aspuru-Guzik, Phoenics: A Bayesian Optimizer for Chemistry, *ACS Cent. Sci.*, 2018, **4**, 1134–1145.
128. T. Fukazawa, Y. Harashima, Z. Hou and T. Miyake, Bayesian optimization of chemical composition: A comprehensive framework and its application to RFe12-type magnet compounds, *Phys. Rev. Mater.*, 2019, **3**(5), 053807.
129. R.-R. Griffiths and J. M. Hernández-Lobato, Constrained Bayesian Optimization for Automatic Chemical Design, *arXiv [stat.ML]*, 2017.
130. L. Chan, G. R. Hutchison and G. M. Morris, Bayesian optimization for conformer generation, *J. Cheminform.*, 2019, **11**, 32.
131. T. P. Lillicrap, J. J. Hunt, A. Pritzel, N. Heess, T. Erez, Y. Tassa, D. Silver, D. Wierstra, Continuous control with deep reinforcement learning, 2015, arXiv preprint arXiv:1509.02971.
132. J. Kober, J. A. Bagnell and J. Peters, Reinforcement learning in robotics: A survey, *Int. J. Rob. Res.*, 2013, **32**, 1238–1274.

133. X. Yang, J. Zhang, K. Yoshizoe, K. Terayama and K. Tsuda, ChemTS: an efficient python library for de novo molecular generation, *Sci. Technol. Adv. Mater.*, 2017, **18**, 972–976.
134. M. Sumita, X. Yang, S. Ishihara, R. Tamura and K. Tsuda, Hunting for Organic Molecules with Artificial Intelligence: Molecules Optimized for Desired Excitation Energies, *ACS Cent. Sci.*, 2018, **4**, 1126–1133.
135. T. Matsumoto, H. Du and J. S. Lindsey, A two-tiered strategy for simplex and multidirectional optimization of reactions with an automated chemistry workstation, *Chemom. Intell. Lab. Syst.*, 2002, **62**, 149–158.
136. R. Matsuda, M. Ishibashi and Y. Takeda, Simplex optimization of reaction conditions with an automated system, *Chem. Pharm. Bull.*, 1988, **36**, 3512–3518.
137. C. Porte, W. Debreuille, F. Draskovic and A. Delacroix, Automation and optimization by simplex methods of 6-chlorohexanol synthesis, *Process Control Qual.*, 1996, **4**, 111–122.
138. J. M. Dixon and J. S. Lindsey, Performance of search algorithms in the examination of chemical reaction spaces with an automated chemistry workstation, *J. Assoc. Lab. Autom.*, 2004, **9**, 364–374.
139. H. Du and J. S. Lindsey, An approach for parallel and adaptive screening of discrete compounds followed by reaction optimization using an automated chemistry workstation, *Chemom. Intell. Lab. Syst.*, 2002, **62**, 159–170.
140. H. Du, W. Shen, P. Y. Kuo and J. S. Lindsey, Decision-tree programs for an adaptive automated chemistry workstation. Application to catalyst screening experiments, *Chemom. Intell. Lab. Syst.*, 1999, **48**, 205–217.
141. P. Y. Kuo, H. Du, L. Andrew Corkan, K. Yang and J. S. Lindsey, A planning module for performing grid search, factorial design, and related combinatorial studies on an automated chemistry workstation, *Chemom. Intell. Lab. Syst.*, 1999, **48**, 219–234.
142. H. Du, S. Jindal and J. S. Lindsey, Implementation of the multidirectional search algorithm on an automated chemistry workstation. A parallel yet adaptive approach for reaction optimization, *Chemom. Intell. Lab. Syst.*, 1999, **48**, 235–256.
143. J.-C. Plouvier, L. Andrew Corkan and J. S. Lindsey, Experiment planner for strategic experimentation with an automated chemistry workstation, *Chemom. Intell. Lab. Syst.*, 1992, **17**, 75–94.
144. T. Matsumoto, H. Du and J. S. Lindsey, A parallel simplex search method for use with an automated chemistry workstation, *Chemom. Intell. Lab. Syst.*, 2002, **62**, 129–147.
145. J. M. Dixon, H. Du, D. G. Cork and J. S. Lindsey, An experiment planner for performing successive focused grid searches with an automated chemistry workstation, *Chemom. Intell. Lab. Syst.*, 2002, **62**, 115–128.
146. F. Häse, L. M. Roch and A. Aspuru-Guzik, Chimera: enabling hierarchy based multi-objective optimization for self-driving laboratories, *Chem. Sci.*, 2018, **9**, 7642–7655.

147. L. M. Roch, *et al.*, ChemOS: An Orchestration Software to Democratize Autonomous Discovery, *ChemRxiv*, 2018, DOI: 10.26434/chemrxiv.5953606.v1.
148. A. Zakutayev, N. Wunder, M. Schwarting, J. D. Perkins, R. White, K. Munch, W. Tumas and C. Phillips, An open experimental database for exploring inorganic materials, *Sci. Data*, 2018, **5**, 180053.
149. J. Li, *et al.*, AIR-Chem: Authentic Intelligent Robotics for Chemistry, *J. Phys. Chem. A*, 2018, **122**, 9142–9148.
150. I. M. Pendleton, *et al.*, Experiment Specification, Capture and Laboratory Automation Technology (ESCALATE): a software pipeline for automated chemical experimentation and data management, *Magn. Reson. Chem.*, 2019, **9**, 1–14.
151. N. Adams and U. S. Schubert, From data to knowledge: chemical data management, data mining, and modeling in polymer science, *J. Comb. Chem.*, 2004, **6**, 12–23.
152. N. Adams and U. S. Schubert, Software Solutions for Combinatorial and High-Throughput Materials and Polymer Research, *Macromol. Rapid Commun.*, 2004, **25**, 48–58.
153. M. Bates, A. J. Berliner, J. Lachoff, P. R. Jaschke and E. S. Groban, Wet Lab Accelerator: A Web-Based Application Democratizing Laboratory Automation for Synthetic Biology, *ACS Synth. Biol.*, 2017, **6**, 167–171.
154. G. Linshiz, *et al.*, PaR-PaR laboratory automation platform, *ACS Synth. Biol.*, 2013, **2**, 216–222.
155. E. Whitehead, F. Rudolf, H.-M. Kaltenbach and J. Stelling, Automated Planning Enables Complex Protocols on Liquid-Handling Robots, *ACS Synth. Biol.*, 2018, **7**, 922–932.
156. J. Keller Vrana, A. Miller, G. Newman and E. Klavins, *Aquarium: The Laboratory Operating System (Version v2.5.0) Zenodo*, 2019.
157. B. Miles and P. L. Lee, Achieving Reproducibility and Closed-Loop Automation in Biological Experimentation with an IoT-Enabled Lab of the Future, *SLAS Technol.*, 2018, **23**, 432–439.
158. G. Schneider and P. Wrede, Artificial neural networks for computer-based molecular design, *Prog. Biophys. Mol. Biol.*, 1998, **70**, 175–222.
159. G. S. Sittampalam, D. D. Rudnicki, D. A. Tagle, A. Simeonov and C. P. Austin, Mapping biologically active chemical space to accelerate drug discovery, *Nat. Rev. Drug Discovery*, 2019, **18**, 83–84.
160. S. Langner, F. Häse, J. D. Perea, T. Stubhan, J. Hauch, L. M. Roch, T. Heumueller, A. Aspuru-Guzik and C. J. Brabec, Beyond Ternary OPV: High-Throughput Experimentation and Self-Driving Laboratories Optimize Multicomponent Systems, *Adv. Mater.*, 2020, **32**(14), 1907801.
161. R. A. Skilton, *et al.*, Remote-controlled experiments with cloud chemistry, *Nat. Chem.*, 2015, **7**, 1–5.
162. B. Maruyama, *et al.*, Autonomous Experimentation Applied to Carbon Nanotube Synthesis, *Microsc. Microanal.*, 2017, **23**, 182.

163. V. Duros, *et al.*, Human versus Robots in the Discovery and Crystallization of Gigantic Polyoxometalates, *Angew. Chem., Int. Ed.*, 2017, **56**, 10815–10820.
164. D. Xue, *et al.*, Accelerated search for materials with targeted properties by adaptive design, *Nat. Commun.*, 2016, **7**, 11241.
165. V. Dragone, V. Sans, A. B. Henson, J. M. Granda and L. Cronin, An autonomous organic reaction search engine for chemical reactivity, *Nat. Commun.*, 2017, **8**, 15733.
166. J. Snoek, K. Swersky, R. Zemel and R. Adams, Input Warping for Bayesian Optimization of Non-Stationary Functions, in *International Conference on Machine Learning 1674–1682*, jmlr.org, 2014.
167. J. Bergstra and Y. Bengio, Random Search for Hyper-Parameter Optimization, *J. Mach. Learn. Res.*, 2012, **13**, 281–305.
168. E. Gamma, R. Helm, R. Johnson and J. Vlissides, *Design Patterns: Elements of Reusable Object-Oriented Software (Adobe Reader)*, Pearson Education, 1994.
169. S. Bird, E. Klein and E. Loper, *Natural Language Processing with Python: Analyzing Text with the Natural Language Toolkit*, 'O'Reilly Media, Inc., 2009.
170. M. Abadi, *et. al.*, TensorFlow: Large-Scale Machine Learning on Heterogeneous Distributed Systems, *arXiv [cs.DC]*, 2016.
171. Materials Innovation Factory – University of Liverpool, University of Liverpool https://www.liverpool.ac.uk/materials-innovation-factory/.
172. High-Throughput Molecular Screening Center, Scripps Research. https://www.scripps.edu/science-and-medicine/cores-and-services/high-throughput-molecular-screening-center/index.html.
173. Centre for Rapid Online Analysis of Reactions (ROAR), Imperial College London, http://www.imperial.ac.uk/a-z-research/rapid-online-analysis-of-reactions/.
174. Homepage, High Throughput Screening Laboratory, https://hts.ku.edu/homepage, 2013.
175. High Throughput Experimentation (HTE), https://coperetgroup.ethz.ch/research/high-throughput-experimentation–hte–.html.
176. T. C. Malig, *et al.*, Real-time HPLC-MS reaction progress monitoring using an automated analytical platform, *React. Chem. Eng.*, 2017, **2**, 309–314.
177. L. M. Roch, F. Häse, C. Kreisbeck, T. Tamayo-Mendoza, L. P. Yunker, J. E. Hein and A. Aspuru-Guzik, ChemOS: an orchestration software to democratize autonomous discovery, *PLoS One*, 2020, **15**(4), e0229862.
178. S. Langner, *et al.*, Beyond Ternary OPV: High-Throughput Experimentation and Self-Driving Laboratories Optimize Multicomponent Systems, *Adv. Mater.*, 2020, 1907801.
179. T. J. Dougherty and M. J. Pucci, *Antibiotic Discovery and Development*, Springer Science & Business Media, 2011.
180. C. A. Lipinski, F. Lombardo, B. W. Dominy and P. J. Feeney, Experimental and computational approaches to estimate solubility and

permeability in drug discovery and development settings, *Adv. Drug Delivery Rev.*, 1997, **23**, 3–25.
181. C. A. Lipinski, Lead- and drug-like compounds: the rule-of-five revolution, *Drug Discovery Today: Technol.*, 2004, **1**, 337–341.
182. G. Sliwoski, S. Kothiwale, J. Meiler and E. W. Jr. Lowe, Computational methods in drug discovery, *Pharmacol. Rev.*, 2014, **66**, 334–395.
183. C. W. Murray and D. C. Rees, The rise of fragment-based drug discovery, *Nat. Chem.*, 2009, **1**, 187–192.
184. M. Williams and T. Daviter, *Protein-Ligand Interactions: Methods and Applications*, Humana Press, 2016.
185. E. O. Pyzer-Knapp, C. Suh, R. Gómez-Bombarelli, J. Aguilera-Iparraguirre and A. Aspuru-Guzik, What Is High-Throughput Virtual Screening? A Perspective from Organic Materials Discovery, *Annu. Rev. Mater. Res.*, 2015, **45**, 195–216.
186. A. Kadurin, S. Nikolenko, K. Khrabrov, A. Aliper and A. Zhavoronkov, druGAN: An Advanced Generative Adversarial Autoencoder Model for de Novo Generation of New Molecules with Desired Molecular Properties in Silico, *Mol. Pharm.*, 2017, **14**, 3098–3104.
187. D. Merk, L. Friedrich, F. Grisoni and G. Schneider, De Novo Design of Bioactive Small Molecules by Artificial Intelligence, *Mol. Inf.*, 2018, **37**, 201700153.
188. R. Gómez-Bombarelli, *et al.*, Automatic Chemical Design Using a Data-Driven Continuous Representation of Molecules, *ACS Cent. Sci.*, 2018, **4**, 268–276.
189. A. Zhavoronkov, *et al.*, Deep learning enables rapid identification of potent DDR1 kinase inhibitors, *Nat. Biotechnol.*, 2019, **37**, 1038–1040.
190. S. V. Ley, R. J. Ingham, M. O'Brien and D. L. Browne, Camera-enabled techniques for organic synthesis, *Beilstein J. Org. Chem.*, 2013, **9**, 1051–1072.
191. Z. Chen and X. Huang, End-to-end learning for lane keeping of self-driving cars, *2017 IEEE Intelligent Vehicles Symposium (IV)*, 10.1109/ivs.2017.7995975, 2017.
192. R. Ramakrishnan, P. O. Dral, M. Rupp and O. A. von Lilienfeld, Big Data Meets Quantum Chemistry Approximations: The Δ-Machine Learning Approach, *J. Chem. Theory Comput.*, 2015, **11**, 2087–2096.
193. R. Gómez-Bombarelli, Reaction: The Near Future of Artificial Intelligence in Materials Discovery, *Chemistry*, 2018, **4**, 1189–1190.

Section 8: Summary and Outlook

CHAPTER 17

Summary and Outlook

NATHAN BROWN

BenevolentAI, 4-8 Maple Street, London W1T 5HD, UK
Email: nathan.brown@benevolent.ai

17.1 Introduction

This book has been an effort to cover the many various aspects of artificial intelligence as it is applied to drug discovery. While many new methods and approaches have been introduced, and their impact or potential impact in drug discovery, has been presented, it is also important to reflect on the various advances that have been made in all fields as a whole to optimising drug discovery and thereby improving success rates in this endeavour.

A grand challenge of artificial intelligence in drug discovery is to anticipate the potential benefits and adverse effects prior to testing these potential drugs experimentally, but realising that judicious use of experimentation is essential to enable these advances.

We are experiencing a step-change in the ability of algorithms to apply available data in helping design the next generation of therapeutics. It is unlikely that any one innovation will be the dominant method going forward, but it is likely however that novel combinations of both new and old methods, with appropriate enhancements will lead to significant advances in these methods in drug discovery efforts.

17.2 Challenges

There are a number of challenges in effectively applying artificial intelligence in different phases of drug discovery to deliver success. Some have been summarised here and potential approaches to overcoming them.

Drug Discovery Series No. 75
Artificial Intelligence in Drug Discovery
Edited by Nathan Brown

Published by the Royal Society of Chemistry, www.rsc.org

17.2.1 Data

As highlighted in many chapters in this book, data is what drives success and successful observations in both drug discovery and the application of artificial intelligence in drug discovery. The generation of data is both expensive and takes time. Not only do we have to use expensive materials to assay compounds, but we also have the cost and time of synthesising those compounds in the first instance. A number of approaches can be predicted to have a significant impact on reducing the time and expense of making and testing new compounds.

Over the years, laboratories have taken advantage of streamlined processes and various laboratory automation techniques to increase throughput and reduce associated costs in making and testing compounds. As we progress with an ever more connected world significant additional improvements will be made in these aspects and drive optimisation of these challenges in the laboratory, but also in logistics of delivering compounds and data where they need to be so they can be taken into account.

The improvement of automated synthesis planning methods will increasingly streamline the process from design idea to experimental realisation of a synthesised compound. In the same way that various literature database searching has improved the ability to quickly identify the most interesting routes for synthesis, new artificial intelligence driven tools will not only be able identify the prior art in the literature but also to summarise these data to distil down to the most salient details, thereby acting as a timesaver and freeing up more expert time to digest the information presented and select the best recommendation.

It is clear that good quality experimental data is essential in building good quality models that can objectively inform decision making. However, it is also possible to use artificial intelligence to augment the existing data with machine learning using data from other related endpoints to simulate additional data. Similarly, machine learning can be used with molecular simulations, as we have seen, to augment datasets with new simulated data that can then be applied with the machine learning algorithms to build better quality models. While work in this area is relatively nascent, it is anticipated that this work will lead to significant advances, although likely still not replace experimental data.

17.2.2 Compute

Bigger computers will allow even more *in silico* experiments to be run allowing a full characterisation of the parameters. While we already have vast computational resources, effective systems to take advantage of these are still being refined for the multitude of applications to which they can be applied.

Hyperparameter optimisation remains a significant challenge in machine learning models, where there may be many different parameters that can be tuned to improve model performance. Indeed, these parameters may even be interdependent which also significantly increases the complexity of the search space in which we are optimising. While the same optimisation algorithms can be applied in optimising the parameter sets as have been discussed, the combinatorics of the search space explodes rapidly as we increase the number and diversity of parameters being optimised.

The increase not only in compute, but also how we use the available computational resources, will permit ever more increasing refinements in artificial intelligence and machine learning algorithm as applied to challenges in drug discovery, which will in turn increase the confidence of these methods in their applications and positive contributions to project progress towards clinical candidates and approved drugs.

17.2.3 Culture

With all the rapid advances in the way that we work with technology and embracing the positive improvements in our lives, it is also important to note that there is also a less positive element to this engagement and adoption of this technology. New technological advances, especially those that move at a fast pace such as what is being observed now with artificial intelligence and machine learning, will necessarily disrupt the *status quo* and potentially lead to some significant reticence in embracing this change. While this reticence can often be seen as negative and backwards, it is an essential component of any radical change in the way we conduct our work and requires all those involved to find the best way of working with the new technologies.

17.3 Summary

It is clear that artificial intelligence and machine learning are enacting a huge change in the world and is significantly disrupting many industries around the world and drug discovery is no exception. While the long term effects on the efficiency gains and how the new technologies alter the performance of drug discovery in general, it is safe to say that the current positive observations indicate a strong improvement in the positive.

While we are also embedding these technologies within our organisations, the field and new research are progressing at a fast pace. The learning from practical application of these new methods will feed further research into improving those practical applications with improved methods. As the theory and application develops further and improves over time, the contribution of artificial intelligence to drug discovery will become increasingly clear.

Subject Index

www.ingramcontent.com/pod-product-compliance
Ingram Content Group UK Ltd.
Pitfield, Milton Keynes, MK11 3LW, UK
UKHW022101170926
8624IPUK00023B/13

9 781788 015479